# DIVERSITY OF LIFE

CECIE STARR / RALPH TAGGART

# BIOLOGY

*The Unity and Diversity of Life*

NINTH EDITION

LISA STARR

*Biological Illustrator*

Brooks/Cole
Thomson Learning™

*Australia • Canada • Mexico • Singapore • Spain • United Kingdom • United States*

BIOLOGY PUBLISHER: Jack C. Carey

PROJECT DEVELOPMENT EDITOR: Kristin Milotich

MEDIA PROJECT MANAGER: Pat Waldo

EDITORIAL ASSISTANT: Daniel Lombardino

DEVELOPMENTAL EDITOR: Mary Arbogast

MARKETING TEAM: Rachel Alvelais, Mandie Houghan,
Laura Hubrich, Carla Martin-Falcone

ASSISTANT EDITOR: Daniel Lombardino

PRODUCTION EDITOR: Jamie Sue Brooks

PRODUCTION DIRECTOR: Mary Douglas, Rogue Valley Publications

EDITORIAL PRODUCTION: Karen Stough, Jamee Rae

PERMISSIONS EDITOR: Mary Kay Hancharick

TEXT AND COVER DESIGN, ART DIRECTION: Gary Head,
Gary Head Design

COVER PHOTO: *Snow leopard, a currently endangered species*
(*Section 28.2*). © Art Wolfe/PhotoResearchers

ART EDITOR: Mary Douglas, Jamie Sue Brooks

INTERIOR ILLUSTRATION: Lisa Starr, Raychel Ciemma,
Precision Graphics, Preface Inc., Robert Demarest, Jan Flessner,
Darwen Hennings, Vally Hennings, Betsy Palay, Nadine Sokol,
Kevin Somerville, Lloyd Townsend

PHOTO RESEARCHER: Myrna Engler

PRINT BUYER: Vena M. Dyer

COMPOSITION: Preface, Inc. (Angela Harris)

COLOR PROCESSING: H&S Graphics (Tom Anderson,
Michelle Kessel, Laurie Riggle, and John Deady)

COVER PRINTING, PRINTING AND BINDING: Von Hoffmann Press

BOOKS IN THE BROOKS/COLE BIOLOGY SERIES

*Biology: The Unity and Diversity of Life*, Ninth, Starr/Taggart
*Biology: Concepts and Applications*, Fourth, Starr
*Laboratory Manual for Biology*, Perry/Morton/Perry
*Human Biology*, Fourth, Starr/McMillan
*Perspectives in Human Biology*, Knapp
*Human Physiology*, Fourth, Sherwood
*Fundamentals of Physiology*, Second, Sherwood
*Psychobiology: The Neuron and Behavior*, Hoyenga/Hoyenga

*Introduction to Cell and Molecular Biology*, Wolfe
*Molecular and Cellular Biology*, Wolfe

*Introduction to Microbiology*, Second, Ingraham/Ingraham

*Genetics: The Continuity of Life*, Fairbanks
*Human Heredity*, Fifth, Cummings
*Introduction to Biotechnology*, Barnum
*Sex, Evolution, and Behavior*, Second, Daly/Wilson

*Plant Biology*, Rost et al.
*Plant Physiology*, Fourth, Salisbury/Ross
*Plant Physiology Laboratory Manual*, Ross

*General Ecology*, Second, Krohne
*Living in the Environment*, Eleventh, Miller
*Environmental Science*, Eighth, Miller
*Sustaining the Earth*, Fourth, Miller
*Environment: Problems and Solutions*, Miller
*Environmental Science*, Fifth, Chiras

*Oceanography: An Invitation to Marine Science*, Third, Garrison
*Essentials of Oceanography*, Second, Garrison
*Introduction to Ocean Sciences*, Segar
*Oceanography: An Introduction*, Fifth, Ingmanson/Wallace
*Marine Life and the Sea*, Milne

Printed in United States of America

10 9 8 7 6 5 4 3 2 1

For permission to use material from this work, contact us by
www.thomsonrights.com
Fax: 1-800-730-2215
Phone: 1-800-730-2214

ISBN 0-534-37943-5

For more information about this or any other Brooks/Cole products,
contact:
BROOKS/COLE
511 Forest Lodge Road
Pacific Grove, CA 93950 USA
www.brookscole.com
1-800-423-0563 (Thomson Learning Academic Resource Center)

# CONTENTS IN BRIEF

*Highlighted chapters are included in DIVERSITY OF LIFE.*

# DETAILED CONTENTS

# Preface

Teachers of introductory biology know all about the Red Queen effect, whereby one runs as fast as one can to stay in the same place. New and modified information from hundreds of fields of inquiry piles up daily, and somehow teachers are expected to distill it into Biology Lite, a one-course zip through the high points that still manages to help students deepen their understanding of a world of unbelievable richness.

We offer this book as a way to find coherence in the diversity and underlying unity of life. With its examples of problem solving and experiments, it shows the power of thinking critically about the natural world. It also highlights key concepts, current understandings, and research trends for major fields of biological inquiry. It explains the structure and function of a broad sampling of organisms in enough detail so students can develop a working vocabulary about life's parts and processes.

**TOPIC SPREADS**  Ongoing feedback from teachers of more than three million students helped us refine our approach to writing. We keep the story line in focus for students by subscribing to the question "How do you eat an elephant?" and its answer, "One bite at a time." We organized descriptions, art, and supporting evidence for each topic on two facing pages, at most. Each *topic spread* starts with a numbered tab and concludes with boldface, summary statements of key points. Students can use the statements to check whether they understand one topic spread's concepts before starting another.

By clearly organizing topics within chapters, we offer teachers more flexibility in assigning text material to fit course requirements. For example, they might or may not assign the enriching detail of the Focus essays. All topic spreads within a chapter flow as parts of the same story, but some spreads clearly offer depth that may be treated as optional.

The topic spreads are not gimmicks. Ongoing feedback guided decisions about when to add depth and when to loosen core material with applications. Within the spreads, headings and subheadings help students keep track of the hierarchy of information. Transitions between spreads help them keep the greater story in focus and discourage memorization for its own sake. To avoid disrupting the basic story line while still attending to interested students, we include some enriching details in optional illustrations.

The clear organization helps students find assigned topics easily and has translated into improved test scores. This is a tangible outcome, but we are more pleased that the clarity helps give students enough confidence to dig deeper into biological science. We also are happy to hear that the story has lured many of them into reading far more than they planned to do.

**BALANCING CONCEPTS WITH APPLICATIONS**  Each chapter starts with a lively or sobering application and an adjoining list of key concepts, the chapter's advance organizer. Strategically placed examples of applications parallel core material, not so many as to be distracting but enough to keep students interested in continuing with the basics. Brief applications are integrated in the text. Focus essays afford more depth on many medical, environmental, and social issues without interrupting the conceptual flow.

**FOUNDATIONS FOR CRITICAL THINKING**  To help students increase their capacity for critical thinking, we walk them through experiments that yielded evidence in favor of or against hypotheses. We use certain chapter introductions as well as entire chapters to show students some productive results of critical thinking. Also, each chapter has a set of *Critical Thinking* questions.

**VISUAL OVERVIEWS OF CONCEPTS**  We simultaneously develop text and art as inseparable parts of the same story. We give visual learners a means to work their way through a visual overview of major processes before reading the corresponding (and possibly intimidating) text. Students repeatedly let us know how much they appreciate this art. Overview illustrations have step-by-step descriptions of biological parts and processes. Instead of "wordless" diagrams, we break down information into a series of illustrated callouts.

Many drawings are integrated overviews of structure and function. Students need not jump back and forth from text, to tables, to illustrations, and back again to see how an organ system is put together and what its parts do. We hierarchically arrange descriptions of parts to reflect a system's structural and functional organization.

**COLOR CODES**  Consistent use of colors for molecules, cell structures, and processes helps students track what is going on. We use these colors throughout the book:

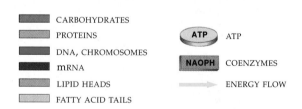

| CARBOHYDRATES | |
| PROTEINS | ATP  ATP |
| DNA, CHROMOSOMES | |
| mRNA | NAOPH  COENZYMES |
| LIPID HEADS | ENERGY FLOW |
| FATTY ACID TAILS | |

**ICONS**  Small, simple diagrams next to an illustration help students relate a topic to the big picture. For instance, a simple diagram of a cell reminds them of the location of the plasma membrane relative to the cytoplasm. Other icons relate reactions and processes to certain locations and to how they tie in with one another. A multimedia icon directs them to art in the CD-ROM packaged in its own envelope at the back of their book. Another directs them to supplemental material on the Web and a third, to InfoTrac.

# THE ORIGIN AND EVOLUTION OF LIFE

## *In The Beginning . . .*

Some clear evening, watch the moon as it rises from the horizon and think of the 380,000 kilometers between it and you. *Five billion trillion times* farther away from you are galaxies—systems of stars—at the boundary of the known universe. Wavelengths traveling through space move faster than anything else—millions of meters per second—yet long wavelengths that originated from faraway galaxies many billions of years ago are only now reaching the Earth.

If we are to accept all known measures, all of the near and distant galaxies in the vast space of the universe are moving away from one another, which means the universe must be expanding. And the prevailing view of how the colossal expansion came about may account for every bit of matter in every living thing.

Think about how you rewind a videotape on a VCR, then imagine "rewinding" the universe. As you do this, the galaxies start moving back together. After 12 to 15 billion years of rewinding, all galaxies, all matter, and all of space are compressed into a hot, dense volume about the size of the sun. You have arrived at time zero.

That incredibly hot, dense state lasted only for an instant. What happened next is known as the **big bang**,

**Figure 21.1** Part of the great Eagle nebula, a hotbed of star formation 7,000 light-years from Earth, in the constellation Serpens. (The Latin *nebula* means mist.) Not shown in this image are a few huge, young stars above the pillars. For the past few million years, intense ultraviolet radiation from the stars has been eroding the less dense surface of the pillars. (By analogy, visualize a strong prevailing wind blowing away sand in a desert and exposing rocks.) Globules of denser gases and dust that have resisted erosion are visible at the surface. Each pillar is wider than our solar system—more than 10 billion miles across! New stars are hatching from the protruding globules; some are shining brightly on the tips of gaseous streamers.

a stupendous, nearly instantaneous distribution of matter and energy throughout the known universe. About a minute later, temperatures dropped to a billion degrees. Fusion reactions produced most of the light elements, including helium, which are still the most abundant elements in the universe. Radio telescopes detect cooled, diluted background radiation that is a relic of the big bang, left over from the beginning of time.

Over the next billion years, uncountable numbers of gaseous particles collided and condensed under gravity's force to become the first stars. When the stars became massive enough, nuclear reactions were ignited in their core, and they gave off tremendous light and heat. Massive stars continued to contract, and many became dense enough to promote the formation of heavier elements.

All stars have a life history, from birth to an often spectacularly explosive death. In what might be called the original stardust memories, the heavier elements released during the explosions became swept up in the gravitational contraction of new stars, and they became raw materials for the formation of even heavier elements. Even as you read this page, the Hubble space telescope is revealing astounding glimpses of such star-forming activity, as in the dust clouds of Orion, Serpens, and other constellations (Figure 21.1).

Now imagine a time long ago, when explosions of dying stars ripped through our galaxy and left behind a dense cloud of dust and gas that extended trillions of kilometers in space. As the cloud cooled, countless bits of matter gravitated toward one another. By 4.6 billion years ago, the cloud had flattened out into a slowly rotating disk. At the dense, hot center of that great disk, the shining star of our solar system—the sun— was born.

The remainder of this chapter is a sweeping slice through time, one that cuts back to the formation of the Earth and the chemical origins of life. It is the starting point for the next seven chapters, which will take us along lines of descent that led to the present range of species diversity.

The story is not complete. Even so, the available evidence from many avenues of research points to a principle that can help us organize bits of information about the past: *Life is a magnificent continuation of the physical and chemical evolution of the universe, of galaxies and stars, and of the planet Earth.*

## KEY CONCEPTS

1. We have evidence that life originated about 3.8 billion years ago. The origin and subsequent evolution of life have been correlated with the physical and chemical evolution of the universe, the stars, and the planet Earth.

2. All inorganic and organic compounds necessary for self-replication, membrane assembly, and metabolism— for the structure and functions of living cells—could have formed spontaneously under conditions that apparently prevailed on the early Earth.

3. The history of life, from its chemical beginnings to the present, spans five intervals of geologic time. It extends through two great eons—the Archean and Proterozoic— and the Paleozoic, Mesozoic, and Cenozoic eras.

4. Not long after life originated, divergences led to two great prokaryotic lineages called the archaebacteria and eubacteria. Soon afterward, the ancestors of eukaryotes diverged from the archaebacterial lineage.

5. Archaebacteria and eubacteria dominated the Archean and Proterozoic eons. Eukaryotic cells originated late in the Proterozoic and became spectacularly diverse. A theory of endosymbiosis helps explain the profusion of specialized organelles that evolved in eukaryotic cells.

6. All six kingdoms of life have a history of persistences, extinctions, and radiations of different lineages.

7. Throughout the history of life, asteroid impacts, drifting and colliding continents, and other environmental insults have had profound impact on the direction of evolution.

## Origin of the Earth

Figure 21.1 gave you a view of one of the vast clouds in the universe. Such clouds consist mostly of hydrogen gas, along with water, iron, silicates, hydrogen cyanide, ammonia, methane, formaldehyde, and other simple inorganic and organic substances. The contracting cloud that became our solar system probably was similar in composition. We assume the edges of that cloud cooled between 4.6 and 4.5 billion years ago. Mineral grains and ice orbiting the new sun started clumping together as a result of electrostatic attraction and gravity's pull (Figure 21.2). In time, larger, faster clumps collided and shattered. Some became more massive by sweeping up asteroids, meteorites, and the other rocky remnants of collisions, and gradually they evolved into planets.

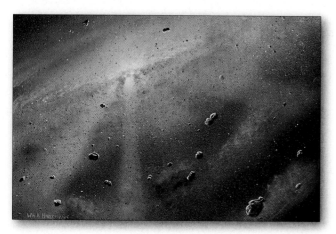

**Figure 21.2** Representation of the cloud of dust, gases, and clumps of rock and ice around the early sun.

As the Earth was forming, much of its inner rocky material melted. Asteroid impacts and the Earth's own internal compression and radioactive decay of minerals could have generated the heat necessary to do this. As rocks melted, nickel, iron, and other heavy materials moved to the Earth's interior; lighter ones floated to the surface. The process produced a crust, mantle, and core. The **crust** is an outer region of basalt, granite, and other low-density rocks. It envelops the intermediate-density rocks of the **mantle**, which wraps around a core of very high-density, partially molten nickel and iron.

Four billion years ago, the Earth was a thin-crusted inferno (Figure 21.3a). Within 200 million years, life had originated on its surface! We have no record of the event. As far as we know, movements in the mantle and crust, volcanic activity, and erosion obliterated all traces of it. Still, we can put together a plausible explanation of how life originated by considering three questions:

*First, can we identify physical and chemical conditions that prevailed on the Earth when life originated?*

*Second, do known physical, chemical, and evolutionary principles support the hypothesis that large organic molecules formed spontaneously and evolved into molecular systems displaying the fundamental properties of life?*

*Third, can we devise experiments to test the hypothesis that living systems emerged by chemical evolution?*

## The First Atmosphere

When the first patches of crust were forming, hot gases blanketed the Earth. Probably this first atmosphere was a mix of gaseous hydrogen ($H_2$), nitrogen ($N_2$), carbon monoxide (CO), and carbon dioxide ($CO_2$). Did it hold gaseous oxygen ($O_2$)? Probably not. Rocks subjected to intense heat, as happens during volcanic eruptions, do release oxygen, but not much. Also, free oxygen would have reacted at once with other elements. An oxygen atom has an electron vacancy in its outermost shell and tends to fill it by bonding with other atoms (Section 2.3).

If the early atmosphere had not been relatively free of oxygen, organic compounds necessary to assemble cells in the first place would not have been able to form spontaneously—*on their own*. Any oxygen would have attacked their structure and disrupted their functioning.

What about liquid water? Dense clouds blanketed the early Earth, but water reaching the molten surface would have evaporated at once. After the crust cooled and solidified, however, rains fell on the parched rocks. For millions of years, the runoff eroded mineral salts and other compounds from the rocks. Salt-laden waters collected in the depressions in the crust and formed the first seas. If liquid water had not accumulated, then cell membranes—which take on their bilayer organization only in water—could not have formed. No membrane, no cell. Life at its most basic level *is* the cell, which has a capacity to survive and reproduce on its own.

## Synthesis of Organic Compounds

Reduce a cell to its lowest common denominator and all that remains are proteins, complex carbohydrates and lipids, and nucleic acids. Existing cells assemble these molecules from smaller organic compounds: the simple sugars, fatty acids, amino acids, and nucleotides. Energy from the environment drives the synthesis reactions. Were small organic compounds also present on the early Earth? Were there sources of energy that spontaneously drove their assembly into the large molecules of life?

Mars, meteorites, the Earth's moon, and the Earth all formed at the same time, from the same cosmic cloud. Rocks collected from Mars, meteorites, and the moon contain precursors of biological molecules, so the same precursors must have been present on the Earth. If this were the case, *then energy from the sun, from lightning, or*

**Figure 21.3** (**a**) Representation of the Earth during its formation, when the moon's orbit was much closer than it is today. If the Earth had condensed into a planet of smaller diameter, its gravitational mass would not have been great enough to hold on to an atmosphere. If it had settled into an orbit closer to the sun, water would have evaporated from its hot surface. If the Earth's orbit had been more distant from the sun, its surface would have been far colder, and water would have been locked up as ice. Without liquid water, life as we know it never would have originated on Earth.

(**b**) Stanley Miller's experimental apparatus, used to study the synthesis of organic compounds under conditions that presumably existed on the early Earth. The condenser cools circulating steam so that water droplets form.

*even from heat being vented from the crust could have been enough to drive their combination into organic molecules.*

Stanley Miller conducted the first experimental test of that prediction. First he mixed methane, hydrogen, ammonia, and water inside a reaction chamber of the sort depicted in Figure 21.3*b*. Then he kept the mixture circulating and bombarded it with a spark discharge to simulate lightning. In less than one week, amino acids and other small organic compounds had formed.

In other experiments that simulated conditions on the early Earth, glucose, ribose, deoxyribose, and other sugars formed spontaneously from formaldehyde, and adenine from hydrogen cyanide. Ribose and adenine occur in ATP, $NAD^+$, and other nucleotides vital to cells.

However, if *complex* organic compounds had formed directly in seawater, could they have lasted long? The spontaneous direction of the required reactions would have been toward hydrolysis, not condensation, in water. How did more lasting bonds form?

By one hypothesis, clay in the rhythmically drained muck of tidal flats and estuaries served as templates (structural patterns) for the spontaneous assembly of proteins and other complex organic compounds. Clay consists of thin, stacked layers of aluminosilicates with metal ions at their surfaces, which attract amino acids. Expose amino acids to some clay, warm the clay with rays from the sun, then alternately moisten and dry it. Condensation reactions will proceed at its surfaces and yield proteins and other complex organic compounds.

By another hypothesis, complex organic compounds formed spontaneously near hydrothermal vents on the seafloor, where species of archaebacteria are thriving today. As experimental tests by Sidney Fox and others show, when amino acids are heated and then placed in water, they spontaneously order themselves into small protein-like molecules, which Fox calls "proteinoids."

However the first proteins formed, their molecular structure dictated how they could interact with other compounds. Suppose some proteins had the structure to function as weak enzymes, and that they hastened bond formation between amino acids. Such enzyme-directed synthesis of proteins would have had selective advantage. In the chemical competition for available amino acids, protein configurations that could promote reactions would win. Proteins would have been favored in still another way—for proteins have the capacity to bind metal ions and other agents of metabolism.

As you will read next, the evolution of metabolism must have been based on such chemical modification. For now, simply reflect on the possibility that selection was at work before the origin of living cells, favoring the chemical evolution of enzymes and other complex organic compounds.

---

**Many diverse experiments yield indirect evidence that the complex organic molecules characteristic of life could have formed under conditions that existed on the early Earth.**

---

# EMERGENCE OF THE FIRST LIVING CELLS

## Origin of Agents of Metabolism

A defining characteristic of life is *metabolism*. The word refers to all the reactions by which cells harness energy and use it to drive their activities, such as biosynthesis. During the first 600 million years or so of Earth history, enzymes, ATP, and other organic compounds may have assembled spontaneously, perhaps in the same physical locations. If they did so, their close association would have naturally promoted chemical interactions and the beginning of metabolic pathways.

Imagine an ancient estuary, rich in clay deposits. It is a coastal region where seawater mixes with mineral-rich water being drained from the land. There, beneath the sun's rays, countless aggregations of organic molecules stick to the clay. At first there are quantities of an amino acid; call it $D$. Throughout the estuary, $D$ molecules get incorporated into new proteins—until the supply of $D$ dwindles. However, suppose a protein that is weakly catalytic also is present in the estuary. It can promote the formation of $D$ by acting on an abundant, simpler substance—call it $C$. By chance, some aggregations of organic molecules include that particular enzyme-like protein, and so they have an advantage in the chemical competition for starting materials.

In time, $C$ molecules become scarce. At that point, the advantage tilts to aggregations that can promote formation of $C$ from even simpler substances $B$ and $A$. Assume $B$ and $A$ are carbon dioxide and water. As you know, carbon dioxide and water occur in essentially unlimited amounts in the atmosphere and in the seas. Chemical selection has favored a synthetic pathway:

$$A + B \longrightarrow C \longrightarrow D$$

Finally, suppose some aggregations are better than others at absorbing and using energy. Which molecules could bestow such an advantage? Think of the energy-trapping pathway that now dominates the world of life: photosynthesis. It starts at pigments called chlorophylls. The portion of a chlorophyll molecule that absorbs light and gives up electrons is a porphyrin ring structure. Porphyrins also are present in cytochromes, which are part of electron transport systems in all photosynthetic and aerobically respiring cells. Porphyrin molecules can spontaneously assemble from formaldehyde—one of the molecular legacies of cosmic clouds (Figure 21.4). Was porphyrin a major electron transporter of some of the early metabolic pathways? Perhaps.

## Origin of Self-Replicating Systems

Another defining characteristic of life is a capacity for reproduction, which now starts with protein-building instructions in DNA. The DNA molecule is fairly stable,

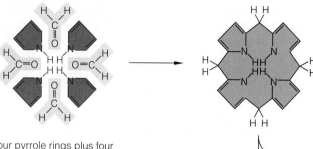

four pyrrole rings plus four
formaldehyde molecules

6H

porphyrin ring system

**Figure 21.4** One hypothetical sequence by which formaldehyde, an organic compound, underwent chemical evolution into porphyrin. Formaldehyde was present when the Earth formed. Porphyrin is the light-trapping and electron-donating component of all existing chlorophyll molecules. It also is a component of cytochrome, which is a protein component of electron transport systems that are part of many metabolic pathways.

chlorophyll *a*, one of the
light-trapping pigments of
photosynthetic plant cells

and it is easily replicated before each cell division. As you know from earlier chapters, arrays of enzymes and RNA molecules operate together and carry out DNA's encoded instructions.

Most existing enzymes get assistance from small organic molecules or metal ions called coenzymes. Intriguingly, some categories of coenzymes have a structure identical to that of RNA nucleotides. Another clue: Mix together and then heat up precursors of ribonucleotides and short chains of phosphate groups, and they self-assemble into single strands of RNA. On the early Earth, energy from the sun's rays or from geothermal events could have driven the spontaneous formation of RNA from such starting materials.

Very simple self-replicating systems of RNA, enzymes, and coenzymes have been created in some laboratories. Did RNA later become the information-

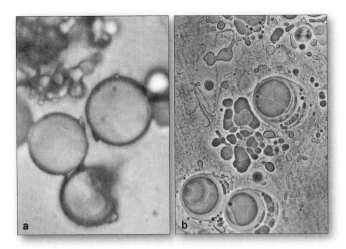

**Figure 21.5** Microscopic spheres of (**a**) proteins and (**b**) lipids that self-assembled under abiotic conditions.

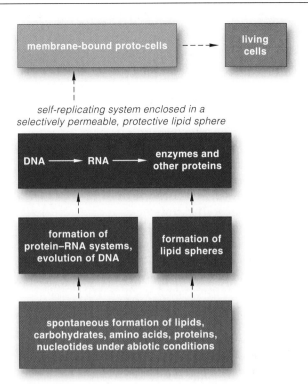

**Figure 21.6** One possible sequence of events that led to the first self-replicating systems, then to the first living cells.

storing templates upon which simple proteins could be synthesized? Perhaps RNA did so initially. We can only speculate at present, because the RNA molecules that we know about are too chemically fragile to serve as templates for protein synthesis.

Yet the use of RNA templates might have set up an **RNA world** that preceded DNA's dominance as the main informational molecule. Whatever the case, DNA eventually assumed this function, probably because it can form long nucleotide chains in more stable fashion.

We still don't know how DNA entered the picture. Until we identify the likely chemical ancestors of RNA and DNA, the story of life's origin will be incomplete. Filling in the details will require imaginative sleuthing. For instance, researchers ran a computer program that incorporated information about natural energy sources and simple inorganic compounds of the sort thought to have been present on the early Earth. They asked their advanced computer to subject the chosen compounds to random chemical competition and natural selection as might have occurred untold billions of times in the past. They ran the program repeatedly. And always the outcome was the same: Simple precursors inevitably evolved—and they organized themselves as interacting systems of large, complex molecules.

## Origin of the First Plasma Membranes

Experimental tests are more revealing of the origin of the plasma membrane—the outermost membrane of all living cells. This cellular component consists of a lipid bilayer, studded with proteins that carry out diverse functions. Its main role is to control which substances move into and out of the cell. Without control, cells can neither exist nor reproduce.

Probably, one avenue of molecular evolution led to **proto-cells**: simple membrane sacs that surrounded and

protected information-storing templates and metabolic agents from the environment. We do know that simple membrane sacs can form spontaneously. For example, for one experiment, Fox heated amino acids until they formed protein-like chains, which he then placed in hot water. After cooling, the chains assembled into small, stable spheres (Figure 21.5*a*). Like membranes of cells, these proteinoid spheres were selectively permeable to various substances. In other experiments, the spheres picked up free lipid molecules from their surroundings, and a lipid–protein film formed at their surface.

In still other experiments, by David Deamer and his coworkers, fatty acids and glycerol combined to form long-tail lipid molecules under laboratory conditions that simulated evaporating tidepools. The lipids self-assembled into small, water-filled sacs. Many were like cell membranes (Figure 21.5*b*).

In short, there are major gaps in the story of life's origins. But there also is strong experimental evidence that chemical evolution probably led to the molecules and structures that are characteristic of life. Figure 21.6 summarizes the milestones in that chemical evolution, which preceded the first cells.

---

**Although the story is not yet complete, many laboratory experiments and computer simulations indirectly show that chemical and molecular evolution could have given rise to proto-cells.**

---

# ORIGIN OF PROKARYOTIC AND EUKARYOTIC CELLS

The first living cells originated in the **Archean** eon, which lasted from 3.9 billion to 2.5 billion years ago. Those cells emerged as molecular extensions of the evolving universe, of our solar system and the Earth. Maybe they originated in tidal flats or seafloor sediments (Section 7.8). Fossils indicate they were like the existing bacteria. Specifically, they were **prokaryotic cells**, without a nucleus. They may have been little more than membrane-bound, self-replicating sacs of DNA and other complex organic molecules. Given the absence of free oxygen, they must have secured energy through anaerobic pathways—fermentation, most likely. Energy was plentiful. Geologic processes already enriched the seas with organic compounds. So "food" was available, predators were absent, and cellular structures were free from oxygen attacks.

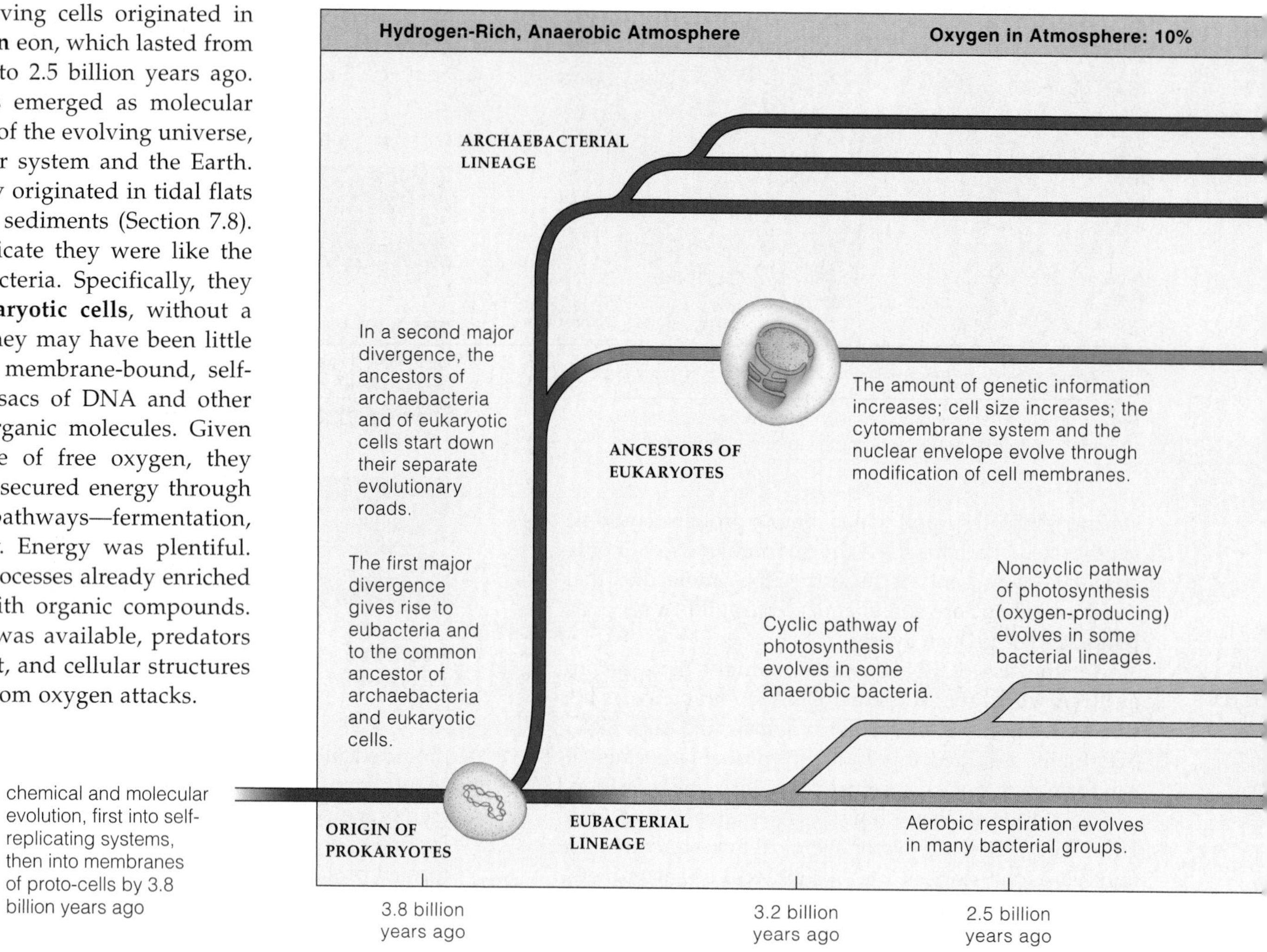

**Hydrogen-Rich, Anaerobic Atmosphere**    **Oxygen in Atmosphere: 10%**

ARCHAEBACTERIAL LINEAGE

In a second major divergence, the ancestors of archaebacteria and of eukaryotic cells start down their separate evolutionary roads.

ANCESTORS OF EUKARYOTES

The amount of genetic information increases; cell size increases; the cytomembrane system and the nuclear envelope evolve through modification of cell membranes.

The first major divergence gives rise to eubacteria and to the common ancestor of archaebacteria and eukaryotic cells.

Cyclic pathway of photosynthesis evolves in some anaerobic bacteria.

Noncyclic pathway of photosynthesis (oxygen-producing) evolves in some bacterial lineages.

chemical and molecular evolution, first into self-replicating systems, then into membranes of proto-cells by 3.8 billion years ago

ORIGIN OF PROKARYOTES

EUBACTERIAL LINEAGE

Aerobic respiration evolves in many bacterial groups.

3.8 billion years ago    3.2 billion years ago    2.5 billion years ago

Some populations of those first prokaryotic cells apparently diverged in two major directions shortly after the time of origin. One lineage gave rise to the **eubacteria**. The other lineage gave rise to the common ancestor of **archaebacteria** and **eukaryotic cells** (Figure 21.7).

Between 3.5 and 3.2 billion years ago, light-trapping pigments, electron transport systems, and other metabolic machinery evolved in some anaerobic eubacteria. With these innovations, the cells became the earliest practitioners of the cyclic pathway of photosynthesis. (You read about this ATP-forming pathway in Section 7.4). Sunlight, an unlimited source of energy, had been tapped. For nearly 2 billion years, photosynthetic descendants of those cells dominated the living world. Their tiny but numerous populations formed very large mats in which sediments had collected. The mats built up, one above the other. Calcium deposits hardened and preserved the mats, which came to be called **stromatolites** (Figure 21.8).

**Figure 21.7** An evolutionary tree of life that reflects mainstream thinking about the connections among major lineages. The diagram incorporates ideas about the origins of some eukaryotic organelles.

**Figure 21.8** Stromatolites exposed at low tide in western Australia's Shark Bay. These mounds started forming 2,000 years ago. Structurally, they are identical with stromatolites that formed more than 3 billion years ago.

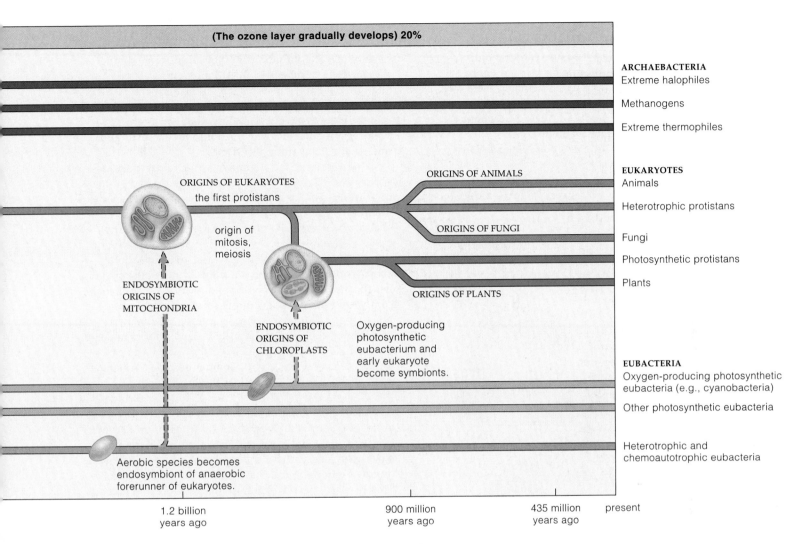

**(The ozone layer gradually develops) 20%**

ARCHAEBACTERIA
Extreme halophiles

Methanogens

Extreme thermophiles

ORIGINS OF EUKARYOTES
the first protistans

ORIGINS OF ANIMALS

EUKARYOTES
Animals

Heterotrophic protistans

origin of
mitosis,
meiosis

ORIGINS OF FUNGI

Fungi

Photosynthetic protistans

Plants

ENDOSYMBIOTIC
ORIGINS OF
MITOCHONDRIA

ORIGINS OF PLANTS

ENDOSYMBIOTIC
ORIGINS OF
CHLOROPLASTS

Oxygen-producing
photosynthetic
eubacterium and
early eukaryote
become symbionts.

EUBACTERIA
Oxygen-producing photosynthetic
eubacteria (e.g., cyanobacteria)

Other photosynthetic eubacteria

Heterotrophic and
chemoautotrophic eubacteria

Aerobic species becomes
endosymbiont of anaerobic
forerunner of eukaryotes.

1.2 billion
years ago

900 million
years ago

435 million
years ago

present

By the dawn of the **Proterozoic** eon, 2.5 billion years ago, photosynthetic machinery had become altered in some eubacterial species, and the noncyclic pathway of photosynthesis emerged. Oxygen, one of the pathway's by-products, started to accumulate. In time, this had two irreversible effects. First, *an oxygen-rich atmosphere stopped the further chemical origin of living cells.* Except in a few anaerobic habitats, complex organic compounds could no longer form spontaneously and resist attack. Second, *aerobic respiration became the dominant energy-releasing pathway.* In many prokaryotic lineages, selection favored metabolic equipment that "neutralized" oxygen by using it as an electron acceptor! This key innovation contributed to the rise of multicelled eukaryotes and their invasion of far-flung environments (Section 8.7).

Eukaryotic cells evolved in the Proterozoic, possibly before 1.2 billion years ago. We have fossils, 900 million years old, of complex algae, fungi, and plant spores. As you know, organelles are the premier defining features of eukaryotic cells. *Where did they come from?* The next section describes a few plausible hypotheses.

By about 800 million years ago, stromatolites along the shores of Laurentia, an early supercontinent, were declining dramatically. Perhaps newly evolved, bacteria-eating animals were using them as a concentrated food source. Fossilized embryos found in China (clusters of cells no wider than a pin) give evidence that the first animals were soft-bodied forerunners of today's sponges and marine worms. Before 570 million years ago, in "Precambrian" times, some of their small descendants started the first adaptive radiation of animals.

---

**The first cells evolved by about 3.8 billion years ago, during the Archean. All were prokaryotic cells, and most, if not all, probably made ATP by fermentation routes.**

**Early on, ancestors of archaebacteria and eukaryotic cells diverged from the lineage that led to modern eubacteria.**

**Oxygen-releasing photosynthetic bacteria evolved. In time, the oxygen-enriched atmosphere put an end to the further spontaneous chemical evolution of life. That atmosphere was a key selection pressure in the evolution of eukaryotic cells.**

---

## 21.4 WHERE DID ORGANELLES COME FROM?

Thanks to Andrew Knoll, William Schopf, Jr., and other globe-hopping microfossil hunters, we have tantalizing evidence of early life. One fossil treasure is a strand of bacterial cells that lived 3.5 billion years ago, not long after life originated. Others, from the Proterozoic, were eukaryotic cells with a few membrane-bound organelles in the cytoplasm (Figure 21.9). Most of their descendants now have a profusion of organelles (Figure 21.10).

*Where did eukaryotic organelles come from?* Speculations abound. Some organelles probably evolved through gene mutations and natural selection. For others, researchers make a good case for evolution by way of endosymbiosis.

**ORIGIN OF THE NUCLEUS AND ER** Prokaryotic cells do not have an abundance of organelles, but some species have interesting infoldings of their plasma membrane (Figure 21.11). Embedded in that membrane are enzymes and other agents of metabolism. In the early forerunners of eukaryotic cells, similar infoldings may have extended into the cytoplasm and served as passageways to the surface. They may have evolved into ER channels and into an envelope around the DNA.

What would be the advantage of such membranous enclosures? Maybe they protected the genes and protein products from "foreigners." Remember, bacterial species often transfer plasmid DNA among themselves. So do yeasts, which are very simple eukaryotic cells. At first, a nuclear envelope may have been favored because it got the cell's genes, replication enzymes, and transcription enzymes out of the cytoplasm. It would have allowed vital genetic messages to be copied and read, free of metabolic competition from what could become an unmanageable hodgepodge of foreign genes. Similarly, ER channels might have kept important proteins and other organic compounds away from metabolically hungry "guests"— foreign cells that one way or another became permanent residents inside the host cell, as described next.

**A THEORY OF ENDOSYMBIOSIS** It appears likely that accidental partnerships between a variety of prokaryotic species arose countless times on the evolutionary road to eukaryotic cells. Perhaps some partnerships resulted in the origin of mitochondria, chloroplasts, and other organelles. This is a story of endosymbiosis, as developed in greatest

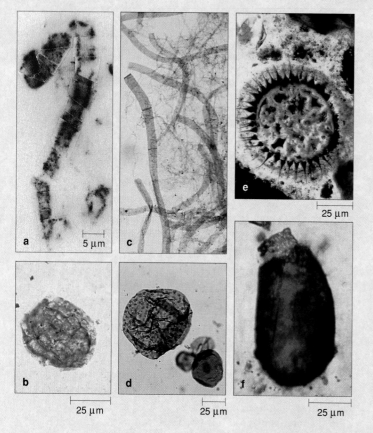

**Figure 21.9** From Australia, (**a**) a strand of walled prokaryotic cells 3.5 billion years old and (**b**) one of the oldest known eukaryotes— a protistan 1.4 billion years old. From Siberia, (**c**) a multicelled alga 900 million to 1 billion years old and (**d**) eukaryotic microplankton 850 million years old. (**e**) From China, a eukaryotic cell that lived 560 million years ago. (**f**) From Spitsbergen, Norway, a protistan that was alive 50 million years before the dawn of the Cambrian.

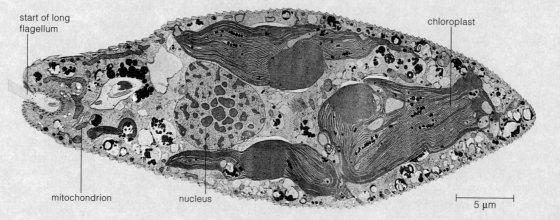

**Figure 21.10** A fine example of the profusion of diverse organelles that are hallmarks of eukaryotic cells: *Euglena*, a single-celled protistan, sliced lengthwise for this micrograph. It also has a long flagellum, which could not fit in the image area at this magnification.

Further reading: Student Guide to InfoTrac on web site →

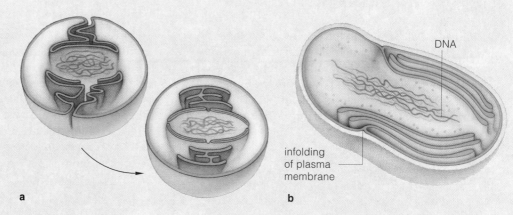

**Figure 21.11** (**a**) Sketch of one idea concerning the origin of endoplasmic reticulum and the nuclear envelope. In prokaryotic ancestors of eukaryotic cells, infoldings of the plasma membrane might have given rise to both cell components. (**b**) Such infoldings are present in the cytoplasm of many kinds of existing bacteria, including *Nitrobacter*, sketched here in cutaway view.

a

DNA

infolding of plasma membrane

b

detail by Lynn Margulis. *Endo-* means within; *symbiosis* means living together. In cases of **endosymbiosis**, one species (the guest) lives permanently inside another species (the host), and the interaction benefits both.

According to one theory, eukaryotic cells evolved by endosymbiosis long after the noncyclic pathway of photosynthesis emerged and oxygen had accumulated to significant levels in the atmosphere. In some bacterial groups, certain electron transport systems in the plasma membrane had become modified and included "extra" cytochromes. The cytochromes could donate electrons to oxygen. Thus the bacteria could extract energy from organic compounds by aerobic respiration.

By 1.2 billion years ago and possibly much earlier, the forerunners of eukaryotes were engulfing aerobic bacteria. Maybe they were like existing soft-bodied, amoebalike cells that weakly tolerate free oxygen. They would have entrapped food by sending out cytoplasmic extensions. Endocytic vesicles could have formed around food and could have delivered it to the cytoplasm for digestion.

A key point of the theory is that some aerobic bacteria resisted digestion and thrived in the protected, nutrient-rich environment. In time, they were releasing extra ATP, which their hosts came to depend on for their growth, increased activity, and assembly of hard parts and other structures. How did the guests benefit? They no longer had to seek food or duplicate the metabolic functions that hosts performed for them. The anaerobic and aerobic cells were now incapable of independent life. The guests had become mitochondria, supreme suppliers of ATP.

**EVIDENCE OF ENDOSYMBIOSIS** Strong evidence favors Margulis's theory. There are plenty of examples of nature continuing to tinker with endosymbionts, including the cell in Figure 21.12. Its mitochondria are like bacteria in size and structure. The inner mitochondrial membrane is like a bacterial plasma membrane. Each mitochondrion replicates its own DNA and divides independently of the host cell's division. A few genetic code words in its DNA and mRNA have unique meanings. They are translated into a few proteins required for special mitochondrial tasks. Compared to the genetic code of living cells, the "mitochondrial code" has a few distinct differences.

Chloroplasts, too, may be stripped-down descendants of oxygen-evolving, photosynthetic bacteria. Perhaps predatory aerobic bacteria engulfed such photosynthetic cells, which escaped digestion, absorbed nutrients from their host's cytoplasm, and continued to function. By providing aerobically respiring hosts with oxygen, they would have promoted their endosymbiotic existence.

In their metabolism and overall nucleic acid sequence, chloroplasts resemble some eubacteria. Their DNA is self-replicating, and they divide independently of the cell's division. Chloroplasts vary in their shape and array of light-absorbing pigments, just as photosynthetic eubacteria do. They might have originated a number of times in a number of different lineages.

Adding to the intrigue are ciliated protistans and marine slugs that "enslave" chloroplasts! The slugs eat algae but retain algal chloroplasts in their gut cells. The chloroplasts draw nutrients from the host tissues, and they continue to photosynthesize and release oxygen for weeks.

However they arose, new kinds of cells did appear on the evolutionary stage. They had become equipped with a nucleus, cytomembranes, and mitochondria, chloroplasts, or both. They were eukaryotic cells, the first **protistans**. With their efficient metabolic strategies, early protistans underwent rapid divergences and adaptive radiations. In no time at all, evolutionarily speaking, some of their descendants gave rise to the great kingdoms of plants, fungi, and animals, as sketched out earlier in Figure 21.7.

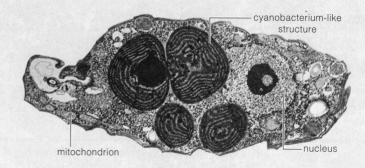

cyanobacterium-like structure

mitochondrion

nucleus

**Figure 21.12** *Cyanophora paradoxa*, a protistan. Its mitochondria resemble bacteria. Its photosynthetic structures look like spherical cyanobacteria (which are photosynthetic) without the cell wall.

We divide the **Paleozoic** into the Cambrian, Ordovician, Silurian, Devonian, Carboniferous, and Permian periods. Before the dawn of that era, gradual rifting had split the supercontinent Laurentia apart. From the Cambrian on into the Silurian, the great fragments straddled the equator, and warm, shallow seas lapped their margins:

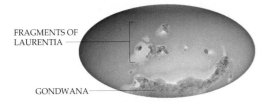

FRAGMENTS OF LAURENTIA

GONDWANA

The global conditions restricted pronounced seasonal changes in prevailing winds, ocean currents, and the upward churning of nutrient deposits on the seafloor. As a result, supplies of nutrients along shorelines at or near the equator were stable but limited.

Most major animal phyla had evolved earlier, in the Precambrian seas. Possibly some of their ancestors were among the **Ediacarans**, peculiar organisms shaped like fronds, disks, and blobs that nearly defy classification (Figure 21.13*a,b*). Like Ediacarans, the early Cambrian animals had flattened bodies, with a good surface-to-volume ratio for taking up nutrients (Figure 21.13*c*). Most existed on or in seafloor sediments, where dead organisms and organic debris settled. They ranged from sponges to simple vertebrates, and they were diverse.

How could so much diversity arise? Possibly genes governing early growth and development were far less intertwined than they are at present, so there might not have been as much selection against mutant alleles and novel traits. Also, warm, shallow waters near the new coasts were vacant adaptive zones with many enticing opportunities for new ways to secure food.

Entombed in sedimentary beds from the Cambrian are fossilized organisms that have punctures, missing chunks, and healed wounds. These are not artifacts of fossilization; the animals were injured while they were alive. Diverse predators and prey with armorlike shells, spines, mouths, and novel feeding structures evolved in short order. Things were starting to get lively!

Later in the Cambrian, temperatures in the shallow seas changed drastically. Trilobites (Figure 21.13*d*), one of the most common invertebrates, almost vanished. In the Ordovician, the supercontinent Gondwana had been drifting south, and parts became submerged in shallow seas. The emergence of vast new marine environments promoted adaptive radiations. Many new reef organisms evolved. Among them were the swift, shelled predators called nautiloids. Their surviving descendants include the chambered nautiluses (Section 26.13).

Later on, Gondwana straddled the South Pole, and huge glaciers formed across its surface. When enormous volumes of water were locked up as ice, shallow seas throughout the world were drained. This was the first ice age that we know about. It may have triggered the first global mass extinction. At the Ordovician–Silurian boundary, reef life everywhere collapsed.

Gondwana drifted north during the Silurian and on into the Devonian. This was a pivotal time of evolution. Reef communities recovered. Armor-plated fishes with massive jaws diversified. And in the wet lowlands, small stalked plants appeared (Figure 21.14*b,c*). So did fungi and many invertebrates, such as segmented worms. In

**Figure 21.13** Representatives from the Precambrian and Cambrian seas. Two Ediacarans, about 600 million years old: (**a**) *Spriggina* and (**b**) *Dickensonia*. The oldest known Ediacarans lived 610 million years ago and the most recent in Cambrian times, 510 million years ago. (**c**) From British Columbia's Burgess Shale, a fossilized marine worm. (**d**) A beautifully preserved fossil of one of the earliest trilobites.

**Figure 21.14** (**a**) Life in the Silurian seas. Some of these shelled animals (nautiloids) were twelve feet long. (**b**) Silurian swamp, dominated by forerunners of modern ferns and club mosses. (**c**) Fossils of a Devonian plant (*Psilophyton*), possibly one of the earliest ancestors of conifers and other seed-bearing plants. (**d**) Reptiles (*Dimetrodon*) of a mostly hotter and drier time—the Permian. Some fossils of these carnivores were found in Texas. Giant club mosses and horsetails had declined, and conifers, cycads, and other gymnosperms replaced them.

As the Permian drew to a close, something caused the greatest of all mass extinctions. More than 50 percent of all families disappeared, and only 5 percent of the known species survived. At the time, all land masses were colliding together. Eventually they formed Pangea, a supercontinent that extended all the way from the North to the South Pole. A single world ocean lapped its margins:

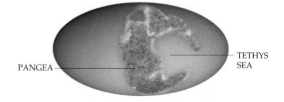

As you will see, the new distribution of oceans, land masses, and land elevations had catastrophic effects on global climates—and on the course of life's evolution.

Devonian times, fishes that would become ancestors to amphibians invaded the land. The fishes sported lobed fins, the forerunners of legs and other limbs. They had simple lungs. As described in Section 27.6, lobed fins and lungs were key innovations that would prove most advantageous for life out of water, in dry land habitats.

Then, as they say, something bad happened. At the boundary between the Devonian and Carboniferous, sea levels swung catastrophically for unknown reasons and caused another mass extinction. Afterward, plants and insects embarked on adaptive radiations on land.

Throughout Carboniferous times, land masses were gradually submerged and drained many times. Organic debris piled up. Then it was compacted and converted to coal, in the manner described in Section 25.4.

Insects, amphibians, and early reptiles flourished in vast swamp forests of Permian times (Figure 21.14*d*). Ancestors of seed-producing plants, including cycads, ginkgos, and conifers, dominated those forests.

Early in the Paleozoic era, organisms of all six kingdoms were flourishing in the seas. By the end of the era, many lineages had successfully invaded the land, including the wet lowlands of the supercontinent Pangea.

# LIFE IN THE MESOZOIC ERA

## Speciation on a Grand Scale

We divide the **Mesozoic** into the Triassic, Jurassic, and Cretaceous periods. It lasted about 175 million years. Early on in the Cretaceous, the supercontinent Pangea started to break up. Its huge fragments slowly drifted apart, and we can assume that the resulting geographic isolation favored divergences and speciation:

This was an era of spectacular expansion in the range of global diversity. In the seas, invertebrates and fishes underwent adaptive radiations. On land, conifers and other seed-bearing **gymnosperms,** insects, and reptiles became the most visibly dominant lineages. Flowering plants—the **angiosperms**—emerged in the late Jurassic or early Cretaceous. It would take them less than 40 million years to displace conifers and related plants in most environments (Figure 21.15 and Section 25.8).

## Rise of the Ruling Reptiles

Early in the Triassic, the first **dinosaurs** evolved from a reptilian lineage. They were not much bigger than a turkey. Possibly most species had high metabolic rates, and maybe they were warm-blooded. Many sprinted about on two legs. Dinosaurs weren't the dominant land animals. Instead, center stage belonged to *Lystrosaurus* and other plant-eating, mammal-like reptiles that were too large to be bothered by most predators.

Adaptive zones may have opened up for dinosaurs in a frightening way. There is a string of five craters in France, Quebec, Manitoba, and North Dakota; one is the size of Rhode Island. About 214 million years ago, according to radioisotope dating, fragments from an asteroid or a comet apparently fell one after another as the Earth spun beneath them, just as a string of huge fragments from comet Shoemaker–Levy 9 hit Jupiter in 1994. The resulting blast waves, global firestorm, lava flows, and earthquakes must have been stupendous. Most of the animals lucky enough to survive that time of mass extinction (as well as later ones) were smaller, had higher rates of metabolism, and were less vulnerable than others to drastic changes in outside temperatures.

Descendants of the surviving dinosaurs became the ruling reptiles; they endured for 140 million years. Some species reached monstrous proportions. Among them were the ultrasaurs, fifteen meters tall.

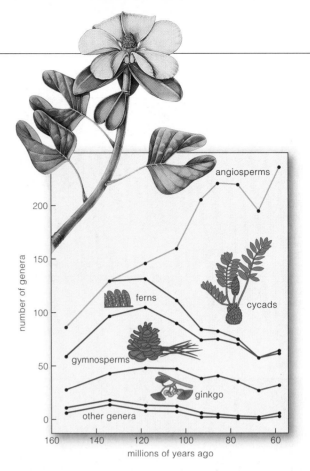

**Figure 21.15** Range of diversity among vascular plants of the Jurassic and Cretaceous. Conifers and other gymnosperms were dominant, then started to decline even before flowering plants began a great adaptive radiation that continued into the Cenozoic. *Upper left:* From the Cretaceous, a floral shoot of *Archaeanthus linnenbergeri*. In many traits, this now-extinct flowering plant resembled living magnolias.

Many dinosaurs perished in another mass extinction at the end of the Jurassic, then in a pulse of extinctions in the Cretaceous. Perhaps plumes of molten material ruptured the crust and triggered changes in the global climate. Perhaps asteroids or comets inflicted the blows. Not all lineages survived those episodes. Yet some did recover, and new ones evolved. Duckbilled dinosaurs appeared in forests and swamps. Tanklike *Triceratops* and other plant eaters flourished in open habitats. They were prey for the fearsomely toothed, agile, and swift *Velociraptor* of motion picture fame.

About 120 million years ago, global temperatures skyrocketed 25 degrees. By one theory, plumes spread out beneath the crust and "greased" the crustal plates into moving twice as fast. A superplume or a rash of them broke through the crust. The crust in what is now the South Atlantic opened like a zipper. Basalt and lava poured from the fissures; volcanoes spewed nutrient-rich ashes. Simultaneously, the plumes released great amounts of carbon dioxide, one of the "greenhouse" gases that absorb some of the heat radiating from the Earth before it escapes into outer space. The nutrient-enriched planet warmed—and remained warm for 20

Further reading: Student Guide to InfoTrac on web site →

**Figure 21.16** If dinosaurs of this sort had not disappeared at the end of the Cretaceous, would then-tiny mammals ever have ventured out from under the shrubbery? Would *you* even be here today?

million years. Photosynthetic organisms flourished on land and in shallow seas. Their remains were slowly buried and converted into the world's vast oil reserves.

About 65 million years ago, the last dinosaurs and many marine organisms vanished in a mass extinction (Figure 21.16). As described in the next section, their disappearance apparently coincided with a direct hit by an asteroid the size of Mount Everest. Over time, the impact site slowly drifted into a position that we now call the northern Yucatán peninsula.

---

**The Mesozoic was a time of major adaptive radiations and of a mass extinction in which the last dinosaurs and many marine organisms disappeared.**

---

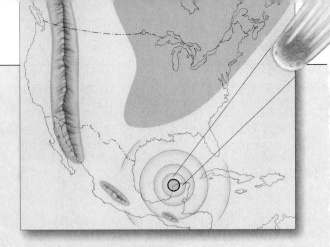

# HORRENDOUS END TO DOMINANCE

It has only been about 55,000 years since the first modern humans walked the Earth. Since then, how many times have people, puffed up with self-importance, set out to conquer neighbors, lands, and seas? Think about it. Then think about the dinosaurs. Were they good at reigning supreme? No question about it; their lineage dominated the land for 140 million years. In the end, did it matter? Not a bit. Sixty-five million years ago, at the Cretaceous–Tertiary (K-T) boundary, nearly all remaining members of their most excellent lineage perished. Why? Bad luck.

A thin layer of iridium-rich rock distributed around the world dates precisely to the K-T boundary. Iridium is rare here but common in asteroids. The **asteroids** are rocky, metallic bodies, 1,000 kilometers to a few meters in diameter, that are hurtling through space. When planets of our solar system were forming, their gravitational pull swept most asteroids from the sky. At least 6,000 still orbit the sun in a belt between Mars and Jupiter (Figure 21.17). The orbits of dozens of others take them across Earth's orbit, like Russian roulette on a cosmic scale.

By analyzing iridium levels in soils, gravity maps, and other evidence, Walter Alvarez and Luis Alvarez hypothesized that an asteroid impact caused the K-T mass extinction. Later, researchers identified the impact site. Massive movements in the crust transported the site to what is now the northern Yucatán peninsula of Mexico (Figure 21.18). The impact crater is 9.6 kilometers deep and 300 kilometers across—wider than Connecticut. This crater as well as other evidence strongly supports what is now called the **K-T asteroid impact theory**.

To make a crater that big, the asteroid had to hit the Earth at 160,000 kilometers (100,000 miles) per hour. At least 200,000 cubic kilometers of debris and dense gases were blasted skyward. The crust itself heaved violently. Monstrous waves, 120 meters high, raced across the ocean, obliterating life on islands and then slamming into coasts of continents. Researchers had hypothesized

**Figure 21.18** Artist's interpretation of what might have happened in the last few minutes of the Cretaceous.

that atmospheric debris blocked out sunlight for months. If so, plants and other producers that sustained the web of life would have withered and died; animals and other consumers would have starved to death. The hypothesis had problems. By some calculations, the volume of debris blasted aloft would not have been enough to have had such significant consequences.

Then comet Shoemaker–Levy 9 slammed into Jupiter. Particles blasted into the Jovian atmosphere triggered an intense heating of an area larger than the Earth. That event supports a **global broiling hypothesis**, proposed first by H. J. Melosh and his colleagues. Briefly stated, energy released at the K-T impact site was equivalent to detonating 100 million nuclear bombs. Trillions of tons of vaporized debris rose in a colossal fireball, then rapidly condensed into particles the size of sand grains. Seconds later, a cooler fireball made of steam, carbon dioxide, and unmelted rock formed. When the debris fell to the Earth, it raised the atmosphere's temperature by thousands of degrees. The sky above the planet must have glowed with heat ten times more intense than the noonday sun above Death Valley in summer. In one horrific hour, nearly all plants erupted in flames and every animal out in the open was broiled alive.

Things haven't settled down much since dinosaurs disappeared. For instance, about 2.3 million years ago, a huge chunk from outer space apparently hit the Pacific Ocean. At about the same time, vast ice sheets started forming abruptly in the Northern Hemisphere. Long-term shifts in climate may have been ushering in this most recent ice age, but the global impact would have accelerated it. Water vaporized by the impact could have contributed to a global cloud cover that limited the amount of sunlight reaching the surface. Ancestors of humans were around when all this happened. The extreme climate shift surely tested their adaptability.

Almost certainly, severe environmental tests await all existing lineages. When we become too smitten with our importance in the world of life, we would do well to step back, from time to time, and reflect on what is going on above and beneath the Earth's surface. Asteroids and superplumes do have a way of leveling the playing field, as they did for gigantic dinosaurs and tiny mammals.

**Figure 21.17** What one of the asteroids looks like, courtesy of the *Galileo* spacecraft that flew past it on the way to Jupiter. This is only a small asteroid; it would extend halfway between Baltimore and Washington, D.C.

Further reading: Student Guide to InfoTrac on web site →

# LIFE IN THE CENOZOIC ERA

The breakup of Pangea triggered events that continued into the present era, the **Cenozoic**. At the dawn of the Cenozoic, land masses were on collision courses:

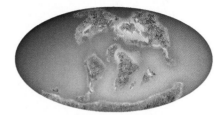

Coastlines fractured. The Cascades, Andes, Himalayas, and Alps rose through volcanic activity, uplifting, and other events at crustal rifts and plate boundaries. These geologic changes brought about major shifts in climate that influenced the further evolution of life.

During the Paleocene epoch, climates were warmer and wetter. Tropical and subtropical forests extended farther north and south than they do today. Woodlands and forests spread even into polar regions. Although their key traits developed before the dinosaurs left the scene, mammals now began a major adaptive radiation.

The global climate warmed even more in the Eocene epoch, and subtropical forests extended north into the polar regions. A variety of mammals, including primates, bats, rhinos, hippos, elephants, horses, and assorted carnivores, emerged. By the late Eocene, climates became cooler and drier, and seasonal changes became pronounced. Woodlands and dry grasslands dominated vast tracts of land. Patterns of vegetational growth changed so much that many mammals were driven to extinction.

From the Oligocene through the Pliocene, an abundance of grazing and browsing animals flourished in the woodlands and grasslands. Among them were camels and the "giraffe rhinoceros," along with some fearsome carnivores that stalked them (Figure 21.19).

At present, the distribution of land masses favors biodiversity. The richest ecosystems are tropical forests of South America, Madagascar, and Southeast Asia, as well as the marine ecosystems of tropical archipelagos in the Pacific. Yet we are in the midst of what may be one of the greatest of all mass extinctions. About 50,000 years ago, nomadic humans followed migrating herds of wild animals around the Northern Hemisphere. By

**Figure 21.19** Examples of Cenozoic species. (**a**) In the early Paleocene, in what is now Wyoming, dense forests of sequoia, laurel, and other plants were home to diverse mammals. On the ground, raccoonlike *Chriacus* faces a tree-climbing rodent (*Ptilodus*). Higher up is *Peradectes*, a marsupial. (**b**) *Indricotherium*, the "giraffe rhinoceros." At 15 tons and 5.5 meters tall at the shoulder, it was the largest mammal known. It browsed in late Eocene and early Miocene woodlands. (**c**) Saber-tooth cats (*Smilodon*) and small horses from the Pleistocene. Fossils were recovered from pitch pools in Rancho LaBrea, California.

a few thousand years later, major groups of mammals were extinct. The pace of extinction has since accelerated as humans hunt for food, fur, feathers, or fun, and as they destroy habitats to clear land for farm animals or crops (Chapters 28 and 51).

---

**Major geologic changes during the Cenozoic triggered shifts in climate. The great adaptive radiation of mammals began, first in tropical forests, then in woodlands and grasslands.**

---

# SUMMARY OF EARTH AND LIFE HISTORY

We conclude this overview chapter with an illustration that correlates milestones in the evolution of life with the evolution of the Earth. As you study Figure 21.20, keep in mind that it is only a generalized summary. For example, it charts the five greatest mass extinctions yet there were numerous others in between. The illustration's diagram of the full range of biodiversity conveys the shrinking and expansion of species through time. It is a combined range for *all* of the major groups on land and in the seas. Remember, each major group has its own evolutionary history. Some of its member species may have persisted to the present; some or all may be extinct. Some of the groups may be represented by only one or a few species; others may be represented by hundreds or thousands or a million.

With these qualifications in mind, you are ready to turn to the next chapters in this unit. They will provide you with far richer detail of the history and the current range of biodiversity for all six kingdoms of life.

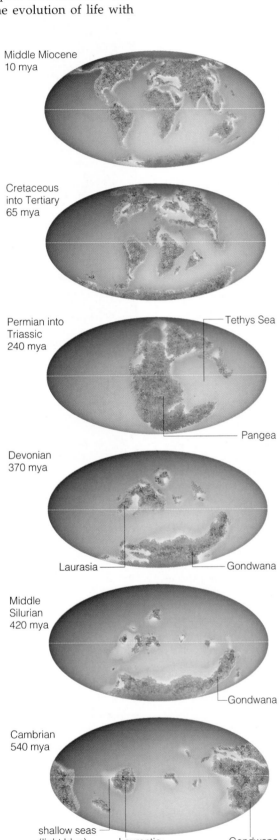

Middle Miocene
10 mya

Cretaceous
into Tertiary
65 mya

Permian into
Triassic
240 mya

Tethys Sea

Pangea

Devonian
370 mya

Laurasia — Gondwana

Middle
Silurian
420 mya

Gondwana

Cambrian
540 mya

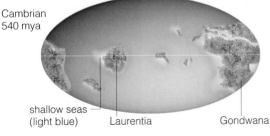

shallow seas
(light blue)   Laurentia   Gondwana

**Figure 21.20**  Summary of major events in the evolution of the Earth and of life. As you read through the chapters to follow, you may wish to return to this illustration now and then to remind yourself of how the details fit into the greater evolutionary picture.

| | Period | Epoch | |
|---|---|---|---|
| CENOZOIC ERA | Quaternary | Recent | |
| | | Pleistocene | |
| | Tertiary | Pliocene | |
| | | Miocene | |
| | | Oligocene | |
| | | Eocene | |
| | | Paleocene | |
| MESOZOIC ERA | Cretaceous | Late | |
| | | Early | |
| | Jurassic | | |
| | Triassic | | |
| PALEOZOIC ERA | Permian | | |
| | Carboniferous | | |
| | Devonian | | |
| | Silurian | | |
| | Ordovician | | |
| | Cambrian | | |
| PROTEROZOIC EON | | | |
| ARCHEAN EON AND EARLIER | | | |

| Millions of Years Ago (mya) | Range of Global Diversity (marine and terrestrial) | Times of Major Geological and Biological Events |
|---|---|---|

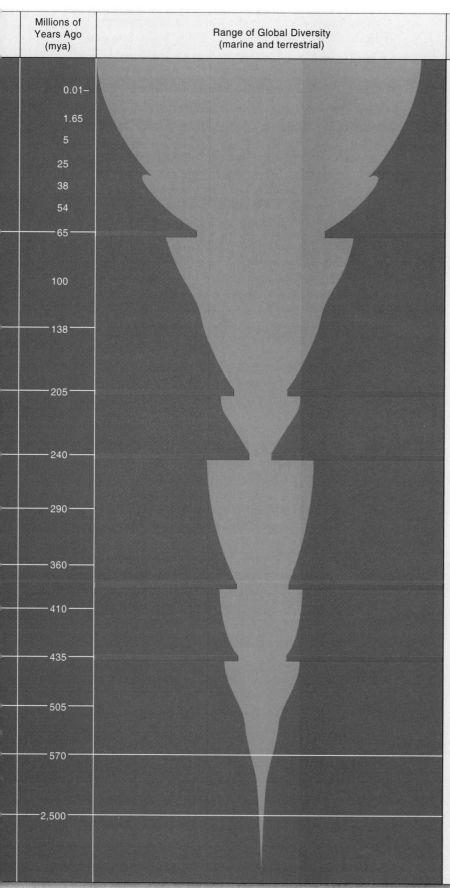

| | |
|---|---|
| 0.01– | |
| 1.65 | |
| 5 | |
| 25 | |
| 38 | |
| 54 | |
| 65 | |
| 100 | |
| 138 | |
| 205 | |
| 240 | |
| 290 | |
| 360 | |
| 410 | |
| 435 | |
| 505 | |
| 570 | |
| 2,500 | |

1.65 mya to present. Major glaciations. Modern humans evolve. Starting with hunters of most recent ice age, their activities set in motion the most recent **mass extinction**.

65–1.65 mya. Colossal mountain building as continents rupture, drift, collide. Major shifts in climate. First tropical and subtropical conditions extend to polar regions. Woodlands, then grasslands emerge as climates get cooler, drier. Major **radiations** of flowering plants, insects, birds, mammals. Origin of earliest human ancestors.

65 mya. Apparent asteroid impact causes **mass extinction** of all dinosaurs and many marine organisms.

135–65 mya. Pangea breakup continues, broad inland seas form. Major **radiations** of marine invertebrates, fishes, insects, and dinosaurs. Origin of angiosperms (flowering plants) late in the Jurassic or early in the Cretaceous.

181–135 mya. Pangea starts to break up. Diverse marine communities. Major **radiations** of dinosaurs.

205 mya. Asteroid impact? **Mass extinction** of many species in seas, some on land; some dinosaurs, mammals survive.

240–205 mya. Recoveries, **radiations** of marine invertebrates, fishes, dinosaurs. Gymnosperms the dominant land plants. Origin of mammals.

240 mya. **Mass extinction**. Nearly all species on land and in seas perish.

280–240 mya. Pangea and worldwide ocean form. Shallow seas squeezed out. **Radiations** of reptiles, gymnosperms.

360–280 mya. Tethys Sea forms. Recurring glaciations. Major **radiations** of insects, amphibians on land. Spore-bearing plants dominate; gymnosperms present. Origin of reptiles.

370 mya. **Mass extinction** of many marine invertebrates, most fishes.

435–360 mya. Laurasia forms. Glaciations as Gondwana crosses South Pole. **Mass extinctions** of many marine species. Gondwana drifts north. Vast swamps, first vascular plants. **Radiations** of fishes continue. Origin of amphibians.

500–435 mya. Gondwana drifts south. Major **radiations** of marine invertebrates and early fishes.

550–500 mya. Land masses dispersed near equator. Simple marine communities. Origin of animals with hard parts.

700–550 mya. Laurentia breaks up; widespread glaciations.

2,500–570 mya. Oxygen accumulating in atmosphere. Origin of aerobic metabolism. Origin of protistans, fungi, animals.

3,800–2,500 mya. Origin of photosynthetic bacteria.

4,600–3,800 mya. Formation of Earth's crust, first atmosphere, and seas. Chemical, molecular evolution leading to origin of life—proto-cells to anaerobic bacteria.

4,600 mya. Origin of early Earth.

## SUMMARY

1. The evolutionary story of life begins with the "big bang," a model for the origin of the universe.

   a. By this model, all matter and all of space were once compressed in a fleeting state of enormous heat and density. Time began with the near-instantaneous distribution of matter and energy through the known universe, which has been expanding ever since.

   b. Helium and the other light elements formed right after the big bang. Heavier elements originated during the formation, evolution, and death of stars.

   c. Every element of the solar system, the Earth, and life itself are products of the physical and chemical evolution of the universe and its stars.

2. Four billion years ago, the Earth was organized as a high-density core, a mantle of intermediate density, and a thin, extremely unstable crust of low-density rocks. Its first atmosphere probably contained gaseous hydrogen, nitrogen, carbon monoxide, and carbon dioxide. Free oxygen and liquid water could not have accumulated at the surface under the prevailing conditions.

3. Water accumulated after the crust cooled off. Over hundreds of millions of years, runoff from rains carried dissolved mineral salts and other compounds to crustal depressions, where the early seas formed. Life could not have originated without salty liquid water.

4. Many diverse studies and experiments have yielded indirect evidence that life originated under conditions that presumably existed on the early Earth.

   a. Comparative investigations of the composition of cosmic clouds, rocks from other planets, and rocks from the Earth's moon suggest that precursors of complex molecules associated with life were available.

   b. In laboratory tests that simulated the primordial conditions, including the absence of free oxygen, those precursors spontaneously assembled into sugars (such as glucose), amino acids, and other organic compounds.

   c. Known chemical principles as well as advanced computer simulations indicate that metabolic pathways could have evolved through chemical competition for limited supplies of organic molecules (which could have accumulated by natural geologic processes in the seas).

   d. Self-replicating systems of RNA, enzymes, and coenzymes have been synthesized in the laboratory. How DNA entered the picture is not yet understood.

   e. In laboratory simulations of conditions thought to have prevailed on the early Earth, lipids as well as lipid–protein membranes having some of the properties of cell membranes have formed spontaneously.

5. Since life originated some 3.8 billion years ago, it has been influenced by major changes in the Earth's crust, atmosphere, and oceans. Forces of change have included asteroid impacts, volcanism, and movements of crustal plates, which altered the distribution of land masses and oceans as well as global and local climates. Another force of change: the activities of organisms—especially the oxygen-releasing photosynthesizers and, currently, the human species.

6. Abrupt discontinuities in the fossil record mark the times of global mass extinctions. They are the boundary markers for five great intervals in a geologic time scale. Radiometric dating has allowed us to assign absolute dates to this time scale:

   a. Archean: 3.9 billion to 2.5 billion years ago
   b. Proterozoic: 2.5 billion to 570 million years ago
   c. Paleozoic: 570 to 240 million years ago
   d. Mesozoic: 240 to 65 million years ago
   e. Cenozoic: 65 million years ago to the present

7. The first living cells were prokaryotic (bacteria). Not long after they arose in the Archean, the first divergence led to eubacteria, and to a shared prokaryotic ancestor of both archaebacteria and eukaryotes. Some eubacteria used a cyclic pathway of photosynthesis.

8. During the Proterozoic, the noncyclic pathway of photosynthesis evolved in some lineages of eubacteria. Oxygen, a by-product of the pathway, slowly started to accumulate in the atmosphere.

   a. Eventually, the atmospheric concentration of free oxygen blocked the further spontaneous formation of organic molecules. From that time on, the spontaneous origin of life was no longer possible on Earth.

   b. The abundance of atmospheric oxygen became a selective pressure that brought about the evolution of aerobic respiration. Aerobic respiration was a key step toward the origin of the first eukaryotic cells.

   c. Mitochondria and chloroplasts, both important eukaryotic organelles, probably evolved as an outcome of endosymbiosis between certain aerobic bacteria and the anaerobic bacterial forerunners of eukaryotes.

   d. The oxygen-rich atmosphere promoted formation of a layer of ozone ($O_3$). In time, that atmospheric shield against destructive ultraviolet radiation allowed some lineages to move out of the seas, into low wetlands.

9. By the early Paleozoic, diverse organisms of all six lineages had become established in the seas. By its end, the invasion of land was under way. From that time on, there have been pulses of mass extinctions and adaptive radiations.

---

### Review Questions

1. Compare the presumed chemical and physical conditions that are thought to have prevailed on the Earth 4 billion years ago with conditions that are now prevailing. *21.1*

2. Describe examples of the kinds of experimental evidence for the spontaneous origin of (a) large organic molecules, (b) the self-assembly of proteins, and (c) the formation of organic membranes and spheres, under laboratory conditions similar to those of the early Earth. *21.1, 21.2*

3. Summarize the key points of the theory of endosymbiotic origins for mitochondria and chloroplasts. Cite evidence that favors this theory. *21.4*

4. Describe the prevailing conditions that probably favored the Cambrian "explosion" of diversity among marine animals, as evidenced by the fossil record. *21.5*

5. During which geologic time spans did plants, fungi, and insects invade the land? What kind of vertebrates first invaded the land, and when? *21.5*

6. What were global conditions like when gymnosperms and dinosaurs originated? *21.6*

7. Briefly explain how an asteroid impact and "global broiling" may have caused the mass extinctions at the K–T boundary. *21.7*

8. Would you expect the Paleozoic, Mesozoic, or Cenozoic to be called "the age of mammals"? As part of your answer, explain the differences between global conditions in each era. *21.8*

## Self-Quiz  (*Answers in Appendix III*)

1. Life originated by _____ .
   a. 4.6 billion years ago       c. 3.8 billion years ago
   b. 2.8 million years ago       d. 3.8 million years ago

2. Through study of the geologic record, we know that the evolution of life has been profoundly influenced by _____ .
   a. tectonic movements of the Earth's crust
   b. bombardment of the Earth by celestial objects
   c. profound shifts in land masses, shorelines, and oceans
   d. physical and chemical evolution of the Earth
   e. all of the above

3. _____ was the first to obtain indirect evidence that organic molecules could have been formed on the early Earth.
   a. Darwin            c. Fox
   b. Miller            d. Margulis

4. An abundance of _____ was conspicuously absent from the Earth's atmosphere 4 billion years ago.
   a. hydrogen          c. carbon monoxide
   b. nitrogen          d. free oxygen

5. Which of the following statements is false?
   a. The first living cells were prokaryotic.
   b. The cyclic pathway of photosynthesis first appeared in some eubacterial species.
   c. Oxygen began accumulating in the atmosphere after the noncyclic pathway of photosynthesis evolved.
   d. In the Proterozoic, increasing levels of atmospheric oxygen enhanced the spontaneous formation of organic molecules.
   e. All are correct.

6. The first eukaryotic cells emerged during the _____ .
   a. Paleozoic         c. Archean        e. Cenozoic
   b. Mesozoic          d. Proterozoic

7. Match the geologic time interval with the events listed.
   ____ Archean     a. major radiations of dinosaurs, origin
   ____ Proterozoic       of flowering plants and mammals
   ____ Paleozoic   b. chemical evolution, origin of life
   ____ Mesozoic    c. major radiations of flowering plants,
   ____ Cenozoic        insects, birds, mammals; emergence
                        of human forms
                    d. oxygen present; origin of aerobic
                        metabolism, protistans, fungi,
                        animals
                    e. rise of early plants on land, origin
                        of amphibians, and origin of reptiles

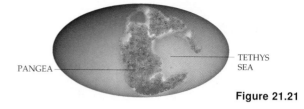

PANGEA ———————— TETHYS
                       SEA

**Figure 21.21**

## Critical Thinking

1. Briefly explain, in terms of hydrophilic and hydrophobic interactions, how proto-cells might have formed in water from aggregations of lipids, proteins, and nucleic acids.

2. Reflect on Figure 21.21. Now reflect on this: the Atlantic Ocean is gradually widening, and the Pacific Ocean and Indian Ocean are closing in on each other. Many millions of years from now, continents will collide and form a second Pangea. Write a short essay on what environmental conditions might be like on that future supercontinent and on what types of species might survive there.

3. According to one estimate, there is a chance that about $10^{20}$ planets have formed in the universe that are capable of sustaining life—but there is only one chance at intelligent life per planet. Given your knowledge of molecular biology and evolutionary processes, do you agree with this estimate? If so, speculate on why the odds are so low.

4. We know of a number of large asteroids that may intersect Earth's orbit in the distant future. There probably are a number we don't know about. Would you use this as an excuse not to worry about polluting the environment, not to take care of your physical health (as by avoiding drugs), and not to care about our cultural evolution? Why or why not?

## Selected Key Terms

| | |
|---|---|
| angiosperm *21.6* | global broiling |
| archaebacterium *21.3* | hypothesis *21.7* |
| Archean *21.3* | gymnosperm *21.6* |
| asteroid *21.7* | K–T asteroid impact theory *21.7* |
| big bang *CI* | mantle, of Earth *21.1* |
| Cenozoic *21.8* | Mesozoic *21.6* |
| crust, of Earth *21.1* | Paleozoic *21.5* |
| dinosaur *21.6* | prokaryotic cell *21.3* |
| Ediacaran *21.5* | Proterozoic *21.3* |
| endosymbiosis | protistan *21.4* |
| (theory of) *21.4* | proto-cell *21.2* |
| eubacterium *21.3* | RNA world *21.2* |
| eukaryotic cell *21.3* | stromatolite *21.3* |

## Readings  *See also www.infotrac-college.com*

de Duve, C. September–October 1995. "The Beginnings of Life on Earth." *American Scientist* 83: 428–437.

————. April 1996. "The Birth of Complex Cells." *Scientific American* 274(4): 50–57.

Dott, R., Jr., and R. Batten. 1994. *Evolution of the Earth*. Fifth edition. New York: McGraw-Hill.

Horgan, J. February 1991. "Trends in Evolution: In the Beginning . . . ." *Scientific American* 264(2): 116–125.

Impey, C. and W. Hartman. 2000. *The Universe Revealed*. Belmont, California: Wadsworth. Paperback

Wright, K. March 1997. "When Life Was Odd." *Discover* 18(3): 52–61. Update on the puzzling Ediacarans.

# BACTERIA AND VIRUSES

## *The Unseen Multitudes*

Did a friend ever mention that you are nearly 1/1,000 of a mile tall? Probably not. What would be the point of measuring people in units as big as miles? Even so, we think this way, in reverse, whenever we measure microorganisms. For the most part, **microorganisms** are single-celled organisms that are too small to be seen without the aid of a microscope.

The bacterial cells shown in Figure 22.1 are a case in point. To measure them, you would have to divide one meter into a thousand units, or millimeters. Next, you would have to divide one of the millimeters into a thousand smaller units, or micrometers. To give you a sense of how small that is, a single millimeter would be about as small as the dot of this "i." And a *thousand* bacteria would fit side by side on top of the dot!

To be sure, viruses are smaller still, as you might deduce after looking at Figure 22.2. We measure them in nanometers (billionths of a meter). But viruses are not alive. We consider them in this chapter because one kind or another infects just about every organism.

Bacteria generally are the smallest microorganisms, but they vastly outnumber the individuals in all other kingdoms combined. Their reproductive potential is staggering. Under ideal conditions, some divide about every twenty minutes. If that rate of reproduction were to hold constant, a single bacterium would have nearly a billion descendants in ten hours!

So why don't the unseen multitudes take over the world? Sooner or later, their burgeoning populations use up available nutrients and pollute the surroundings with their own metabolic wastes. In other words, they alter the very conditions that initially favored their reproduction. Besides this, certain kinds of viruses attack just about every type of bacterium and help keep their population sizes in check. So do seasonal changes in living conditions.

Of course, you probably do not find much comfort in this when you serve as host for one of the pathogenic types. **Pathogens** are infectious, disease-causing agents that invade target organisms and multiply inside or on

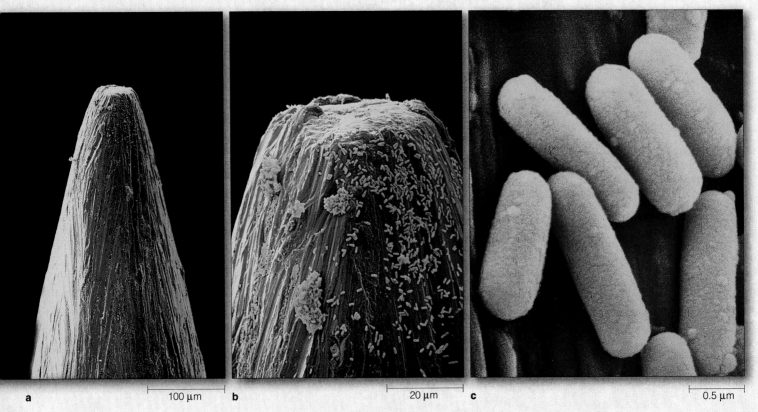

a    100 µm    b    20 µm    c    0.5 µm

**Figure 22.1** **(a–c)** How small are bacteria? Shown here, *Bacillus* cells peppering the tip of a pin. The cells in **(c)** are magnified more than 16,600 times.

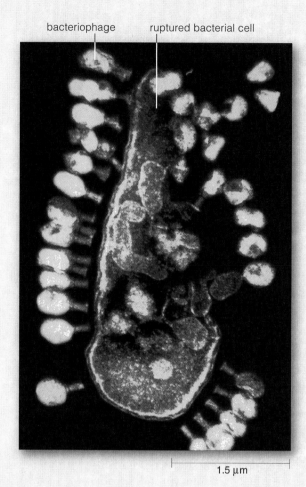

bacteriophage     ruptured bacterial cell

1.5 μm

**Figure 22.2** How small are the viruses? Shown here, bacteriophage particles, each about 225 nanometers tall, that infected a bacterial cell. The cell has ruptured open.

## KEY CONCEPTS

**1.** In structural terms, we find the simplest forms of life among the bacteria. Most bacteria are microscopically small. Smaller still are the viruses.

**2.** Bacteria alone are prokaryotic cells. They do not have a profusion of internal, membrane-bound organelles, as eukaryotic cells do. As a group, bacteria show great metabolic diversity, and many kinds display complex behavior.

**3.** Most bacteria reproduce by prokaryotic fission. This cell division mechanism follows DNA replication and results in two genetically equivalent daughter cells.

**4.** Bacteria were the first living organisms on Earth. Not long after they originated, they diverged into lineages that gave rise to eubacteria and to the common ancestor of archaebacteria and eukaryotic cells.

**5.** A virus, a noncellular infectious particle, consists of nucleic acid (either DNA or RNA), a protein coat, and sometimes an outer envelope. It cannot replicate without pirating the metabolic machinery of a specific type of host cell.

**6.** Nearly all viral multiplication cycles proceed through five steps: attachment to a host cell, penetration of its plasma membrane, replication of viral DNA or RNA and synthesis of viral proteins, then assembly of new viral particles, and finally release from the infected cell.

**7.** Most of us tend to judge microorganisms through the prism of human interests. Yet their lineages are the most ancient, their adaptations are diverse, and they are simply surviving and reproducing like the rest of us.

them. Disease follows when the metabolic activities of pathogenic cells damage the tissues and interfere with the body's normal functioning.

Certain pathogens can indeed make you suffer, but they should not give every microorganism a bad name. For example, think back on the uncountable numbers of photosynthetic bacterial cells in the seas (Section 7.9). Collectively, they help provide food and oxygen for entire communities and have major roles in the global cycling of carbon. Or think about the kinds of bacterial species that feed on organic debris. Together with other decomposers, they help cycle nutrients that sustain entire communities.

From the human perspective, microorganisms are good or bad, even dangerous. Basically, however, they are simply surviving and reproducing like the rest of us, in ways that are the topics of this chapter.

# CHARACTERISTICS OF BACTERIA

Of all organisms, bacteria are the most abundant and far-flung. Thousands of species inhabit diverse places, such as deserts, hot springs, glaciers, and seas. Some have been carrying on for millions of years 2,780 meters (9,121 feet) below the Earth's surface! Billions may live in a handful of rich soil. The ones in your gut and on your skin outnumber your own cells. Bacteria have the longest evolutionary history. Trace any lineage back far enough and you come across bacterial ancestors. From *Escherichia coli* to amoebas, elephants, clams, and coast redwoods, all organisms interconnect, regardless of size, numbers, and evolutionary distance. Table 22.1 and Figure 22.3 introduce features that help characterize the remarkable bacteria.

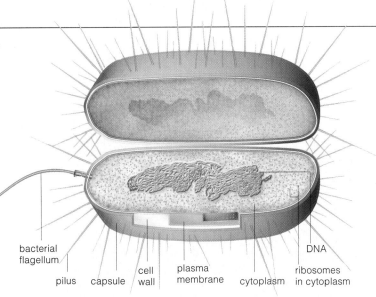

bacterial flagellum · pilus · capsule · cell wall · plasma membrane · cytoplasm · DNA · ribosomes in cytoplasm

**Figure 22.3** Generalized body plan of a bacterium.

## Splendid Metabolic Diversity

All organisms take in energy and carbon to meet their nutritional requirements. Compared with other species, however, bacteria display the greatest diversity in their means of securing resources.

Like plants, *photoautotrophic* bacteria build organic compounds by photosynthesis; they are "self-feeders." They tap sunlight for energy and use carbon dioxide as their carbon source. Their plasma membrane contains the photosynthetic machinery. Some photosynthesizers use electrons and hydrogen from water molecules for the synthesis reactions, and release oxygen. Anaerobic photosynthesizers (which die in the presence of oxygen) get electrons and hydrogen from inorganic compounds, such as gaseous hydrogen and hydrogen sulfide.

*Chemoautotrophic* bacteria are self-feeders. Carbon dioxide is the usual carbon source. Some species strip organic compounds for electrons and hydrogen. Others use inorganic substances, such as gaseous hydrogen, sulfur, nitrogen compounds, and a form of iron.

*Photoheterotrophic* bacteria are not self-feeders. They use energy from the sun for photosynthesis, but their carbon sources are fatty acids, complex carbohydrates, and other compounds that various organisms produce.

*Chemoheterotrophic* bacteria are parasites or saprobes, not self-feeders. The parasites live on or in a living host and draw glucose and other nutrients from it. Saprobic types get nutrients from the organic products, wastes, or remains of other organisms.

## Bacterial Sizes and Shapes

So far, you have a general sense of the microscopically small sizes of bacteria. The width and length of these cells typically fall between 1 and 10 micrometers.

Three basic shapes are common among bacteria. A spherical shape is a **coccus** (plural, cocci; from a word that means berries). A rod shape is a **bacillus** (plural, bacilli, meaning small staffs). A cell body with one or more twists is a **spirillum** (plural, spirilla):

coccus · bacillus · spirillum

Don't let the simple categories fool you. Cocci may also be oval or flattened. Bacilli may be skinny (like straws) or tapered (like cigars). Surface extensions make some bacteria look starlike. Square bacteria live in salt ponds in Egypt. Also, after daughter cells divide, they may stick together in chains, sheets, and other aggregations, as in Figure 22.4. Some spiral species are curved like a comma, and others are flexible or like stiff corkscrews.

## Structural Features

Bacteria alone are **prokaryotic cells**, meaning they were around before the evolution of the nucleated cell (*pro-*, before; *karyon*, nucleus). Few have membrane-bounded compartments of any sort for isolating metabolic events. Reactions take place in the cytoplasm or at the plasma membrane. For example, protein synthesis proceeds at

---

**Table 22.1    Characteristics of Bacterial Cells**

1. All bacterial cells are prokaryotic; they do not have a membrane-bound nucleus.

2. Bacterial cells in general have a single chromosome (a circular DNA molecule); many species also have plasmids.

3. Most bacteria have a cell wall of peptidoglycan.

4. Most bacteria reproduce by prokaryotic fission.

5. Collectively, bacteria show great metabolic diversity.

---

Further reading: Student Guide to InfoTrac on web site →

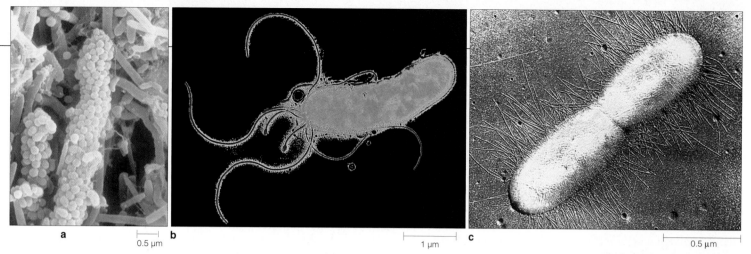

a

0.5 µm

b

c

1 µm

0.5 µm

**Figure 22.4** (**a**) Surface view of numerous bacilli and cocci attached to a human tooth. (**b**) *Helicobacter pylori* cell, with its tuft of flagella. This pathogen can colonize the stomach lining and trigger inflammation. If untreated, an infection can lead to gastritis, peptic ulcers, and possibly stomach cancer. *H. pylori* may contaminate water and food, especially unpasteurized milk. Currently, a combination of antibiotics and an antacid rids the body of the pathogen, after which ulcers heal. (**c**) Example of pili, filamentous structures that project from the surface of many bacterial cells. This *Escherichia coli* cell is dividing in two.

**Figure 22.5** Scanning electron micrographs of bacterial cell walls. (**a**) Dividing *Bacillus subtilis* cell. Its wall has a smooth surface texture. (**b**) A dividing *E. coli* cell. Its wrinkled cell wall is characteristic of Gram-negative bacteria.

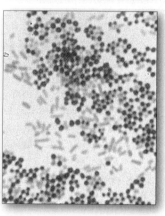

ribosomes that are distributed through the cytoplasm or attached to the side of the plasma membrane facing the interior. This does not mean bacteria are somehow inferior to eukaryotic cells. Being tiny, fast reproducers, they do not require great internal complexity.

A wall usually surrounds the plasma membrane. A **cell wall** is a semirigid, permeable structure that helps a cell maintain shape and resist rupturing when internal fluid pressure rises (Figure 22.5 and Section 5.5). The cell wall of eubacteria is composed of peptidoglycan molecules. In such molecules, peptide groups crosslink numerous polysaccharide strands to one another.

Clinicians can identify many bacterial species on the basis of their wall structure and composition. Consider a staining reaction called the **Gram stain**. A sample of unknown bacterial cells is exposed to a purple dye, then to iodine, an alcohol wash, and a counterstain. The cell wall of *Gram-positive* species remains purple. The wall of *Gram-negative* species loses color after the wash, but the counterstain turns it pink (Figure 22.6).

A sticky mesh, or **glycocalyx**, often encloses the cell wall. It consists of polysaccharides, polypeptides, or both. When highly organized and attached firmly to the wall, we call the mesh a capsule. When less organized and loosely attached to the wall, we call it a slime layer. The mesh helps a bacterial cell attach to teeth, mucous membranes, rocks in streambeds, and other interesting surfaces (Figure 22.4a). It also helps some encapsulated species avoid being engulfed by phagocytic, infection-fighting cells of their host organism.

Some bacteria have one or more **bacterial flagella** (singular, flagellum), which are used in motility. These don't have the same structure as a eukaryotic flagellum,

**Figure 22.6** Gram staining. Cocci and rods smeared on a slide are stained with a purple dye (such as crystal violet), washed off, then stained with iodine. All cells are now purple. The slide is washed with alcohol, which renders the Gram-negative cells colorless. Now the slide is counterstained (with safranin), washed, and dried. Gram-positive cells (*Staphylococcus aureus* in this case) remain purple. The counterstain, however, colors the Gram-negative cells (*Escherichia coli*) pink.

■ stain with purple dye
□ stain with iodine
□ wash with alcohol
▨ counterstain with safranin

and they do not operate the same way. They move the cell by rotating like a propeller. Many species also have **pili** (singular, pilus). These short, filamentous proteins project above the cell wall, as in Figure 22.4c. Some pili help cells adhere to surfaces. Others help them attach to one another as a prelude to conjugation, an interaction that is described in the next section.

Bacteria are microscopic, prokaryotic cells. They generally have one circular bacterial chromosome, and often they have extra DNA in the form of plasmids.

Nearly all bacteria have a wall around the plasma membrane. Many have a capsule or slime layer around the cell wall.

# BACTERIAL GROWTH AND REPRODUCTION

## The Nature of Bacterial Growth

Between cell divisions, bacteria grow through increases in their component parts. We measure the growth of a large, multicelled organism in terms of increases in size, but doing this for a microscopically small bacterium would be a bit pointless. Instead, we measure bacterial growth as an increase in the number of cells in a given population. Under ideal conditions, each cell divides in two, the division of two cells results in four, four result in eight, and so forth. Many types can divide every half hour; a few can do so every ten minutes. Such rates of increase lead to large population sizes in short order.

Natural conditions that promote the growth of some bacterial populations might not sustain the growth of others. Most organisms could not become established in the most extreme environments—say, Antarctica, the Negev Desert, or deep inside the Earth. There, the few species that do endure, inside rocks, rarely reproduce. And few can live in the highly acidic wastewater from mining operations. K-12, the strain of *E. coli* originally isolated from the human gut, has been cultivated for such a long time in the laboratory that it no longer is able to grow when reintroduced into the gut. Through microevolutionary processes, it became adapted to the conditions in its artificial environment.

## Prokaryotic Fission

When a bacterium has nearly doubled in size, it divides in two. Each daughter cell inherits a single **bacterial chromosome**, which is a circularized, double-stranded DNA molecule that has only a few proteins attached to it. In some species, the daughter cell merely buds from the parent cell. Most often, however, bacteria reproduce by a division mechanism called **prokaryotic fission**.

As prokaryotic fission begins, a parent cell replicates its DNA (Figure 22.7). The two molecules of DNA that result are attached to the plasma membrane at adjacent sites. The cell synthesizes lipid and protein molecules, which become incorporated in the membrane between the attachment sites. Membrane growth moves the two DNA molecules apart. New wall material is deposited above the membrane. The membrane and wall grow through the cell midsection and divide the cytoplasm. The result is two genetically equivalent daughter cells.

Especially in microbiology, you may hear someone refer to this division mechanism as *binary* fission, but such usage can cause confusion. The same term applies to a form of asexual reproduction among flatworms and some other multicelled animals. It refers to growth by way of mitotic cell divisions, then division of the whole body into two parts of the same or different sizes.

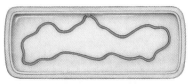

**a** Bacterium (cutaway view) before DNA replication. The bacterial chromosome is attached to the plasma membrane.

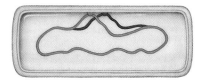

**b** DNA replication starts. It proceeds in two directions away from the same site in the bacterial chromosome.

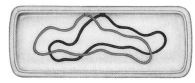

**c** The new copy of DNA is attached at a membrane site near the attachment site of the parent DNA molecule.

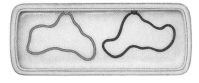

**d** New membrane grows between the two attachment sites. As it increases, it moves the two DNA molecules apart.

**e** At the cell midsection, deposits of new membrane and new wall material extend down into the cytoplasm.

**f** The ongoing, organized deposition of membrane and wall material at the cell midsection divides the cell in two.

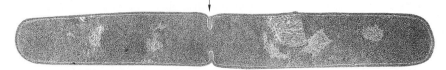

**Figure 22.7** Bacterial reproduction by way of prokaryotic fission, a cell division mechanism. The micrograph shows cytoplasmic division of *Bacillus cereus*, as brought about by the formation of new membrane and wall material. This cell has been magnified 13,000 times its actual size.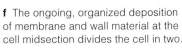

nicked plasmid    conjugation tube

**a** A conjugation tube has already formed between a donor and a recipient cell. An enzyme has nicked the donor's plasmid.

**b** DNA replication starts on the nicked plasmid. The displaced DNA strand moves through the tube and enters the recipient cell.

**c** In the recipient cell, replication starts on the transferred DNA.

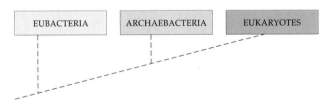

**d** The cells separate from each other; the plasmids circularize.

**Figure 22.8** Simplified sketch of bacterial conjugation. For clarity, the plasmid's size is greatly increased and the bacterial chromosome is not shown.

## Bacterial Conjugation

Daughter bacterial cells also may inherit one or more plasmids. A **plasmid**, recall, is a small, self-replicating circle of extra DNA that has a few genes (Section 16.1). The F (Fertility) plasmid has certain genes that confer the means to engage in **bacterial conjugation**. By this mechanism, a donor cell transfers plasmid DNA to a recipient cell. The transfers are known to occur among many bacterial species, such as *Salmonella*, *Streptococcus*, and *E. coli*—even between *E. coli* and yeast cells, in the laboratory. The F plasmid carries genetic instructions for synthesizing a structure called a sex pilus. Sex pili at the surface of a donor cell can hook onto a recipient cell and pull it right next to the donor. Shortly after the two cells make contact, a conjugation tube develops between them. After this, plasmid DNA is transferred through the tube, in the manner shown in Figure 22.8.

---

Most bacteria reproduce by way of prokaryotic fission, a cell division mechanism that follows DNA replication. Each daughter cell inherits a single bacterial chromosome (one DNA molecule). Many species also transfer plasmid DNA.

---

Just a few decades ago, reconstructing the evolutionary history of bacteria appeared to be an impossible task. Except for stromatolites (Section 21.3), the most ancient groups of bacteria are not well represented in the fossil record. Most groups are not represented at all.

Given their elusive histories, the many thousands of known species of prokaryotic cells traditionally have been classified by **numerical taxonomy**. By this practice, traits of an unidentified cell are compared with those of a known bacterial group. Such traits typically include cell shape, motility, staining attributes of the cell wall, nutritional requirements, metabolic patterns, and the presence or absence of endospores. The greater the total number of traits that the cell has in common with the known group, the closer is the inferred relatedness.

Since the 1970s, nucleic acid hybridization studies, gene sequencing, and other methods of comparative biochemistry have been revealing compelling evidence of bacterial phylogenies (Section 20.4). Comparisons of ribosomal RNAs are notably useful. Remember, rRNAs are indispensable to protein synthesis, and they cannot undergo drastic changes in their overall base sequence without loss of function. The numerous small changes that accumulated in the rRNAs of different lineages can

| EUBACTERIA | ARCHAEBACTERIA | EUKARYOTES |

**Figure 22.9** Evolutionary tree diagram showing a relationship between archaebacteria, eubacteria, and eukaryotes, based on evidence from comparative biochemistry.

be directly measured. Surprisingly, the measurements are uniting some groups that did not even appear to be related on the basis of other tests. For example, it now seems a key divergence began shortly after prokaryotic cells first appeared on Earth (Section 21.3). One branch led to **eubacteria**, the most common prokaryotic cells. (Here, the *eu–* signifies "typical.") The other branch led to the common ancestor of both the **archaebacteria** and the first eukaryotic cells. Figure 22.9 is a tree diagram of these evolutionary relationships.

---

Prokaryotic cells are classified by numerical taxonomy: the total percentage of observable traits they have in common with a known bacterial group. They also are classified more directly by comparisons at the biochemical level.

Prokaryotic cells are now assigned to one of two lineages, called the eubacteria and archaebacteria.

---

# MAJOR BACTERIAL GROUPS

# ARCHAEBACTERIA

By now, you probably have sensed that *species* is the basic unit in bacterial classification schemes. However, the definition of species that fits sexually reproducing organisms does not fit bacteria—which do not form reproductively isolated populations of interbreeding individuals. Each bacterial cell generally does its own thing; genetic recombination is infrequent. Besides this, bacteria do not display spectacular variation in traits, and variations that do occur are dictated by relatively few genes. If two types show only minor differences, one may be classified as a **strain**, not a new species.

These are just a few of the problems we face when attempting to bring more order to our understanding of bacterial kingdoms. Until evolutionary relationships are sorted out, the bacteria will continue to be grouped mainly according to numerical taxonomy, as described earlier. Table 22.2 summarizes some major groupings that are touched upon in this book. Appendix I lists their habitats and characteristics.

**Table 22.2  Representative Archaebacteria and Eubacteria**

| Some Major Groups | Representatives |
| --- | --- |
| **ARCHAEBACTERIA** | |
| Methanogens | *Methanobacterium* |
| Extreme halophiles | *Halobacterium, Halococcus* |
| Extreme thermophiles | *Sulfolobus, Thermoplasma* |
| **EUBACTERIA** | |
| *Photoautotrophs:* | |
| Cyanobacteria, green sulfur bacteria, and purple sulfur bacteria | *Anabaena, Nostoc, Rhodopseudomonas, Chloroflexus* |
| *Photoheterotrophs:* | |
| Purple nonsulfur and green nonsulfur bacteria | *Rhodospirillum, Chlorobium* |
| *Chemoautotrophs:* | |
| Nitrifying, sulfur-oxidizing, and iron-oxidizing bacteria | *Nitrosomonas, Nitrobacter, Thiobacillus* |
| *Chemoheterotrophs:* | |
| Spirochetes | *Spirochaeta, Treponema* |
| Gram-negative aerobic rods and cocci | *Pseudomonas, Neisseria, Rhizobium, Agrobacterium* |
| Gram-negative facultative anaerobic rods | *Salmonella, Escherichia, Proteus, Photobacterium* |
| Rickettsias and chlamydias | *Rickettsia, Chlamydia* |
| Myxobacteria | *Myxococcus* |
| Gram-positive cocci | *Staphylococcus, Streptococcus, Deinococcus* |
| Endospore-forming rods and cocci | *Bacillus, Clostridium* |
| Gram-positive nonsporulating rods | *Lactobacillus, Listeria* |
| Actinomycetes | *Actinomyces, Streptomyces* |

We divide the kingdom of archaebacteria into three major groups: methanogens, extreme halophiles, and extreme thermophiles. In many respects, archaebacteria are unique in their composition, structure, metabolism, and nucleic acid sequences. They differ as much from other bacteria as they do from eukaryotic cells. Many investigators believe that the existing archaebacteria, which can withstand conditions as hostile as those on the early Earth, resemble the first living cells. Hence the name of the kingdom (*archae–* means "beginning").

**METHANOGENS**  The **methanogens** (methane makers) live in swamps, stockyards, the termite and mammalian gut, and other oxygen-free habitats (Figure 22.10). They are strict anaerobes; they die in the presence of oxygen.

Methanogens produce ATP by anaerobic electron transport, a pathway described in Section 8.5. Usually, they use hydrogen gas ($H_2$) as their source of electrons, although some groups get them from ethanol and other alcohols. Nearly all use carbon dioxide as their carbon source and as a final electron acceptor for the reactions, which end with the formation of methane ($CH_4$). As a group, the methanogens produce about 2 billion tons of methane every year. Most of the methane production is released from wetlands, termites, ruminants, landfills, and rice paddies. Collectively their activities influence the levels of carbon dioxide in the atmosphere and the cycling of carbon through ecosystems.

Long ago, huge methane deposits accumulated on the seafloor. For example, geologists recently found 35

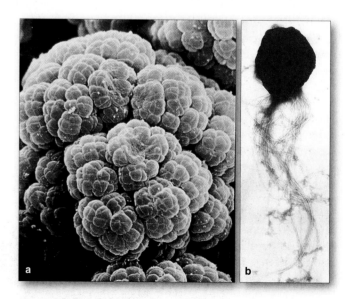

**Figure 22.10**  Representative archaebacteria. (**a**) Scanning electron micrograph of a dense population of *Methanosarcina* cells. Each of these methanogens has a thick polysaccharide wall. (**b**) Transmission electron micrograph of *Methanococcus jannaschii*, a heat-loving methane producer.

**Figure 22.11** (**a**) Representative habitat for methanogens. The stomach chambers of cattle and other ruminants house methanogen populations. Cattle belch—a lot—and release quantities of methane. You might have noticed the resulting exceptionally pungent air around stockyards.

Representative habitats for extreme halophiles. (**b**) Great Salt Lake, Utah. Halophilic bacteria make purplish pigments and a green alga (*Dunaliella salina*) makes beta-carotene pigments when salinity and light intensity are high, and pH and nutrient levels are low. The pigments impart a reddish-purple color to the water. (**c**) Commercial seawater evaporating ponds, San Francisco Bay, another home of *Halobacterium* and *D. salina*. Both organisms thrive in hypersaline conditions, where salt levels greatly exceed that of seawater (for example, 300 versus 35 grams per liter).

(**d**) Representative habitat for extreme thermophiles: hot, sulfur-rich water in Emerald Pool, Yellowstone National Park. Microbes with an abundance of carotenoids impart an orange color to the rim of this hot spring.

billion tons of it 400 kilometers off the South Carolina coast. If the ocean circulation were to change, as it has in the past, then all of that gas could move abruptly and explosively to the surface. The rapid release of so much methane would drastically change the atmosphere and therefore the global climate (Section 49.9).

**EXTREME HALOPHILES**   There are many salty habitats around the world, but **extreme halophiles** (salt lovers) thrive in exceptionally salty ones. Great Salt Lake, the Dead Sea, and seawater evaporation ponds are like this (Figure 22.11*b,c*). Extreme halophiles spoil commercial sea salt, salted fish, and animal hides. Most make ATP by aerobic pathways. When oxygen levels are low, some make ATP by photosynthesis that starts with a unique pigment, bacteriorhodopsin, in the plasma membrane. Absorption of light energy triggers a metabolic process that leads to an increase in an $H^+$ gradient across the plasma membrane. ATP forms when these ions follow the gradient and flow through the interior of proteins that span the membrane (compare Section 7.5).

**EXTREME THERMOPHILES**   Geothermally heated soils, sulfur-rich hot springs, and wastes from coal mines are

habitats of the "heat lovers," or **extreme thermophiles** (Figure 22.11*d*). These archaebacteria are the most heat-tolerant prokaryotes known. Some populations grow at above-boiling temperatures; all do best at temperatures above 80°C! Nearly all are strict anaerobes that require sulfur as an electron acceptor or electron donor.

*Solfolobus*, the first extreme thermophile discovered, grows in acidic hot springs. So does *Thermus aquaticus*, which biotechnologists employ as a source of extremely heat-stable DNA polymerases (Section 16.2).

Certain extreme thermophiles live in shallow water around volcanoes. We find some of these off the coast of Italy where geothermally heated water spews from openings in the sediments. Other species are the basis of food webs in sediments around hydrothermal vents, where the surrounding water can exceed 110°C. They use hydrogen sulfide spewing from these vents as a source of electrons. Their existence at vents is cited as evidence that life may have originated on the seafloor.

Like the first cells on Earth, archaebacteria live in extremely inhospitable habitats.  In their properties, they differ as much from eubacteria as from eukaryotic cells.

## EUBACTERIA

We recognize more than 400 genera of prokaryotes. By far, most of them are eubacteria. Unlike other bacterial cells, eubacteria have fatty acids incorporated into their plasma membrane. When a species has a cell wall—as nearly all do—the wall incorporates peptidoglycan. We still have not learned enough about the evolutionary histories of eubacteria to move much beyond taxonomic classification. Here we focus on modes of nutrition to give you a sense of the biodiversity in this kingdom.

### A Sampling of Biodiversity

PHOTOAUTOTROPHIC EUBACTERIA  The cyanobacteria, once called blue-green algae, are the classic example of photoautotrophic eubacteria. They are also among the most common photoautotrophs. All cyanobacteria are aerobic cells that engage in photosynthesis. Most types live in ponds and other freshwater habitats. They may grow as mucus-sheathed chains of cells, which often form thick, dense, slimy mats at the surface of nutrient-enriched water (Figure 22.12a).

*Anabaena* and other types also convert nitrogen gas ($N_2$) to ammonia, which they use in biosynthesis. When nitrogen compounds are scarce, some cells develop into **heterocysts**. These modified cells make a nitrogen-fixing enzyme (Figure 22.12b,c). They produce and then share nitrogen compounds with photosynthetic cells; in return they receive carbohydrates. The shared substances move freely across cytoplasmic junctions between cells.

Anaerobic photoautotrophs, such as green bacteria, get electrons from hydrogen sulfide and hydrogen gas, not from water. They may resemble anaerobic bacteria in which the cyclic pathway of photosynthesis evolved.

CHEMOAUTOTROPHIC EUBACTERIA  Nearly all eubacteria in this category influence the global cycling of nitrogen, sulfur, and other nutrients. For example, as you know, nitrogen is a major building block for amino acids and proteins. Without it, there would be no life. Nitrifying bacteria in soil strip electrons from ammonia. Plants use the end product, nitrate, as a nitrogen source.

CHEMOHETEROTROPHIC EUBACTERIA  Nearly all bacteria fall in this category. Many, including pseudomonads, are decomposers; their enzymes can break down organic compounds and even pesticides in soil. Other "good" species, at least by our standards, are *Lactobacillus* (used in the manufacture of pickles, sauerkraut, buttermilk, and yogurt) and actinomycetes (sources of antibiotics). *E. coli* produces vitamin K and compounds that help us digest fat, it helps newborn mammals digest milk, and its activities help stop many food-borne pathogens from colonizing the gut. Sugarcane and corn benefit from a symbiont, the nitrogen-fixing spirochete *Azospirillum*.

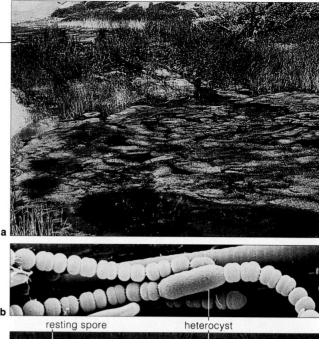

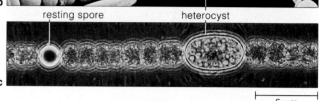

resting spore          heterocyst

5 μm

**Figure 22.12**  Cyanobacteria—common photoautotrophs. (**a**) A cyanobacterial population near the surface of a nutrient-enriched pond. (**b,c**) Resting spores form when conditions do not favor growth. A nitrogen-fixing heterocyst is also shown.

The plants use some of the nitrogen initially fixed by the bacterium and give up some sugars to it. *Rhizobium* is a fine symbiont with peas, beans, and other legumes. *Deinococcus radiodurans* resists high radiation doses that would kill other organisms. A genetically engineered strain holds promise for bioremediation; its *E. coli* genes allow it to break down toxic nuclear wastes, including extremely nasty mercury compounds. At present, there is no other mechanism in place to rid the environment of the alarming accumulation of nuclear wastes.

Also in this category are most pathogenic bacteria. We admire pseudomonads as decomposers in soil, not when they grow on our soaps, antiseptics, and other carbon-rich goods. They are especially bad because they can transfer plasmids with antibiotic-resistance genes.

Some *E. coli* strains cause a form of diarrhea that is the main cause of infant death in developing countries. *Clostridium botulinum* can taint fermented grain as well as food in improperly sterilized or sealed cans and jars. Its toxins cause *botulism*, a form of poisoning that can disrupt breathing and lead to death. *C. tetani*, one of its relatives, causes the disease *tetanus* (Section 34.5).

Like many other bacteria that typically live in soil, *C. tetani* can form an **endospore**. This resting structure forms inside the cell, around a copy of the bacterial chromosome and part of the cytoplasm (Figure 22.13). Endospores form when a depletion of nitrogen or some other nutrient arrests cell growth. They are released as

Further reading: Student Guide to InfoTrac on web site →

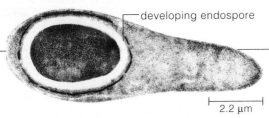

developing endospore

2.2 μm

**Figure 22.13** Developing endospore in *Clostridium tetani*, one of the dangerous pathogens.

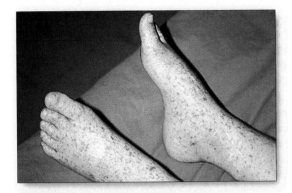

**Figure 22.14** Rash typical of Rocky Mountain spotted fever, a disease caused by *Rickettsia rickettsii*. Infected ticks transmit the bacterium to humans with their bites.

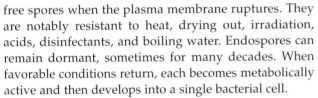

a

0.25 μm

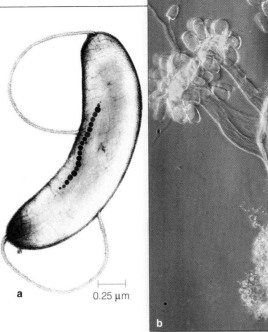

b

**Figure 22.15** (**a**) Micrograph of a magnetotactic bacterium. Inside the cytoplasm is a chain of magnetite particles that acts like a compass when the bacterium moves about. (**b**) A gathering of myxobacteria (*Chondromyces crocatus*). Cells of this species aggregate and form spore-bearing structures.

free spores when the plasma membrane ruptures. They are notably resistant to heat, drying out, irradiation, acids, disinfectants, and boiling water. Endospores can remain dormant, sometimes for many decades. When favorable conditions return, each becomes metabolically active and then develops into a single bacterial cell.

Many chemoautotrophic eubacteria taxi from host to host inside the gut of insects. For example, bites from infected ticks transmit *Borrelia burgdorferi* from deer and some other wild animals to humans, who develop *Lyme disease*. A "bull's-eye" rash often develops around the bite (Section 26.20). Severe headaches, backaches, chills, and fatigue follow. Without prompt treatment, the symptoms worsen. Another tick-traveling pathogen, *Rickettsia rickettsii*, causes *Rocky Mountain spotted fever*. After the bacterium enters a host through a tick bite, it penetrates the cytoplasm and the nucleus of host cells. Three to twelve days after this, a high fever and severe headache develop. Three to five days after that, a rash develops on the hands and feet (Figure 22.14). Diarrhea and gastrointestinal cramps are common. If untreated, the symptoms can persist for more than two weeks.

### Regarding the "Simple" Bacteria

Bacteria are small. Their insides are not elaborate. *But bacteria are not simple.* A brief look at their behavior will reinforce this point. Bacteria move toward nutrient-rich regions. Aerobes move toward oxygen; anaerobes avoid it. Photosynthetic types move into light and away from light that is too intense. Many species can tumble away from toxins. Such behaviors often start with membrane receptors that are stimulated by changes in chemical

conditions or in the intensity of light coming in from a particular direction. Such a change triggers a shift in metabolic activities inside the cell, which might then lead to an adjustment in the direction of movement.

Magnetotactic bacteria contain a chain of magnetite particles that serves as a tiny compass (Figure 22.15*a*). The compass helps them sense which way is north and also down. These bacteria swim toward the bottom of a body of water, where oxygen concentrations are lower and therefore more suitable for their growth.

Some species even show *collective* behavior, as when millions of *Myxococcus xanthus* cells form a "predatory" colony. These cells secrete enzymes that digest "prey," such as cyanobacteria, that become stuck to the colony. Then the cells absorb the breakdown products. What's more, the cells migrate, change direction, and move as a single unit toward what may be food!

Many myxobacteria colonies form **fruiting bodies** (spore-bearing structures). Under suitable conditions, some cells in the colony differentiate and form a slime stalk, others form branchings, and others form clusters of spores (Figure 22.15*b*). The spores disperse when a cluster bursts open; each may give rise to a new colony. As you will see in the next chapter, certain eukaryotic species also form spore-bearing structures.

Eubacteria are the most common and diverse prokaryotic cells. They are adapted to nearly all environments.

### Defining Characteristics

In ancient Rome, *virus* meant "poison" or "venomous secretion." In the late 1800s, this rather nasty word was bestowed on newly discovered pathogens, smaller than the bacteria being studied by Louis Pasteur and others.

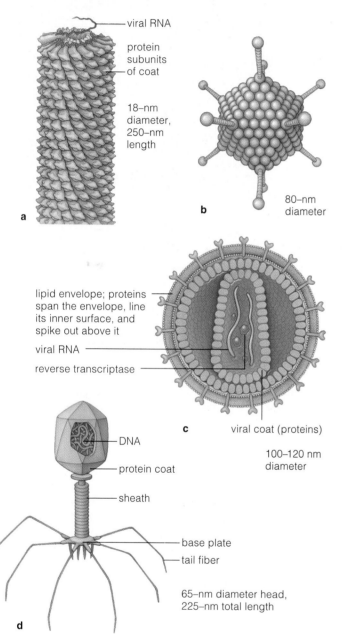

**Figure 22.16** Body plans of viruses. (**a**) *Helical* viruses have a rod-shaped coat of protein subunits, coiled helically around their nucleic acid. The upper subunits of the coat have been removed from this portion of a tobacco mosaic virus to reveal the RNA. (**b**) *Polyhedral* viruses, including this adenovirus, have a many-sided coat. (**c**) *Enveloped* viruses, including HIV, have an envelope around a helical or polyhedral coat. (**d**) *Complex* viruses, such as T-even bacteriophages, have additional structures attached to the coat.

Many viruses deserve the name. They attack humans, cats, cattle, birds, insects, plants, fungi, protistans, and bacteria. You name it, there are viruses that can infect it.

Today we define a **virus** as a noncellular infectious agent that has two characteristics. First, a viral particle consists of a protein coat wrapped around a nucleic acid core—that is, around its genetic material. Second, the virus cannot reproduce itself. It can be reproduced only after its genetic material and a few enzymes enter a host cell and subvert the cell's biosynthetic machinery.

The genetic material of a virus is DNA *or* RNA. The coat consists of one or more types of protein subunits organized into a rodlike or polyhedral (many-sided) shape, as in Figure 22.16. The coat protects the genetic material during the journey to a new host cell. It also contains proteins that can bind with specific receptors on host cells. An envelope, made mostly of membrane remnants from a previously infected cell, encloses the coat of some viruses. The envelope bristles with spikes of glycoproteins. Coats of complex viruses have sheaths, tail fibers, and other accessory structures attached.

The vertebrate immune system detects certain viral proteins. The problem is, genes for many viral proteins mutate at high frequencies, so a virus often can elude the immune fighters. For example, people susceptible to lung infections get new "flu shots" each year because envelope spikes on influenza viruses keep changing.

### Examples of Viruses

Each kind of virus can multiply only in certain hosts. It cannot be studied easily unless the investigator cultures living host cells. This is why much of our knowledge of viruses comes from **bacteriophages**, a group of viruses that infect bacterial cells. Unlike cells of humans and other complex, multicelled species, bacterial cells can be cultured easily and rapidly. This is also why bacteria and bacteriophages were used in early experiments to determine the function of DNA (Section 13.1). They are still used as research tools in genetic engineering.

Table 22.3 lists some major groups of animal viruses. These viruses contain double- or single-stranded DNA or RNA, which is replicated in various ways. Animal viruses range in size from parvoviruses (18 nanometers) to brick-shaped poxviruses (350 nanometers). Many kinds cause diseases, such as the common cold, certain cancers, warts, herpes, and influenza (Figure 22.17). One, HIV, is the trigger for *AIDS*. By attacking certain white blood cells, HIV weakens the immune system's ability to fight infections that might not otherwise be life threatening. Researchers attempting to develop drugs against HIV and other diseases use HeLa cells and other immortal cell lineages for early experiments (Section 9.6). Later they must use laboratory animals,

Further reading: Student Guide to InfoTrac on web site →

## Table 22.3  Classification of Some Major Animal Viruses

| DNA VIRUSES | Some Diseases/Consequences |
| --- | --- |
| Parvoviruses | Gastroenteritis; roseola (fever, rash) in small children; aggravation of symptoms of sickle-cell anemia |
| Adenoviruses | Respiratory infections (fever, cough, sore throat, rash), diarrhea in infants, conjunctivitis (inflamed, pebbly eye membranes); some cause tumors |
| Papovaviruses | Benign and malignant warts |
| Orthopoxviruses | Smallpox, cowpox, monkeypox |
| Herpesviruses: | |
| *H. simplex* type I | Oral herpes, cold sores |
| *H. simplex* type II | Genital herpes (Section 45.14) |
| Varicella–zoster | Chicken pox, shingles |
| Epstein–Barr | Infectious mononucleosis; cancers of skin, liver, cervix, pharynx; Burkitt's lymphoma (malignant tumor of jaw, face) |
| Cytomegalovirus | Hearing loss, mental retardation |
| Hepadnavirus | Hepatitis B (severe liver infection) |

| RNA VIRUSES | Some Diseases/Consequences |
| --- | --- |
| Picornaviruses: | |
| Enteroviruses | Polio, hemorrhagic eye disease, hepatitis A (infectious hepatitis) |
| Rhinoviruses | Common cold |
| Hepatitis A virus | Inflammation of liver, kidneys, spleen |
| Togaviruses | Forms of encephalitis (inflammation of the brain), rubella |
| Flaviviruses | Yellow fever (fever, chills, jaundice), dengue (fever, severe muscle pain) St. Louis encephalitis |
| Coronaviruses | Upper respiratory infections, colds |
| Rhabdoviruses | Rabies, other animal diseases |
| Filoviruses | Hemorrhagic fevers, as by *Ebola* virus (Section 22.10) |
| Paramyxoviruses | Measles, mumps, respiratory ailments |
| Orthomyxoviruses | Influenza |
| Bunyaviruses | |
| Bunyamwera virus | California encephalitis |
| Phlebovirus | Hemorrhagic fever, encephalitis |
| Hantavirus | Hemorrhagic fever, kidney failure |
| Arenaviruses | Hemorrhagic fevers |
| Retroviruses: | |
| HTLV-I, HTLV-II* | Adult T-cell leukemia |
| HIV | AIDS |
| Reoviruses | Respiratory and intestinal infections |

\* Human T-cell leukemia virus.

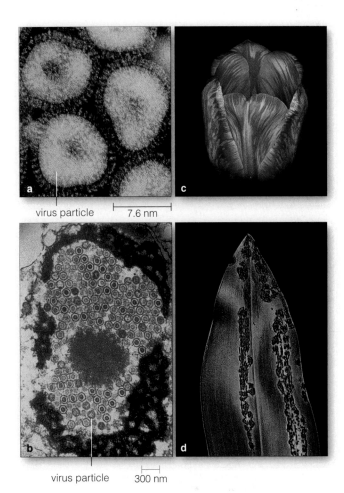

virus particle    7.6 nm

virus particle    300 nm

**Figure 22.17**  Some viruses and their effects. (**a**) Particles of a DNA virus that causes a herpes infection in humans. (**b**) Particles of an enveloped RNA virus that causes influenza in humans. Spikes project from the lipid envelope. (**c**) Streaking of a tulip blossom. A harmless virus infected pigment-forming cells in the colorless parts. (**d**) An orchid leaf infected by a rhabdovirus.

then human volunteers, to test a new drug for toxicity and effectiveness. Why? It takes a functioning immune system to test responses.

Plant viruses must breach plant cell walls to cause diseases. They typically hitch rides on the piercing or sucking devices of insects that feed on plant juices. Some RNA viruses infect tobacco plants (the tobacco mosaic virus), barley, potatoes, and other major crop plants. Certain DNA viruses infect such valuable crops as cauliflower and corn. Figure 22.17*c,d* shows visible effects of two viral infections.

**A virus is a nonliving infectious particle that consists of nucleic acid enclosed in a protein coat and sometimes an outer envelope. It cannot multiply without pirating the metabolic machinery of a specific type of host cell.**

**Diverse viruses infect organisms in all kingdoms.**

# VIRAL MULTIPLICATION CYCLES

Viruses multiply in a variety of ways. Even so, nearly all of their multiplication cycles proceed through five basic steps, as outlined here:

1. *Attachment.* A virus attaches to a host cell. Any cell is a suitable host if molecular groups on a virus particle are able to chemically recognize and lock on to specific molecular groups at the cell surface.

2. *Penetration.* Either the whole virus or its genetic material alone penetrates the cell's cytoplasm.

3. *Replication and synthesis.* In an act of molecular piracy, the viral DNA or RNA directs the host cell into producing many copies of viral nucleic acids and proteins, including enzymes.

4. *Assembly.* The viral nucleic acids and viral proteins are put together to form new infectious particles.

5. *Release.* New virus particles are released from the cell.

Consider this list with respect to some bacteriophages. Lytic and lysogenic pathways are common among their replication cycles. In a **lytic pathway**, steps 1 through 4 proceed rapidly, and new particles are released when a host cell undergoes lysis (Figure 22.18). **Lysis** means the plasma membrane, cell wall, or both are damaged, so that cytoplasm leaks out and the cell dies. Late into most lytic pathways, a viral enzyme that causes swift destruction of the bacterial cell wall is synthesized.

In a **lysogenic pathway**, a latent period extends the cycle. The virus does not kill its host outright. Instead, a viral enzyme cuts a host chromosome, then integrates viral genes into it. Thus, when an infected cell prepares to divide, it replicates the recombinant molecule. As a result of that single instance of genetic recombination, miniature time bombs are passed on to all of that cell's descendants. Later on, a molecular signal or some other stimulus may reactivate the cycle.

Latency occurs in the multiplication cycles of many viruses, not just among bacteriophages. Type I *Herpes simplex*, which causes *cold sores*, is an example. Nearly everybody harbors this virus. It remains latent inside a ganglion (an aggregation of neuron cell bodies) in facial tissue. Stress factors such as sunburn can reactivate the virus. Then, virus particles move down the neurons to their tips near the skin. There they infect epithelial cells and cause painful skin eruptions.

Like other enveloped viruses, the herpesviruses can enter a host cell by their own version of endocytosis, then leave it by budding from the plasma membrane. Figure 22.19 shows how they accomplish this.

The multiplication cycle of the RNA viruses has an interesting twist to it. In the host cell's cytoplasm, their

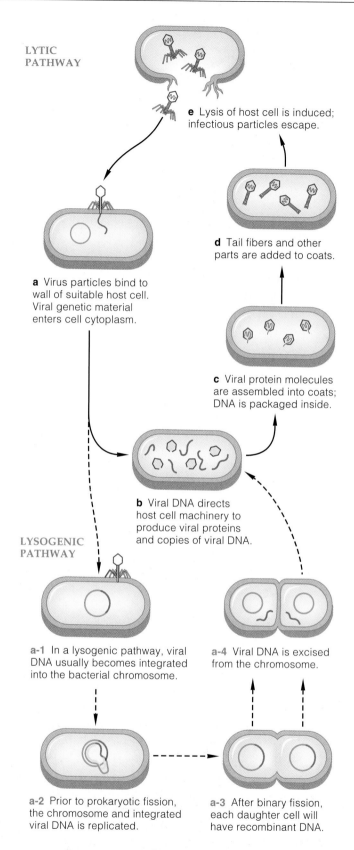

**LYTIC PATHWAY**

**e** Lysis of host cell is induced; infectious particles escape.

**a** Virus particles bind to wall of suitable host cell. Viral genetic material enters cell cytoplasm.

**d** Tail fibers and other parts are added to coats.

**c** Viral protein molecules are assembled into coats; DNA is packaged inside.

**b** Viral DNA directs host cell machinery to produce viral proteins and copies of viral DNA.

**LYSOGENIC PATHWAY**

**a-1** In a lysogenic pathway, viral DNA usually becomes integrated into the bacterial chromosome.

**a-4** Viral DNA is excised from the chromosome.

**a-2** Prior to prokaryotic fission, the chromosome and integrated viral DNA is replicated.

**a-3** After binary fission, each daughter cell will have recombinant DNA.

**Figure 22.18** Generalized multiplication cycle for some bacteriophages. New viral particles may be produced and released by a lytic pathway. For certain viruses, the lytic pathway may expand to include a lysogenic pathway.

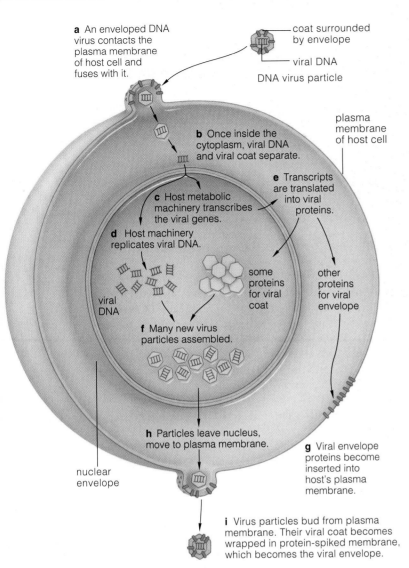

**a** An enveloped DNA virus contacts the plasma membrane of host cell and fuses with it.

coat surrounded by envelope

viral DNA

DNA virus particle

plasma membrane of host cell

**b** Once inside the cytoplasm, viral DNA and viral coat separate.

**e** Transcripts are translated into viral proteins.

**c** Host metabolic machinery transcribes the viral genes.

**d** Host machinery replicates viral DNA.

viral DNA

some proteins for viral coat

other proteins for viral envelope

**f** Many new virus particles assembled.

**h** Particles leave nucleus, move to plasma membrane.

nuclear envelope

**g** Viral envelope proteins become inserted into host's plasma membrane.

**i** Virus particles bud from plasma membrane. Their viral coat becomes wrapped in protein-spiked membrane, which becomes the viral envelope.

**j** The finished particle is equipped to infect a new potential host cell.

**Figure 22.19** Multiplication cycle of one type of enveloped DNA virus infecting an animal cell.

RNA serves as a template for synthesizing either DNA or mRNA. For example, HIV, a retrovirus, carts its own enzymes into cells. These assemble DNA on viral RNA by reverse transcription (Sections 16.1 and 40.11).

The multiplication cycles of nearly all viruses include five basic steps: attachment to a suitable host cell, penetration into the cell, viral DNA or RNA replication and synthesis of viral proteins, assembly of new viral particles, and then release from the infected cell.

Bacteriophage replication cycles commonly follow a rapid, lytic pathway and an extended, lysogenic pathway.

Replication cycles of RNA viruses involve the use of viral RNA as a template for synthesizing DNA or mRNA.

**VIROIDS** You may have trouble visualizing this, but some infectious agents are more stripped down than viruses. **Viroids** are tightly folded strands or circles of RNA, smaller than anything in viruses. They contain no protein-coding genes whatsoever. In this they resemble the noncoding portions of eukaryotic DNA called introns (Section 14.2). They might be self-splicing introns that escaped from DNA; they might even be remnants left over from an RNA world (Section 21.2). Viroids have no protein coat, but their tight folding might help protect them from a host's enzymes. These bits of "naked" RNA are known to cause a number of diseases in plants. Each year, they destroy many millions of dollars' worth of potatoes, citrus, and other important crop plants.

**PRIONS** Eight rare, fatal degenerative diseases of the nervous system are linked to small proteins known as **prions**. The proteins are altered products of a gene that is present in unaffected as well as infected individuals. Unaltered and altered forms of the protein molecule are found at the surface of neurons, the communication cells of the nervous system. You may have heard of *kuru* and *Creutzfeldt–Jakob* diseases (CJD). They slowly destroy muscle coordination and brain functions in humans. *Scrapie*, a disease of sheep, is so named because infected animals rub against trees or posts until they scrape off most of their wool.

In 1996, more than 150,000 cattle in Great Britain staggered about, drooled, and showed other symptoms of BSE (bovine spongiform encephalopathy), or *mad cow disease*. Prions had caused abnormal changes in proteins of neurons, and spongy holes and amyloid deposits had formed in the cattle brains. BSE is fatal.

How did it happen? Ground-up tissues of sheep that had died of scrapie had been used as a supplement in cattle feed. The practice was banned in 1988, but prions were already in the food chain.

Medical researchers already knew that 1 in 1 million people around the world develops Creutzfeldt–Jakob disease each year. Prions also are linked to this fatal condition. Symptoms, which include loss of vision and speech, rapid mental deterioration, and spastic paralysis, may take decades to develop. At the same time as the major outbreak of BSE in Great Britain, ten people were diagnosed with a variant of CJD; four had already died, probably as a result of eating meat of infected animals. Genetic analysis and interviews linked all the cases to exposure to BSE.

Ten other countries also had reported cases of BSE, possibly as a result of importing cattle feed from Great Britain. The incidence of BSE in Great Britain is declining since strict precautionary measures have been instituted. In the United States, importation of live cattle or meat products from BSE-affected countries has been banned since 1989.

# 22.10 THE NATURE OF INFECTIOUS DISEASES

**CATEGORIES OF DISEASES**   Just by being human, you are a potential host for a great many pathogenic bacteria, viruses, fungi, protozoans, and parasitic worms. When a pathogen has invaded your body and is multiplying in host cells and tissues, this is an **infection**. Its outcome, **disease**, results if the body's defenses cannot be mobilized fast enough to prevent the pathogen's activities from interfering with normal body functions. You may have heard of *contagious* diseases. This simply means that the pathogenic agents can be transmitted by direct contact with body fluids secreted or otherwise released from infected individuals, as by explosive wet sneezes.

During an **epidemic**, a disease spreads rapidly through part of a population for a limited time, then the outbreak subsides. What happens when an epidemic breaks out in several countries around the world at the same time? We call this a **pandemic**. AIDS, an incurable disease, is an example. Millions are infected by the causative agent, HIV (short for the *Human Immunodeficiency Virus*).

*Sporadic* diseases such as whooping cough break out irregularly and affect just a few people. *Endemic* diseases pop up more or less continuously, but they don't spread far in large populations. Tuberculosis is like this. So is impetigo, a highly contagious bacterial infection that often spreads no further than, say, a single day-care center.

**AN EVOLUTIONARY VIEW**   Now look at disease in terms of a pathogen's prospects for survival. Like you, *a pathogen lives only for as long as it has access to outside sources of energy and raw materials*. To a microscopic organism or viral particle, a human host is a veritable treasurehouse of both. With such a bountiful supply of resources, the pathogen can multiply or replicate itself to amazing population sizes. And, evolutionarily speaking, the ones that leave the most descendants win.

There are two significant barriers to complete world dominance by pathogens. First, any species with a history of being attacked by a particular pathogen has coevolved with it and has built-in defenses against it. The premier example is the vertebrate immune system, as described in Chapter 40. Second, if a pathogen kills its host too quickly, it may vanish before having time to infect a new one. This is one reason why most pathogens have less-than-fatal effects on a host. After all, an infected individual that lives

longer spreads more of the next generation of a pathogen and contributes to its reproductive or replicative success. Usually, a host dies only if a pathogen enters its body in overwhelming numbers, if it is a novel host (one with no coevolved defenses against that pathogen), or if a mutant strain of the pathogen emerges and can surmount the host's current array of defenses.

Being equipped with the evolutionary perspective, you probably can work out the numbers on your own. To wit: *The greater the population density of host individuals, the greater will be the kinds and frequencies of infectious diseases transmitted among them.* This brings us to the bad news.

***EBOLA* AND OTHER EMERGING PATHOGENS**   Thanks to planes, trains, and automobiles, people travel often and in droves all around the world. Among their more exotic destinations are virgin tropical forests and other remote environments where the human body was an infrequent (or nonexistent) opportunity for pathogens. At present, strange and often dangerous pathogens are opportunistic about the novel, two-legged treasurehouses of metabolic machinery and nutrients entering their habitats. Human travelers can become infected within hours and taxi such pathogens far away, and eventually back home.

We know little about the deadly, **emerging pathogens**. Some have been around for a long time and only now are taking great advantage of the increased presence of the novel human hosts. Others are newly mutated strains of existing species.

Consider *Ebola*, one of the viruses that cause a deadly, hemorrhagic fever. This RNA virus appears filamentous or circular in electron micrographs (Figure 22.20). It may have coevolved with monkeys in Africa's tropical forests. By 1976, it was infecting humans. It kills between 70 and 90 percent of its victims. There is no vaccine or treatment for the disease, which starts with high fever and flu-like aches. Within a few days, nausea, vomiting, and diarrhea begin. Blood vessels are destroyed. Blood seeps from the circulatory system, into the surrounding tissues and out through all the body's orifices. The liver and kidneys may rapidly turn to mush. Patients often become deranged and soon die of circulatory shock. There have been four *Ebola* epidemics since 1976. Each time, governments mobilized agencies around the world; quarantine procedures were implemented to limit the spread of the disease.

Or consider the *monkeypox* virus, a relative of the smallpox virus. After infecting a host, this DNA virus replicates in the lymph nodes, spleen, and bone marrow. The bloodstream delivers new viral particles to the skin, where they produce large, hard, painful, pus-filled sores. When the virus overwhelms the immune system, death follows from secondary infections, especially pneumonia.

Like smallpox, one of the most deadly human diseases, monkeypox kills as many as one in ten. Unlike smallpox, which is highly contagious, monkeypox was not much of a threat at first. Its few victims were mainly children who

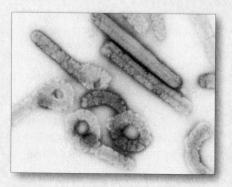

**Figure 22.20** Transmission electron micrograph of *Ebola* virus particles.

| Table 22.4 | The World's Eight Most Deadly Infectious Diseases | | |
|---|---|---|---|
| Disease | Cause | Estimated New Cases per Year | Estimated Deaths per Year |
| Acute respiratory infections* | Bacteria, viruses | 1 billion | 4.7 million |
| Diarrheas** | Bacteria, viruses, parasites | 1.8 billion | 3.1 million |
| Tuberculosis | Bacteria | 9 million | 3.1 million |
| Malaria | Sporozoans | 110 million | 2.5–2.7 million |
| AIDS | HIV | 5.6 million | 2.6 million |
| Measles | Viruses | 200 million | 1 million |
| Hepatitis B | Virus | 200 million | 1 million |
| Tetanus | Bacteria | 1 million | 500,000 |

\* Includes pneumonia, influenza, and whooping cough.
\*\* Includes amoebic dysentery, cryptosporidiosis, and gastroenteritis.

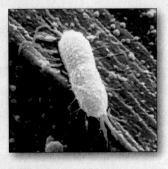

**Figure 22.21** Bits of food and *Pseudomonas* on a kitchen knife blade. The cell attached itself to the blade by pili at its surface. Think about *that* the next time you use an unwashed kitchen knife previously used on raw poultry, beef, or seafood.

had trapped, skinned, and eaten infected monkeys and other small animals in the African forests. Virus particles slipped into the body through small cracks in the hands.

A mutant monkeypox virus appears to have entered villages. More than *Ebola* and other attention-grabbing pathogens, it has the potential to break out in pandemics. The virus easily infects and replicates itself in human hosts, and it resists breakdown in the environment. Paul Ewald, an evolutionary biologist, says it has the right starting material for making a very nasty pathogen.

The smallpox vaccine works against monkeypox virus. But the World Health Organization essentially eradicated smallpox decades ago. There are few stores of the vaccine, and there are no plans to manufacture more.

**ABOUT THE DRUG-RESISTANT STRAINS** There is an old saying that, when you attack nature, it comes back at you with a pitchfork. We have already considered antibiotic resistance in several contexts, in Sections 1.4, 1.6, and 18.4. Here we reinforce the consequences in terms of the upsurge in infectious diseases.

One of the resistant pathogens is having a field day close to home. Because of economic pressures over the past few decades, the number of working mothers has skyrocketed in the United States. So has the number of preschoolers enrolled in day-care centers. In effect, each center is a population of hosts whose immune systems are still developing and vulnerable to contagious diseases. Now think of *Streptococcus pneumoniae*. This bacterium causes pneumonia, meningitis, and middle-ear infections in people of all ages; about 40,000 to 50,000 cases end in death every year. The risk of infection is 36 times greater in large day-care centers than it is for children cared for at home. As recently as 1988, drug-resistant strains of *S. pneumoniae* were practically unheard of in the United States. They are becoming the rule, not the exception.

A mere twenty-five years ago, antibiotics and vaccines were considered invincible, and the Surgeon General of the United States announced we could "close the book on infectious diseases." There are more than 6 billion people.

Most live in crowded cities. During any specified interval, as many as 50 million people are on the move within and between countries in search of a better life (Section 46.7). Even putting aside emerging pathogens, is it any wonder that the pathogens responsible for cholera, tuberculosis, and other familiar diseases are striking with a vengeance? Table 22.4 lists some sobering numbers.

**PATHOGENS LURKING IN THE KITCHEN** Each year, food poisoning hits more than 80 million people in the United States alone. It kills about 9,000 of them, mainly the very young, the very old, or those with compromised immune systems. Yet the cases reported to health agencies number only in the thousands; people often dismiss their misery as "the 24-hour flu." *Salmonella enteriditis* as well as certain strains of *E. coli* are prominent among the culprits. They typically are ingested with undercooked beef or poultry, contaminated water, unpasteurized milk or cider, and vegetables grown in fields fertilized with manure.

A few examples: In 1966, *S. enteriditis* slipped into ingredients used to make a popular ice cream. In a single outbreak, possibly as many as 224,000 people suffered from stomach cramps, diarrhea, and other symptoms of the disease *salmonella*. In 1993 in Washington State, more than 500 people became sick and three children died after eating restaurant hamburgers tainted with a pathogenic strain of *E. coli*. A recent outbreak of food poisoning was traced to unpasteurized apple juice.

Outbreaks such as these certainly grab our attention, yet pathogens lurking in the kitchen probably account for at least half of all cases of food poisoning. Examine wood or plastic cutting boards or sleek, stainless steel knives with electron microscopes and you see nooks and crannies where microbes can lurk (Figure 22.21). Carlos Enriquez and his colleagues at the University of Arizona sampled 75 dishrags and 325 sponges in several homes. Most harbored colonies of *Salmonella* and *E. coli*, as well as *Pseudomonas* and *Staphylococcus*. Bacteria can live two weeks in a wet sponge. Use the sponge to wipe down the kitchen, and you spread them about. Antibacterial soaps, detergents, and weak solutions of household bleach get rid of them. You can sanitize sponges simply by putting them through a dishwasher cycle.

Do you think food poisoning is of small concern? The annual cost of treating the infections known to be caused by food-borne pathogens ranges between 5 billion and 22 billion dollars.

## SUMMARY

1. A major divergence occurred soon after the origin of life. One lineage gave rise to eubacteria, the other to common ancestors of archaebacteria and eukaryotes. Archaebacteria (methanogens, extreme halophiles, and extreme thermophiles) live in extreme environments, like those in which life probably originated. They are unique in wall structure and other features, and they share some features with eukaryotic cells. By far, the most common existing prokaryotic cells are eubacteria.

2. All bacteria are prokaryotic cells; they do not have a nucleus. Some species have membrane infoldings and other structures in the cytoplasm. None has a profusion of organelles, which is typical of eukaryotic cells.

   a. Three basic bacterial shapes are common: cocci (spheres), bacilli (rods), and spirilla (spirals).

   b. Nearly all eubacteria have a cell wall consisting of peptidoglycan. It protects the plasma membrane and helps it resist rupturing. The composition and structure of the cell wall help identify particular bacterial species.

   c. A sticky mesh of polysaccharides (a glycocalyx) may surround the wall as a capsule or slime layer. It helps the bacterium attach to substrates and sometimes resist a host organism's infection-fighting mechanisms.

   d. Some species have one or more bacterial flagella, structures that rotate a bit like a propeller and serve in motility. Many have pili, filamentous proteins that help cells adhere to a surface or facilitate conjugation.

3. Prokaryotic fission is a cell division mechanism used only by bacteria. The mechanism involves replication of the bacterial chromosome and division of a parent cell into two genetically equivalent daughter cells.

4. Many species have plasmids, small circles of DNA that are replicated independently of the single, circular bacterial chromosome. Plasmids may be transmitted to daughter cells and may be transferred to cells of the same species or a different species by bacterial conjugation.

5. Bacteria as a group show great metabolic diversity, as in their modes of acquiring energy and carbon.

   a. *Photoautotrophs* use sunlight and carbon dioxide during photosynthesis. They include cyanobacteria and other oxygen-releasing species of eubacteria. They also include green nonsulfur and purple nonsulfur bacteria that do not produce oxygen; these get electrons from sulfur and other inorganic compounds, not from water.

   b. *Photoheterotrophs* use sunlight energy and organic compounds as sources of carbon. They include certain archaebacteria and eubacteria.

   c. *Chemoautotrophs*, including the nitrifying bacteria, use carbon dioxide but not sunlight. Different kinds get energy by stripping electrons from a variety of organic or inorganic substances.

   d. *Chemoheterotrophs* are parasites (which get carbon and energy from living hosts) or saprobes (which feed on the organic products, wastes, or remains of other organisms). Most bacterial species are in this category. They include major decomposers and pathogens.

6. Bacteria make behavioral responses to stimuli, as when they move toward regions with more nutrients.

7. Viruses are nonliving, noncellular agents that infect particular species of nearly all organisms.

   a. Each virus particle consists of a core of DNA or RNA and a protein coat that sometimes is enclosed in a lipid envelope. Many glycoproteins project, spikelike, from the envelopes. The coats of complex viruses have sheaths, tail fibers, and other accessory structures.

   b. A virus particle cannot reproduce on its own. Its genetic material must enter a host cell and direct the cellular machinery to synthesize materials necessary to produce new virus particles.

8. Nearly all viral multiplication cycles include five steps: attachment to a suitable host cell, penetration of it, viral DNA or RNA replication and protein synthesis, assembly of new viral particles, and release.

9. Two pathways are common in the multiplication cycle of bacteriophages (bacteria-infecting viruses). In a lytic pathway, multiplication is rapid and new viral particles are released by lysis. In a lysogenic pathway, the infection enters a latent period. The host cell is not killed outright, and the viral nucleic acid may undergo genetic recombination with a host cell chromosome.

10. Multiplication cycles of viruses are diverse. They are rapid or they enter a latent phase. Penetration and release of most enveloped types occur by endocytosis and budding. DNA viruses spend part of the cycle in the nucleus of a host cell. RNA viruses complete the cycle in the cytoplasm. The viral RNA is the template for mRNA synthesis and for protein synthesis.

---

### Review Questions

1. Describe the key metabolic and structural features of bacteria. Make sketches of the three basic shapes of bacterial cells. *22.1*

2. Compared to your own bodily growth, how is bacterial growth measured? *22.2*

3. With respect to bacterial classification, what are some of the pitfalls of numerical taxonomy? How are comparisons of rRNA sequences assisting classification efforts? *22.3*

4. Name a few of the photoautotrophic, chemoautotrophic, and chemoheterotrophic eubacteria. Describe some that are likely to give humans the most trouble, medically speaking. *22.4, 22.6*

5. What is a virus? Why is a virus considered to be no more alive than a chromosome? *22.7*

6. Distinguish between:
   a. microorganism and pathogen *CI*
   b. infection and disease *22.10*
   c. epidemic and pandemic *22.10*

7. Label the components of the following viruses: 22.7

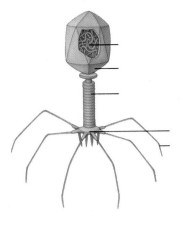 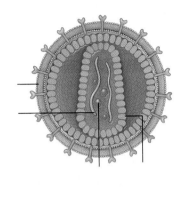

## Self-Quiz (Answers in Appendix III)

1. Nondividing bacteria have _____ chromosome(s) and may have extra circles of _____ called plasmids.
   a. one; RNA
   b. two; RNA
   c. one; DNA
   d. two; DNA

2. _____ live in habitats much like those of the early Earth.
   a. Cyanobacteria
   b. Eubacteria
   c. Archaebacteria
   d. Protozoans

3. Which of the following is not grouped with archaebacteria?
   a. halophiles
   b. cyanobacteria
   c. thermophiles
   d. methanogens

4. Bacteria reproduce by _____ .
   a. mitosis
   b. meiosis
   c. prokaryotic fission
   d. longitudinal fission

5. Eubacterial cell walls are composed of _____ ; and a sticky mesh of polysaccharides, a _____ , often surrounds the wall.
   a. peptidoglycan; plasma membrane
   b. cellulose; glycocalyx
   c. cellulose; plasma membrane
   d. peptidoglycan; glycocalyx

6. Most bacteria are _____ and include major decomposers and pathogens.
   a. photoautotrophs
   b. photoheterotrophs
   c. chemoautotrophs
   d. chemoheterotrophs

7. Viruses are _____ .
   a. the simplest living organisms
   b. infectious particles
   c. nonliving
   d. both a and b
   e. both b and c

8. Viruses have a _____ and a _____ .
   a. DNA core; carbohydrate coat
   b. DNA or RNA core; plasma membrane
   c. DNA-containing nucleus; lipid envelope
   d. DNA or RNA core; protein coat

9. Match the terms with their most suitable description.
   ____ archaebacteria
   ____ eubacteria
   ____ virus
   ____ plasmid
   ____ prokaryotic fission
   ____ methanogen
   ____ prion
   a. infectious small protein
   b. nonliving infectious particle with nucleic acid core and protein coat
   c. at home in stockyards
   d. methanogens, extreme halophiles, extreme thermophiles
   e. most common prokaryotic cells
   f. small circle of bacterial DNA
   g. bacterial cell division mechanism

## Critical Thinking

1. *Salmonella* bacteria, which cause a form of food poisoning, often live in poultry and eggs, but not in newly hatched chicks until chicks eat bacteria-laden feces of healthy adult chickens. Harmless bacteria ingested this way colonize the surface of intestinal cells, leaving no place for *Salmonella* to take hold. Some farmers raise thousands of chicks in confined quarters with no adult chickens. Should they consider feeding the chicks a known mixture of bacteria from a lab or a mixture of unknown bacteria from healthy adult chickens? Devise an experiment to test which approach may be more effective.

2. Reflect on the description of mad cow disease in Section 22.9. The FDA is moving to prohibit all protein supplements for sheep, cattle, and other ruminants. Investigate the extent of this practice in the United States and how it has been monitored.

3. Water seeping from the earth at boiling springs is above 100°C; at hot springs (Figure 22.11*d*) it approaches the boiling point. The water gradually cools as it spills out through shallow channels. Using the following list as a guide, predict which organisms you might find (a) in boiling spring water, (b) near the edges of a hot spring, and (c) along the edges of an outflow channel where the upper temperature is 72°C. (Not all species of the groups listed grow at these temperatures.)

| | |
|---|---|
| Fishes 38°C | Fungi 60°–62°C |
| Insects 45°–50°C | Cyanobacteria 70°–74°C |
| Vascular plants 45°C | Methanogens 110°C |
| Algae 55–60°C | Extreme thermophiles 113°C |

## Selected Key Terms

| | |
|---|---|
| archaebacterium 22.3 | infection 22.10 |
| bacillus (bacilli) 22.1 | lysis 22.8 |
| bacterial chromosome 22.2 | lysogenic pathway 22.8 |
| bacterial conjugation 22.2 | lytic pathway 22.8 |
| bacterial flagellum 22.1 | methanogen 22.5 |
| bacteriophage 22.7 | microorganism CI |
| cell wall 22.1 | numerical taxonomy 22.3 |
| coccus (cocci) 22.1 | pandemic 22.10 |
| disease 22.10 | pathogen CI |
| emerging pathogen 22.10 | pilus (pili) 22.1 |
| endospore 22.6 | plasmid 22.2 |
| epidemic 22.10 | prion 22.9 |
| eubacterium 22.3 | prokaryotic cell 22.1 |
| extreme thermophile 22.5 | prokaryotic fission 22.2 |
| fruiting body 22.6 | spirillum (spirilla) 22.1 |
| glycocalyx 22.1 | strain (bacterial) 22.4 |
| Gram stain 22.1 | viroid 22.9 |
| halophile 22.5 | virus 22.7 |
| heterocyst 22.6 | |

## Readings  See also www.infotrac-college.com

Brock, T., M. Madigan, J. Martinko, and J. Parker. 1997. *Biology of Microorganisms*. Eighth edition. Englewood Cliffs, New Jersey: Prentice-Hall. Exceptionally lucid, well-illustrated survey.

Daszak, P., A. Cunningham, and A. Hyatt. 21 January 2000. "Emerging Infectious Diseases of Wildlife—Threats to Biodiversity and Human Health." *Science,* 287:443–449

Hively, W. May 1997. "Looking for Life in All the Wrong Places." *Discover,* 76–85. Account of microbes that live in the most extreme environments on and in the earth.

Oernt, W. October 1999. "Killer Pox in the Congo." *Discover,* 74–79.

# 23

## PROTISTANS

### *Kingdom at the Crossroads*

More than 2 billion years ago, in tidal flats and soils, in estuaries, lagoons, lakes, and streams, prokaryotic cells were inconspicuously changing the world. Ever since the time of life's origin, Earth's atmosphere had been free of oxygen, and anaerobic bacteria had reigned supreme. Their realm was about to shrink, drastically.

An oxygen-releasing pathway of photosynthesis was operating in vast populations of bacterial cells. At first the free oxygen combined with iron in rocks. When iron-rich deposits were rusted out—oxidized— free oxygen started to accumulate. With nothing much to combine with, it slowly became more concentrated in water, then in air.

The oxygen-enriched atmosphere was a selection pressure of global dimensions. Thereafter, anaerobic species that could not neutralize the highly reactive, potentially lethal gas were restricted to black muds, stagnant waters, and other anaerobic habitats. Others developed oxygen tolerance.

It must have been a short evolutionary step from having a capacity to neutralize oxygen to using it in metabolism. Why? Aerobic species soon emerged in nearly all bacterial groups.

The metabolic innovation was the start of rampant competition for resources. In the presence of so much oxygen, energy-rich organic compounds no longer could accumulate by geochemical processes. Organic compounds formed by living cells became the premier source of carbon and energy. Novel ways of acquiring and using those organic compounds developed. Many diverse bacteria engaged in new kinds of partnerships, predations, and parasitic interactions. As outlined in Section 21.4, some of those evolutionary experiments gave rise to eukaryotic cells.

Of all existing organisms, **protistans** are the ones most like those earliest, structurally simple eukaryotes. They differ from their bacterial ancestors in several respects. At the least, protistan cells have a nucleus, large ribosomes, mitochondria, endoplasmic reticulum, and Golgi bodies. Their chromosomes consist of DNA molecules closely associated with many histones and other proteins. They assemble microtubules for use in a cytoskeleton, in spindles that move chromosomes, and in a 9 + 2 core of flagella or cilia. Many species contain chloroplasts. And, depending on the species, protistans divide by way of mitosis, meiosis, or both.

Photosynthetic types range from microscopic single cells to giant seaweeds (Figure 23.1). Saprobes resemble certain bacteria and fungi. Some of the predators and

**Figure 23.1** A tiny sampling of protistan diversity. (**a**) Along a rocky shoreline, a stand of *Postelsia palmaeformis*, one of the multicelled brown algae. (**b**) *Micrasterias*, a single-celled freshwater green alga, here dividing in two. (**c**) *Physarum*, a plasmodial slime mold. This mass of yellow-gold cells is migrating on a rotting log. (**d**) Mealtime for *Didinium*, a ciliated protozoan with a big mouth. *Paramecium*, a different ciliated protozoan, is poised at the mouth (*left*) and swallowed (*right*).

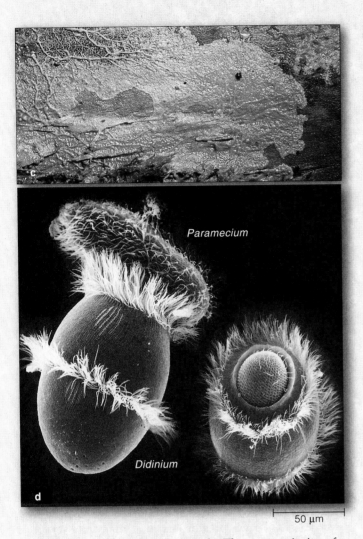

*Paramecium*

*Didinium*

50 µm

## KEY CONCEPTS

**1.** Protistans are easily distinguishable from prokaryotes (the bacteria) but are difficult to classify with respect to other eukaryotes. They also differ enormously from one another in their morphology and life-styles.

**2.** Among the fungus-like protistans are parasitic and predatory molds that produce spores. The majority of the chytrids and water molds are single-celled decomposers in aquatic habitats. The phagocytic slime molds live as single amoeboid cells and as aggregations of cells that migrate together and form spore-producing structures.

**3.** The animal-like protistans include nonphotosynthetic flagellated protozoans, sporozoans, and ciliates. Among their ranks are single-celled predators, grazers, and parasites. In any given year, hundreds of millions of people are affected by infections caused by a few dozen species of flagellated protozoans and sporozoans.

**4.** Among the plant-like protistans are single-celled, photosynthetic flagellates. Astounding numbers of them live freely or as colonies in the open ocean and other bodies of water. These protistans are important members of phytoplankton, the food producers that are the start of nearly all food webs of aquatic habitats.

**5.** The protistans most like plants are the red, brown, and green algae; indeed, many botanists prefer placing them in the plant kingdom. Most species are photosynthetic. They range from single-celled to multicelled forms that show great diversity in size, morphology, life-styles, reproductive modes, and habitats.

parasites even resemble animals. The vast majority of protistans are single-celled, but nearly every lineage also includes multicelled forms. Some groups have notably close evolutionary ties to other kingdoms.

We turn now to the major lineages. Chytrids, water molds, slime molds, protozoans, and sporozoans are heterotrophs. Among the euglenoids, chrysophytes, and dinoflagellates are species that are photosynthetic, heterotrophic, or both. Most red, brown, and green algae are evolutionarily committed to photosynthesis.

Opinions differ on how to classify many of these confoundingly diverse organisms. For instance, many biologists view the multicelled algae as protistans, and about as many view them as plants. Don't worry about memorizing the classification schemes. Simply become familiar with the various groups. Ongoing studies are bringing the evolutionary picture into sharper focus.

# PARASITIC AND PREDATORY MOLDS

Three groups of protistans, the **chytrids**, **water molds**, and **slime molds**, have members that resemble fungi in certain respects. Like fungi, all produce spore-bearing structures and are heterotrophs. Like fungi, many are saprobic decomposers or parasites. A **saprobe** secretes digestive enzymes that break down organic compounds made by other organisms, then absorbs the breakdown products. Unlike fungi, the slime molds are phagocytic predators. Also, members of all three groups differ from fungi in producing motile cells during the life cycle.

## Chytrids

The chytrids (Chytridiomycota) are common in marine and freshwater habitats, where they absorb nutrients from plant debris or from necrotic plant tissues. Most of the 575 species are saprobic decomposers, and some are parasites.

In damp environments, chytrid population growth results in fuzzy, pale masses that resemble fungal molds. In some biochemical respects, also, chytrids are similar to fungi. Chitin reinforces the cell wall of some species, as it does in fungi.

The single-celled species produce flagellated asexual spores. After spores are released, they may settle onto a host cell, then germinate and develop into globular cells (Figure 23.2*a*). These cells have rootlike absorptive structures of a type called rhizoids. When the cells are mature, they become spore-producing structures.

Like many fungi, complex multicelled species have a **mycelium** (plural, mycelia), a mesh of fine absorptive filaments called **hyphae** (singular, hypha). Individual cell walls may or may not cut across hyphae. Either way, cytoplasmic streaming through a mycelium distributes enzymes and nutrients from digested food (Section 4.9).

## Water Molds

Water molds (Oomycota) might be distantly related to yellow-green and brown algae. Most of the 580 known types are key saprobic decomposers in aquatic habitats. Of these, many are free-living species that get nutrients from plant debris in ponds, lakes, and streams. Others live inside necrotic tissues of living plants. Some water molds parasitize aquatic organisms. The pale, cottony growths you may have seen on some aquarium fish are mycelia of a parasitic type, *Saprolegnia* (Figure 23.2*b*).

Most water molds produce an extensive mycelium. Some of the hyphae differentiate into gamete-producing structures. (An *antheridium* produces male gametes; an *oogonium* produces female gametes, as in Figure 23.2*b*.) At fertilization, a male and female gamete fuse to form a diploid zygote, which develops into a thick-walled resting spore. A new mycelium develops from the spore

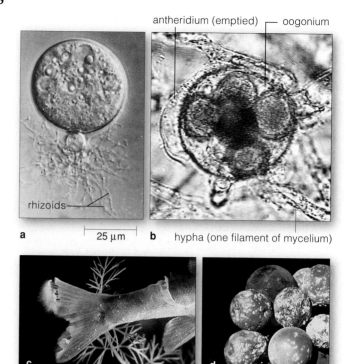

Figure 23.2 (**a**) *Chytridium confervale*, a common chytrid. This globe-shaped cell will develop into a spore-producing structure. (**b**) *Saprolegnia*, one of the parasitic water molds. (**c**) Observable outcome of a *Saprolegnia* attack on the tissues of an aquarium fish. (**d**) Grapes with downy mildew, an outcome of an attack by *Plasmopara viticola*.

after it germinates. Water molds reproduce asexually also, by way of flagellated, asexual spores.

Certain water molds are major plant pathogens. For example, grapes with *downy mildew* have been attacked by *Plasmopara viticola* (Figure 23.2*d*). Another pathogen seriously influenced human affairs. Over a century ago, Irish peasants grew potatoes as their main food crop. Between 1845 and 1860, the growing seasons were cool and damp, year after year, and conditions encouraged the rapid spread of *Phytophthora infestans*. This water mold causes *late blight*, a rotting of potato (and tomato) plants. Its abundant spores were dispersed unimpeded through the watery film on the plants. Destruction was rampant. During a fifteen-year period, one-third of the population in Ireland starved to death, died during the outbreak of typhoid fever that followed as a secondary effect, or fled to the United States and other countries.

Today, potato late blight is the single most costly biological constraint on global food production. Efforts to control it are also the cause of one of the largest uses of pesticides. Researchers at the University of Maryland and elsewhere are utilizing expressed sequence tagging and other methods of recombinant DNA technology to develop genetic databases for *P. infestans*. Their goal is

Further reading: Student Guide to InfoTrac on web site →

**Figure 23.3** A cellular slime mold, *Dictyostelium discoideum*. (**a**) Its life cycle includes a spore-producing stage. Spores give rise to free-living amoebas, which grow and divide until food (soil bacteria) dwindles. In response to cyclic AMP, a chemical signal that they secrete, amoebas stream toward one another. As the amoebas aggregate, their plasma membranes become sticky and they adhere to one another. A cellulose sheath forms around the aggregation, which starts crawling like a slug (**b–d**). Some slugs may incorporate 100,000 amoebas.

As a slug migrates, amoebas develop and differentiate into prestalk (*red*), prespore (*white*), and anteriorlike cells (*brown dots*). Prestalk cells secrete ammonia in amounts that vary with temperature and light intensity. The slug moves most rapidly in response to intermediate levels of ammonia (not too little, not too much), which correspond to warm, moist conditions (not too cold or hot, not soppy or dry). Where such conditions occur at the surface of a substrate, prestalk cells and prespore cells differentiate and form a stalked, spore-bearing structure (**e–g**). The anteriorlike cells, which sort into two groups, may function in elevating the nonmotile spores for dispersal from the top of the spore-bearing structure.

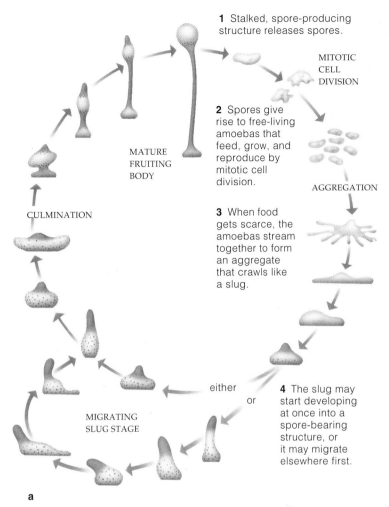

1 Stalked, spore-producing structure releases spores.

MITOTIC CELL DIVISION

2 Spores give rise to free-living amoebas that feed, grow, and reproduce by mitotic cell division.

AGGREGATION

3 When food gets scarce, the amoebas stream together to form an aggregate that crawls like a slug.

either or

4 The slug may start developing at once into a spore-bearing structure, or it may migrate elsewhere first.

MATURE FRUITING BODY

CULMINATION

MIGRATING SLUG STAGE

a

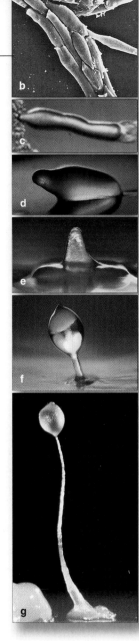

b

c

d

e

f

g

to help ensure an adequate global food supply as world population increases and resources dwindle.

## Slime Molds

All slime molds produce free-living, amoebalike cells during part of the life cycle. The group includes *cellular* slime molds (Acrasiomycota) and *plasmodial* slime molds (Myxomycota). Their cells crawl on rotting plant parts, such as decaying leaves and bark. Like true amoebas, they are phagocytic predators. They engulf bacteria and yeasts, spores, and various organic compounds. When nutrients dwindle, many of the starving cells aggregate and form a slimy mass. The mass may migrate to a new location where conditions favor growth. Collectively, the contractions of single cells move the mass. Later in the life cycle, the amoebalike cells develop into a few cell types that differentiate and form a spore-bearing structure. After the spores are released, they germinate on warm, damp surfaces. Each spore gives rise to one amoebalike cell. Sexual reproduction (by gametes) also is common among slime molds.

*Dictyostelium discoideum*, one of 70 species of cellular slime molds, is often used for laboratory studies of development. Figure 23.3 shows its life cycle. Figure 23.1*c* shows one of the 500 species of plasmodial slime molds. When the amoebalike cells of members of this phylum aggregate, each cell's plasma membrane breaks down. Cytoplasm flows unimpeded and so distributes nutrients and oxygen through the mass.

Commonly, a streaming mass will occupy several square meters and migrate if food runs out. The mass is the plasmodium. You may have observed one crossing a lawn or a road, or even climbing a tree.

Most chytrids and water molds are saprobic decomposers of aquatic habitats. Some are single cells. Like slime molds, many are free-living predators some of the time and simple experiments in multicellularity at other times.

During a slime mold life cycle, amoeboid cells aggregate to form a migrating mass. Cells in the mass differentiate, then they form reproductive structures and spores or gametes.

## THE ANIMAL-LIKE PROTISTANS

About 65,000 named species of protistans are known informally as **protozoans** ("first animals"), because they may resemble single-celled, heterotrophic protistans that gave rise to animals. They include amoeboid protozoans, ciliated protozoans, and animal-like flagellates. Table 23.1, which lists all of the groups, provides a preview of their key features.

| Table 23.1   Major Groups of Animal-Like Protistans |
| --- |
| SARCODINA<br>Amoeboid protozoans. Soft or shelled bodies; locomotion by pseudopods. Free-living or endosymbiotic heterotrophs.<br>  RHIZOPODS. Naked amoebas, foraminiferans.<br>  ACTINOPODS. Radiolarians, heliozoans<br><br>CILIOPHORA<br>Diverse free-living, sessile, and motile heterotrophs with numerous cilia. Predators or symbionts, some parasitic.<br><br>MASTIGOPHORA<br>Animal-like flagellates. Includes free-living and parasitic heterotrophs. Locomotion by one or more flagella.<br><br>APICOMPLEXA<br>Parasitic heterotrophs with a complex of structures, such as rings, tubules, and cones, at the head end. The most familiar members of a group commonly called sporozoans. |

Different protozoans are predators, parasites, and grazers that actively move through their surroundings to secure a meal. All four groups include species that house endosymbiotic algae. Free-living species live in damp soil, in a variety of freshwater habitats such as ponds and lakes, and in marine habitats. Symbiotic and parasitic species live inside or on the moist tissues of other organisms. Some species are major pathogens.

Protozoans reproduce either asexually or sexually. Many species alternate between reproductive modes in response to environmental conditions. Most commonly, asexual reproduction is by **binary fission**, a process by which the body of the individual divides in two. The division plane is random for amoebas, longitudinal for flagellates, and transverse for ciliates. Budding from the parent organism also occurs. Some species undergo multiple fission. (More than two nuclei form, then each nucleus and a bit of cytoplasm separate as a daughter cell.) Many parasitic types go through an encysted stage. The **cyst**, a resistant body covering made of their own secretions, helps them wait out stressful conditions.

**Like animals, protozoans include predatory, parasitic, and grazing species. They reproduce sexually or asexually, as by binary fission. Some species are major pathogens.**

## AMOEBOID PROTOZOANS

Let's first look at **amoeboid protozoans** (Sarcodina): the naked amoebas, foraminiferans, heliozoans, and radiolarians. None has permanent motile structures. Each moves by a combination of cytoplasmic streaming and the formation of **pseudopods** ("false feet"). As you read in Section 4.9, pseudopods are dynamic, reversible cytoplasmic extensions of a cell body. Figure 23.4 shows the rather thick pseudopods of a naked amoeba.

### Rhizopods

Naked amoebas and foraminiferans are members of a group known as **rhizopods**. The constantly changing cytoskeleton lends structural support to the soft body of a naked amoeba, which never shows any symmetry. You find these protistans in damp soil, fresh water, and saltwater. Most species are free-living phagocytes that engulf other protozoans and bacteria. Some live in the gut of invertebrates or vertebrates and normally do no harm. A few are opportunistic parasites. So are certain free-living species that happen to enter the gut along with food or water. When conditions in the gut fan their population growth, they cause intestinal problems.

*Entamoeba histolytica*, a pathogenic type, completes its life cycle in human intestines. The infective stage becomes encysted and departs from the body in feces. The encysted stage is inactive in water or soil. But after it enters a new host, the trophozoite—the active, motile feeding stage—emerges from it. Trophozoites multiply rapidly by binary fission and release lysozymes, which destroy the cells that make up the lining of the host's large intestine (colon) and rectum. Painful abdominal cramps and diarrhea are symptoms of this infectious disease, which is called *amoebic dysentery*.

Amoebic dysentery is most prevalent in subtropical and tropical regions with contaminated water supplies

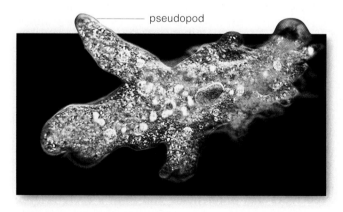

**Figure 23.4** *Amoeba proteus*, one of the naked amoebas. Compared with most of the other amoeboid protozoans, it has stubbier pseudopods. This species is a favorite for laboratory experiments in biology classes.

— pseudopod

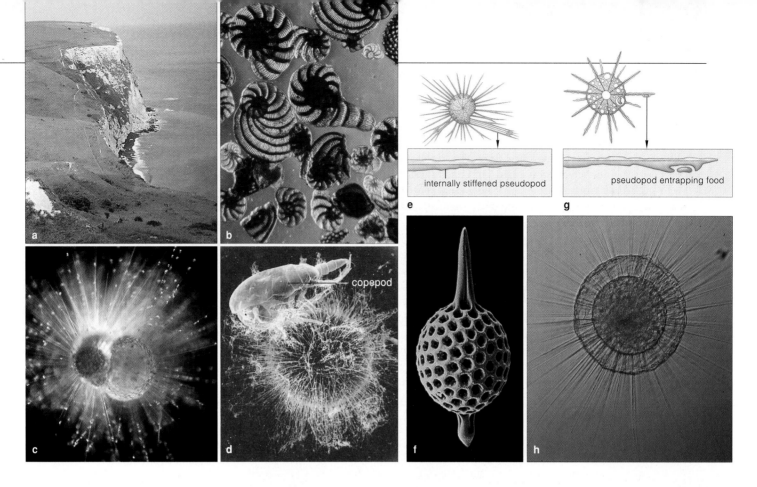

internally stiffened pseudopod

pseudopod entrapping food

copepod

**Figure 23.5** Foraminiferans: a look at how tiny cells often make grand structures. (**a**) The white cliffs of Dover, England, formed more than 200 million years ago by fossilization and compression of countless foraminiferan shells (**b**). Living species usually host symbionts, such as the golden algae visible in (**c**). Copepods and other prey become trapped in the sticky pseudopods (**d**). (**e**) Foraminiferan body plan. A core of microtubules reinforces the pseudopods. (**f**) Radiolarian skeleton. (**g**,**h**) Body plan and micrograph of a heliozoan, with its reinforced pseudopods.

and poor sanitation. Where cyst-harboring water, fruits, or vegetables abound, between 10 and 50 percent of the human population may be infected. The severity of an infection depends on the virulence of the *E. histolytica* strain and the state of an individual's immune system. People who recover show full or partial resistance to subsequent infection. Worldwide, amoebic dysentery is a leading cause of death among infants and toddlers, whose immune systems are not yet well developed.

Unlike the naked amoebas, foraminiferans have a highly perforated external "shell" of secreted organic substances, which are hardened with calcium carbonate (Figure 23.5*a–e*). Thin pseudopods extend through the perforations, and prey becomes trapped in mucus that covers them. Most foraminiferans live on the seafloor. All but 1 percent of the named species are extinct. The fossilized, compacted shells of foraminiferans make up great limestone beds, including the famed white cliffs of Dover, England (Figure 23.5*a*). We use their calcified remains in cement and blackboard chalk.

### Actinopods

**Actinopods** ("ray feet") include the radiolarians and heliozoans. Their name refers to the numerous slender, reinforced pseudopods that radiate from the body. Like foraminiferans, radiolarians are well represented in the fossil record; their ornate, silica-hardened parts resist degradation (Figure 23.5*f*). Many skeletal parts are thin and project outward with stiffened pseudopods. Nearly

all radiolarians drift with ocean currents as members of **plankton**. The word refers to aquatic communities of drifting or motile organisms that are mostly microscopic in size. A few species of radiolarians form colonies in which numerous cells are cemented together.

Heliozoans ("sun animals") have many pseudopods radiating like the sun's rays (Figure 23.5*g*,*h*). Most are free-floating or bottom-dwelling species of freshwater habitats. As with radiolarians, a membrane or capsule divides the single-celled body into two biochemically distinct zones. The inner zone contains the nucleus. The outer zone contains so many digestive vacuoles that the cell appears almost frothy in micrographs. All those vacuoles impart buoyancy to the cell, which remains suspended in the water.

---

Amoeboid protozoans are soft-bodied single cells, some with hardened skeletal elements. All form pseudopods for use in motility and prey capture. Free-living and gut-dwelling species, some parasitic, belong to this group.

---

## CILIATED PROTOZOANS

**Ciliated protozoans** (Ciliophora) typically have profuse arrays of cilia at their surface. You read about these fine motile structures in Section 4.9. Most ciliates use them for swimming through freshwater and marine habitats, where they prey on bacteria, tiny algae, and one another, as Figure 23.1*d* so aptly suggested. Their cilia beat in such a synchronized pattern over the body surface, they call to mind a soft wind through a field of tall grasses.

*Paramecium* is typical of the group (Figure 23.6). A fully grown cell is about 150 to 200 micrometers long. Outer membranes form a **pellicle**, a body covering that may be rigid or quite flexible, depending on how those membranes are organized. Inside the gullet, which starts at an oral depression at the body surface, some cilia sweep food-laden water into the cell body. There, the food becomes enclosed in enzyme-filled vesicles and is digested. As with amoeboid protozoans, *Paramecium*'s internal solute concentrations are greater than those of its surroundings. Therefore, it must continually counter

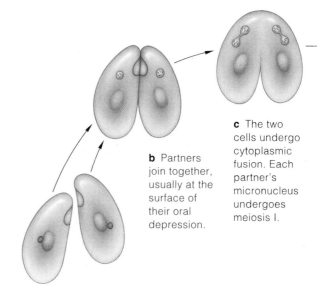

**c** The two cells undergo cytoplasmic fusion. Each partner's micronucleus undergoes meiosis I.

**b** Partners join together, usually at the surface of their oral depression.

**a** Prospective partners meet up.

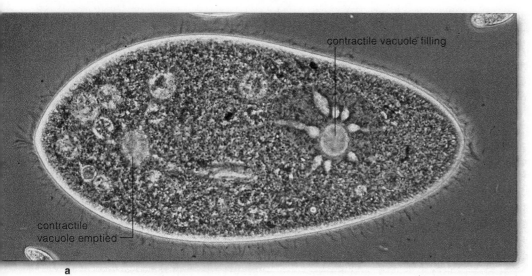

contractile vacuole filling

contractile vacuole emptied

**a**

20 µm

**Figure 23.6** Representatives of two groups of ciliated protozoans.

*Left and below:* Paramecia. (**a**) Light micrograph of a living paramecium. (**b**) Generalized diagram of the body plan for the genus *Paramecium*. (**c**) Components of the pellicle. Trichocysts, which span the pellicle, are organelles that discharge threads when the cell is irritated. They might be a defense against predators.

*Facing page:* (**d**) Scanning electron micrograph of the arrays of cilia at the surface of these single-celled predators. (**e**) Hypotrichs. This free-living species lives in the Bahamas. The hypotrichs are the most animal-like of the ciliates. They run around on leglike tufts of cilia. Some have a "head" end with modified sensory cilia.

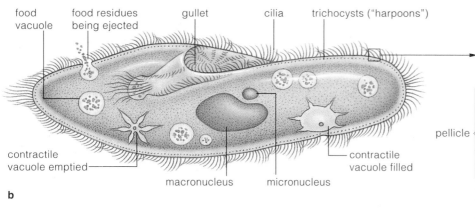

food vacuole    food residues being ejected    gullet    cilia    trichocysts ("harpoons")

contractile vacuole emptied

macronucleus    micronucleus

contractile vacuole filled

**b**

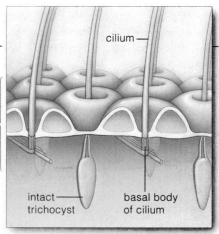

cilium

pellicle

intact trichocyst

basal body of cilium

**c**

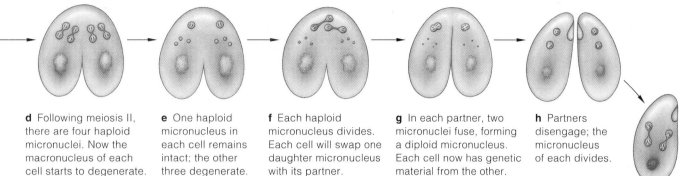

**d** Following meiosis II, there are four haploid micronuclei. Now the macronucleus of each cell starts to degenerate.

**e** One haploid micronucleus in each cell remains intact; the other three degenerate.

**f** Each haploid micronucleus divides. Each cell will swap one daughter micronucleus with its partner.

**g** In each partner, two micronuclei fuse, forming a diploid micronucleus. Each cell now has genetic material from the other.

**h** Partners disengage; the micronucleus of each divides.

**Figure 23.7** Generalized diagram of protozoan conjugation, an unusual form of sexual reproduction, as demonstrated by the two ciliates that first encounter each other in (**a**).

**i** Micronuclei divide again in each cell; the original macronucleus degenerates.

water's tendency to diffuse inward by osmosis. (Here you may wish to refer to Section 5.5.) Like amoebas, it uses **contractile vacuoles**. Tiny tubes extending from the center of these organelles collect excess water that moves osmotically into the cell body. Filled vacuoles contract and force water through a pore to the outside.

As is the case for other protistan groups, the ciliates show diversity in life-styles. Like *Paramecium*, about 65 percent of the known species are free-living and motile. Others attach themselves permanently or temporarily to substrates, often by means of a stalked component. Some form colonies. About 30 percent live as symbionts in or on other organisms. *Balantidium coli* is the largest protozoan parasite of humans, with effects like those of *Entamoeba histolytica* infections.

Ciliates can reproduce sexually and asexually, and things get interesting because each cell commonly has

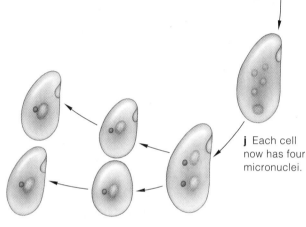

**j** Each cell now has four micronuclei.

**m** Two daughter cells result. Each has a micronucleus and a macronucleus.

**l** Cytoplasmic division now begins.

**k** Two of the micronuclei develop into macronuclei.

two types of nuclei. When a cell reproduces asexually by binary fission, a small, diploid *micro*nucleus divides by mitosis. And the large *macro*nucleus lengthens and splits in two a bit sloppily; some of the DNA may spill out. Most often, sexual reproduction occurs by a unique form of **conjugation**, as shown in Figure 23.7. In brief, the partner ciliates repeatedly divide their micronuclei, swap two daughter micronuclei, and allow two others to fuse and form a diploid macronucleus to replace the one that disappears. And you probably thought sex among the single-celled critters was simple!

Ciliated protozoans bear a profusion of cilia. These motile structures are used for swimming and for beating food into an oral cavity. Each ciliate usually has two types of nuclei, which are distributed to daughter cells in unique ways during asexual and sexual reproduction.

Like other protistans, the ciliates show great diversity in life-styles and diverse variations on the basic body plan.

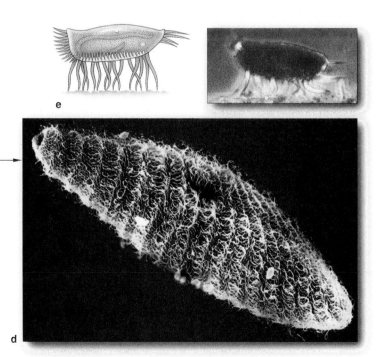

e

d

## ANIMAL-LIKE FLAGELLATES

**Animal-like flagellates** (Mastigophora) are free-living predators and parasites bearing one to several flagella. Members of one group are equipped with flagella *and* pseudopods, which is evidence of close evolutionary links with the amoebas. The free-living types abound in freshwater or marine habitats. Parasitic types live in the moist tissues of plants and animals.

*Giardia lamblia*, for example, is an internal parasite of humans, foraging cattle, and numerous wild animals such as beavers. Cells (trophozoites) survive outside the body in cysts, usually in feces-contaminated water. Ingesting the cysts is a prelude to infection. Stomach acid stimulates the cysts into releasing the trophozoites, which attach to the epithelial lining of the small intestine (Figure 23.8). After cells reproduce, they move into the large intestine (colon), then are expelled with feces. They infect new hosts who ingest contaminated water or food. The parasite also may be transmitted by way of anal intercourse with an infected partner.

Often the parasite causes only mild intestinal upsets. But it also causes a severe form of gastroenteritis called *Giardiasis*. Disease symptoms include bloating, nausea, intestinal cramps, and explosive, foul-smelling, watery diarrhea. The illness may recur over months or years.

Worldwide, 20 percent of the human population may be infected at a given time. Giardiasis is prevalent in developing countries, in overcrowded regions with poor sanitation and water quality control. In the United States, waterborne outbreaks have occurred mainly in mountains from coast to coast. For safety, hikers and

**Figure 23.8**
Scanning electron micrographs of *Giardia lamblia*, an animal-like flagellate that causes intestinal disturbances.
(**a**) This cell is adhering to epithelium by means of its sucking disk.
(**b**) When a cell detaches, its disk often leaves an impression on the epithelial surface that is distinctive when viewed with the aid of electron microscopes.

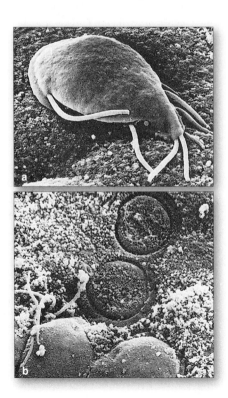

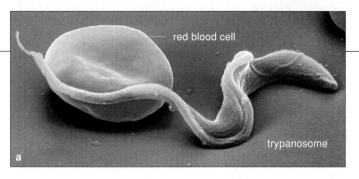

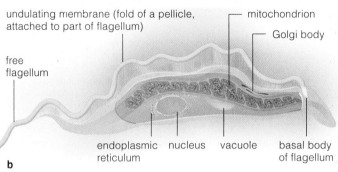

undulating membrane (fold of a pellicle, attached to part of flagellum) — mitochondrion — Golgi body

free flagellum

endoplasmic reticulum — nucleus — vacuole — basal body of flagellum

b

**Figure 23.9** Light micrograph and diagram of *Trypanosoma brucei*. This animal-like flagellate causes African sleeping sickness.

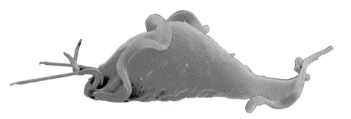

**Figure 23.10** Scanning electron micrograph of *Trichomonas vaginalis*. This pathogenic species causes trichomoniasis, a highly prevalent, sexually transmitted disease.

campers should boil any water drawn from wilderness sources, including remote streams, before drinking it.

Trypanosomes also are serious parasites in many regions. *Trypanosoma brucei* is a type that causes *African sleeping sickness*, a severe disease of the nervous system (Figure 23.9). The tsetse fly is its vector between hosts. Another type, *T. cruzi*, is prevalent in South America and Mexico. It causes *Chagas disease*. Bugs pick up the parasite when they feed on infected humans and other animals. The parasite multiplies in the insect gut, then may be excreted onto a host. Any scratches in the skin invite infection. The liver and spleen enlarge, eyelids and the face swell up, then the brain and heart become severely damaged. There is no treatment or cure.

Many of the trichomonads are among the parasitic types. *Trichomonas vaginalis*, a worldwide nuisance, can infect human hosts during sexual intercourse (Figure 23.10). Unless trichomonad infections are treated, they can damage the urinary and reproductive tracts.

**Animal-like flagellates include internal parasites that are responsible for serious disease throughout the world.**

## SPOROZOANS

**Sporozoan** is an informal name for parasitic protistans that complete part of the life cycle *inside specific cells of host organisms*. These parasites form sporozoites, a type of motile infective stage. Some have encysted stages. At one end of the cell body is a complex of distinctive structures that function in penetrating host cells. The sporozoans are often grouped as phylum Apicomplexa.

Many sporozoans cause serious human diseases. For example, *cryptosporidiosis* is a waterborne disease caused by *Cryptosporidium*. The parasite invades epithelial cells of the small intestine or the respiratory system (Figure 23.11). Two to ten days after the infection, people suffer stomach cramps, watery diarrhea, and a slight fever. The symptoms may recur, often months or years later in people with weak immune systems. Infected people who do not show symptoms can still infect others.

The parasite can survive outside the body in lakes, rivers, streams, swimming pools, jacuzzis, chlorinated drinking water, and ice. (*Cryptosporidium*, unlike most pathogens, does not succumb to chlorine.) It may now contaminate most water supplies, and the only way to get rid of it is to pass water through a reverse osmosis filter or bring it to a rolling boil for a full minute. It should not be present in distilled water, and apparently it cannot survive in hot coffee or tea.

Another sporozoan, *Pneumocystis carinii*, lives in the lungs of humans and many domestic and wild animals. Malnutrition or a weakened immune system give it the opportunity to threaten its host. For example, it causes a deadly form of pneumonia in about two-thirds of all AIDS cases. A resistant cyst forms during its life cycle. Sporozoites released during an asexual phase develop into a stage that lives in interstitial tissues of the lungs (Figure 23.12). Tiny air sacs in the lungs fill with foamy material teeming with parasites. Fever, coughing, rapid breathing, and blue skin around the mouth and eyes follow. Untreated patients die from asphyxia; they stop breathing. Even with treatment, death rates are high.

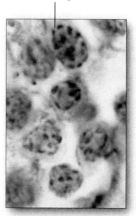

Encysted infective bodies in lung tissue

**Figure 23.12** *Above:* An encysted stage of *Pneumocystis carinii*, the agent of a form of pneumonia that can kill people with compromised immune systems, as in AIDS.

**Figure 23.13** *Right:* Standoff between a well-informed mother-to-be and a possible reservoir for the sporozoan *Toxoplasma*.

The sporozoan *Toxoplasma* uses domestic and wild cats as definitive hosts. Its intermediate hosts are other domestic and wild animals—and humans. Its cysts may be present in raw or undercooked meat. Cockroaches and flies also can move cysts from cat feces onto food. The disease *toxoplasmosis* has flu-like symptoms. It is not common, but small epidemics do occur. The disease is dangerous for immune-compromised people and for embryos of pregnant women. It can cause miscarriages or birth defects if the parasite infects the unborn child. Any cat, no matter how coddled, can pass cysts. That is why all pregnant women should avoid stray cats and never empty litterboxes, clean up housecat "accidents," or clean out children's sandboxes (Figure 23.13).

The next section takes a close look at *Plasmodium*, the sporozoan agent of malaria. Why have we focused so far on the most notorious animal-like flagellates and sporozoans? After all, fewer than two dozen species cause serious diseases in humans. The reason is that they infect hundreds of millions of people every year. There are no effective vaccines against them.

**Sporozoans are internal parasites that produce infective, motile stages. Many species form encysted stages.**

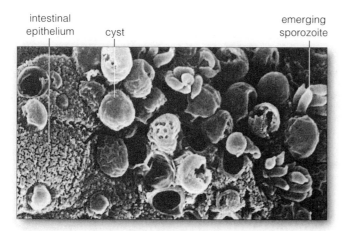

intestinal epithelium   cyst   emerging sporozoite

**Figure 23.11** *Cryptosporidium* in intestinal epithelium.

## 23.7  MALARIA AND THE NIGHT-FEEDING MOSQUITOES

Each year in Africa alone, about a million people die of *malaria*, a long-lasting disease that four different parasitic species of *Plasmodium* can cause. This type of sporozoan currently has infected more than 100 million people.

Only female mosquitoes of the genus *Anopheles* can transmit the parasites to human hosts. Their bite delivers sporozoites (an infective, motile stage) to the host's blood. The bloodstream transports sporozoites to the liver, where they undergo multiple fission. Some of the resulting cells, called merozoites, penetrate and asexually reproduce in red blood cells, which they destroy. Others develop into male and female gametocytes, which eventually mature into gametes (Figure 23.14).

Symptoms of malaria begin after infected cells abruptly rupture and release merozoites, metabolic wastes, and cellular debris into the individual's bloodstream. Shaking, chills, a burning fever, and drenching sweats are classic symptoms. After one episode, symptoms subside for a few weeks or months. Infected individuals might even feel normal, but relapses inevitably recur. In time, anemia and gross enlargement of the liver and spleen may result.

Amazingly, the *Plasmodium* life cycle is sensitive to the body temperature and oxygen levels inside humans and mosquitoes. The gametocytes cannot mature in humans, who are warm-bodied and have little free oxygen in their blood (most is bound to hemoglobin). They are induced to mature inside the mosquito, which has a lower body temperature and which incidentally slurps in oxygen from the air along with gametocyte-containing blood.

Mature gametes fuse to form zygotes. These repeatedly divide and form many sporozoites, which migrate to the female mosquito's salivary glands and await the next bite.

Malaria has been around for a long time. People were describing its symptoms more than 2,000 years ago. It got its name in the seventeenth century, when Italians made a connection between the disease and noxious gases from swamps near Rome, where mosquitoes flourished (*mal*, bad; *aria*, air). By severely incapacitating so many people, malaria contributed to the decline of the ancient Greek and Roman empires. Much later in time, it incapacitated soldiers during the United States Civil War, World War II, and the Korean and Vietnam conflicts.

Throughout human history, malaria has been most prevalent in tropical and subtropical parts of Africa. Now, however, the numbers of cases reported in North America and elsewhere are increasing dramatically, owing to the hordes of globe-hopping travelers and unprecedented levels of human immigration.

Travelers who intend to visit countries with high rates of malaria are advised to use antimalarial drugs such as chloroquine. But certain strains of *Plasmodium* are now resistant to the drugs, and a vaccine has been difficult to develop. *Vaccines* are preparations that induce the body to build up resistance to a specific pathogen. Experimental vaccines for malaria are not equally effective against all the stages that develop during sporozoan life cycles. This is generally the case for vaccines that researchers hope to develop against most parasites with complex life cycles.

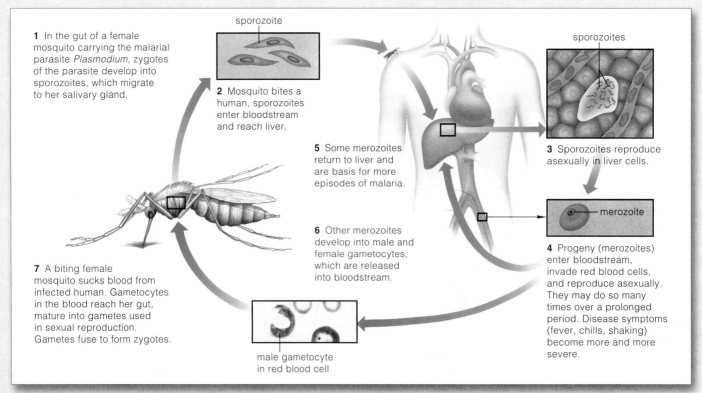

**1** In the gut of a female mosquito carrying the malarial parasite *Plasmodium*, zygotes of the parasite develop into sporozoites, which migrate to her salivary gland.

sporozoite

**2** Mosquito bites a human, sporozoites enter bloodstream and reach liver.

**5** Some merozoites return to liver and are basis for more episodes of malaria.

**6** Other merozoites develop into male and female gametocytes, which are released into bloodstream.

**7** A biting female mosquito sucks blood from infected human. Gametocytes in the blood reach her gut, mature into gametes used in sexual reproduction. Gametes fuse to form zygotes.

male gametocyte in red blood cell

sporozoites

**3** Sporozoites reproduce asexually in liver cells.

merozoite

**4** Progeny (merozoites) enter bloodstream, invade red blood cells, and reproduce asexually. They may do so many times over a prolonged period. Disease symptoms (fever, chills, shaking) become more and more severe.

**Figure 23.14**  Life cycle of one of the sporozoans (*Plasmodium*) that causes malaria.

# REGARDING "THE ALGAE"

The remainder of this chapter surveys the protistans, largely photosynthetic, that are informally known as "the algae." Table 23.2 lists six groups, although no one yet knows how many there really are.

First we will consider the euglenoids, chrysophytes, and dinoflagellates. Single-celled species dominate all three groups, which is not the case for the red, brown, and green algae. The majority belong to **phytoplankton**: communities of aquatic photosynthetic species, mainly microscopic, that drift or swim weakly through water. As food producers, these algae are major contributors to nearly all food webs in aquatic habitats. They form impressively large populations in freshwater, brackish water, and seawater (Section 7.9).

| Table 23.2 | The Mostly Photosynthetic Protistans |
|---|---|

*Mostly free-living cells. Mostly photoautotrophs, the majority being components of phytoplankton*

| EUGLENOPHYTA | Euglenoids. Single cells. Photoautotrophs, heterotrophs, or both |
|---|---|
| CHRYSOPHYTA | Chrysophytes. Mostly single cells, some colonial. Mostly photoautotrophs, some heterotrophs. Differences in pigment arrays, cell wall, and type of flagellated cell |
| | Golden algae |
| | Yellow-green algae |
| | Diatoms |
| | Coccolithophores |
| PYRRHOPHYTA | Dinoflagellates. Single cells; some symbionts. Photoautotrophs, heterotrophs |

*Mostly multicelled, aquatic, photoautotrophs with ecological roles comparable to those of land plants*

| RHODOPHYTA | Red algae |
|---|---|
| PHAEOPHYTA | Brown algae |
| CHLOROPHYTA | Green algae |

Most red, brown, and green algae are multicelled photoautotrophic species, some as tall as trees. Reflect on the variation and you might wonder how one kingdom can accommodate multicelled species in addition to all the one-celled forms. The answer is that considerable research has revealed strong resemblances in cellular structures, metabolism, chromosomal organization, life cycles, and genetics among these groups.

---

Many of the single-celled and multicelled protistans are photosynthetic species informally called "the algae."

Most algae are aquatic. Many single-celled species are key components of phytoplankton. Multicelled species play ecological roles comparable to those of land plants.

---

# EUGLENOIDS

The **euglenoids** (Euglenophyta) are a classic example of evolutionary experimentation. Euglenoids are free-living, flagellated cells that abound in freshwater and in stagnant ponds and lakes. Most of the 1,000 known species are *photoautotrophs*. The rest are *heterotrophs* that subsist on organic compounds dissolved in the water.

A *Euglena* cell has a profusion of organelles (Figure 23.15). Among these are chloroplasts with chlorophylls *a* and *b*, and carotenoids—the same pigments that occur in plant chloroplasts. *Euglena* cells also sport flagella (one long, one short) and a contractile vacuole—just as animal-like protozoans do. Moreover, like some protozoans, *Euglena* has a pellicle, this one being a flexible cover that has many spiral strips of a translucent, protein-rich material. Some of the pigments form an "eyespot" that partly shields a light-sensitive receptor. By moving the long flagellum, this single cell can keep the receptor exposed to light and so stays where the light is most suitable for its photosynthetic activities.

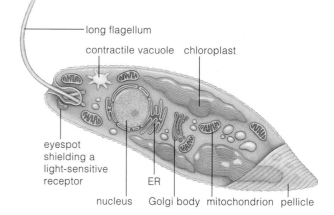

**Figure 23.15** *Euglena* body plan. Compare this diagram with the electron micrograph in Section 21.4.

Generally, "self-feeders" make their own vitamins, which are required for growth. But all photoautotrophic euglenoids must get vitamin $B_{12}$ from their surroundings (most cannot make vitamin $B_1$, either). Probably their chloroplasts originated from endosymbiotic green algae, in the manner described in Section 21.4.

Given their array of traits, could euglenoids be very close relatives of some existing protozoans? Maybe, but probably not. Researchers believe their similarities are more likely an outcome of convergent evolution.

---

The euglenoids are free-living, flagellated cells in fresh and stagnant bodies of water. Most are photoautotrophs; others are heterotrophs that feed on dissolved organic compounds.

---

# CHRYSOPHYTES AND DINOFLAGELLATES

## Chrysophytes

Most of the **chrysophytes** (Chrysophyta) are free-living photosynthetic cells that contain chlorophylls $a$, $c_1$, and $c_2$. The golden algae, yellow-green algae, diatoms, and coccolithophores belong to this large group. From time to time, some species undergo stupendous increases in population size, an event called an **algal bloom**.

GOLDEN AND YELLOW-GREEN ALGAE   Silica scales or other hard parts cover cells of many of the 500 species of the chrysophytes called **golden algae** (Figure 23.16*a*). Chlorophylls of these cells are masked by fucoxanthin, a golden-brown carotenoid. Except for the chlorophylls, some amoeboid forms closely resemble true amoebas.

**Yellow-green algae** are common chrysophytes in many aquatic habitats. Fucoxanthin is absent from the 600 known species. Most are not motile, but they have flagellated gametes. Figure 23.16*b* shows an example.

DIATOMS   The 5,600 existing species of **diatoms** have a silica "shell" of two perforated parts that overlap like a pillbox (Figure 23.16*c*). For 100 million years, finely crushed shells of at least 35,000 extinct species of these mostly photosynthetic cells accumulated at the bottom of lakes and seas. Many sediments contain the deposits, which we quarry for use in insulation, abrasives, and filters. Each year, for instance, more than 270,000 metric tons are quarried near Lompoc, California.

COCCOLITHOPHORES   Calcium carbonate plates encase the **coccolithophores** (Figure 23.16*d*). Most of the 500 or so existing species are single-celled photosynthesizers of marine habitats, especially in the tropics. Accumulations of the plates in the past helped form ocean sediments, and chalk and limestone deposits (including Dover's white cliffs). During algal blooms, mucous secretions around these cells clog fish gills. Also, a by-product of their metabolism (dimethyl sulfide) is noxious enough to make migratory fishes deviate from normal routes.

## Dinoflagellates

We know of more than 1,200 species of **dinoflagellates** (Pyrrhophyta). Most are single photosynthetic cells that are major producers in marine phytoplankton. A few are symbionts with corals. Some bioluminescent types make warm seawater shimmer at night.

Dinoflagellates bear flagella and plates of cellulose (Figure 23.17*a*). When they reproduce, the cell and its cellulose plates divide into two daughter cells. The cells are yellow-green, green, blue, brown, or red, depending on the pigments and endosymbiotic history.

**Red tides** form near coasts when certain species of dinoflagellates undergo blooms that turn the water rust-red or brown (Figure 23.17*b*). Then, the concentration of dinoflagellates may briefly reach 6 to 8 million cells per liter of water. A few species produce toxins that kill fish and other animals. In one year, for example, 150,000 grebes and 5,000 brown pelicans died at the Salton Sea in southern California. People who eat seafood tainted with some of the toxins can suffer brain damage.

## Algal Blooms and the Cell From Hell

Since 1991, dinoflagellate blooms have destroyed more than a billion fish near the coasts of North Carolina, Virginia, and Maryland. In 1999, for example, close to a million fish died in tributaries of the Pocomoke River. Atlantic menhaden piled up on banks of the Bullbeggar Creek. The agent of their death was *Pfiesteria piscicida*, known for good reason as the "cell from hell."

The life cycle of *P. piscicida* proceeds through at least twenty-four encysted, flagellated, and amoeboid forms

**Figure 23.16**   A sampling of chrysophytes. (**a**) *Synura*, a golden alga with a fishy odor, forms colonies in phytoplankton. (**b**) This colonial yellow-green alga, *Mischococcus*, occurs in phytoplankton. (**c**) Diatom shells, which fit together like a pillbox. The four small shells formed by cell division. New walls form inside a parent cell wall, so diatoms become smaller with each division. When too small, a spore forms inside the shell. It germinates and grows into a full-size diatom. (**d**) One of the coccolithophores, which have calcium carbonate plates.

**a** Perforated plates of a dinoflagellate

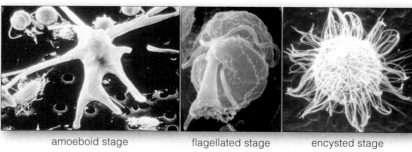

amoeboid stage          flagellated stage          encysted stage

**c** A few stages in the life cycle of *Pfiesteria*

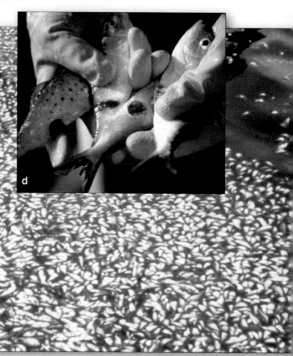

**Figure 23.17** (**a**) Scanning electron micrograph of a dinoflagellate. (**b**) *Left:* Red tide near the coast of central California. *Right:* Small portion of a fish kill resulting from a dinoflagellate bloom. (**c**) Stages in the life cycle of *Pfiesteria piscicida*. (**d**) The kind of damage inflicted by the cell from hell.

(Figure 23.17*c*). *P. piscicida* waits out adverse conditions as cysts in sediments of estuaries. Its flagellated forms (sexual and asexual) usually feed by attaching to prey and sucking out fluids. When a large school of fish (such as oily menhaden) linger to feed, their excretions incite encysted cells to emerge and release toxins, which slow down the fish and stop them from escaping. The toxins also eat away patches of fish skin. *P. piscicida* feeds on the sloughed epidermis, blood, and tissue juices oozing from the open sores (Figure 23.17*d*). Now the flagellates give rise to amoeboids that feast on dead fish. All of the changes may proceed within a matter of hours.

Nitrogen, phosphorus, and other nutrients in fertilizer runoff and raw sewage fan such algal blooms around the world. The blooms usually occur in shallow, warm water that has become enriched with nutrients. JoAnn Burkholder, a North Carolina State University botanist, correlated *Pfiesteria* blooms with hundreds of millions of gallons of raw sewage from hog and chicken farms. More than 16 million hogs are being raised in 3,500 industrial-scale hog farms in eastern North Carolina. Untreated wastes in holding lagoons often spill into rivers that drain into the poorly flushed Chesapeake Bay estuaries and into rivers in Alabama, Delaware, Florida, and Virginia. For instance, after one heavy rain, a hog farm spill exceeded by three times the volume of oil spilled from the *Valdez* (Section 51.8). In 1997 alone, *Pfiesteria* inflicted a loss of 60 million dollars on fishing and tourism industries. The huge loss finally prodded health, agricultural, and environmental agencies into starting coordinated research to address the problem.

---

**Chrysophytes and dinoflagellates help form the food base for marine communities. Human activities that fan their explosive population growth can cause enormous damage.**

---

# RED ALGAE

Of 4,100 known species of **red algae** (Rhodophyta), nearly all are marine; only 200 live in freshwater. Red algae are most abundant in warm currents and tropical seas, often at surprising depths (265 meters below the surface) in clear water. A few occur in phytoplankton. Some encrusting types contribute to the formation of coral reefs and banks (Sections 28.3 and 50.12).

Cells of some red algae have walls hardened with calcium carbonate. Different species appear red, green, purple, or nearly black, depending on which accessory pigments (phycobilins, mostly) mask the chlorophyll *a*. Phycobilins are good at absorbing green and blue-green wavelengths, which are able to penetrate deep waters. (Chlorophylls are more efficient at absorbing red and blue wavelengths, which may not reach far below the water's surface.) The chloroplasts of red algae resemble cyanobacteria, which suggests endosymbiotic origins.

The life cycle of most species includes multicelled stages, but these lack tissues or organs (Figure 23.18). Reproductive modes are diverse, with complex asexual and sexual phases. Figure 23.19 shows an example.

Red algae have a flexible, slippery texture because of mucous material in their cell walls. Agar is made of extracts of wall material from several species. The inert, gelatinous substance is used as a moisture-preserving agent in baked goods and cosmetics, as a setting agent

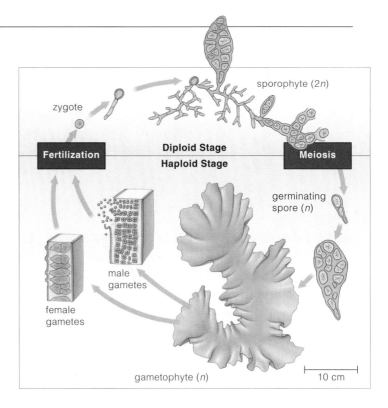

**Figure 23.19** Life cycle of *Porphyra*. From 1623 to the 1950s, Japanese fishermen cultivated and harvested a species of this red alga in early fall. The rest of the year, it seemed to vanish. Kathleen Drew-Baker studied the sheetlike form of *P. umbilicus* in the laboratory. She observed gametes forming in packets interspersed between vegetative cells near the sheet margins; the sheet was a gametophyte (gamete-producing body). She also observed gametes in a petri dish. Zygotes that formed after fertilization grew on bits of shell in the dish before developing into a tiny branching, filamentous form. This was how *Porphyra* spent most of the year! People already recognized the filamentous form as common pinkish growths on shells. But it was so different from the sheetlike form that it was viewed as a separate species. It is the alga's diploid sporophyte (spore-producing body).

Drew-Baker's discovery of alternating stages in the alga's life history, and the realization that the diploid stage could be grown on shells or other calcium-rich surfaces, revolutionized the nori industry. Within a few years, researchers worked out the life cycle of *P. tenera*, the harvested species. By 1960, *nori cultivation* was a billion-dollar industry.

**Figure 23.18** (**a**) A red alga (*Bonnemaisonia hamifera*). Such a growth pattern (branching, filamentous) is the most common. (**b**) From a tropical reef, a red alga showing sheetlike growth.

for jellies, and as culture gels. We also shape it into soft capsules for delivery of drugs and food supplements. Carrageenan, extracted from *Eucheuma*, is a stabilizer in paints, dairy products, and many other emulsions.

Humans find different species of *Porphyra* tasty as well as nutritious. You may know it from sushi bars as *nori*, a wrapping for rice and fish (Figure 23.19).

---

Red algae are photosynthetic protistans with phycobilins and other accessory pigments that mask their chlorophyll *a*. Usually, multicelled stages develop during the life cycles. Most species are aquatic. They show great diversity in size, morphology, life-styles, reproductive modes, and habitats.

---

# BROWN ALGAE

Walk along a rocky shore at low tide and you may come across olive-green or brown seaweeds, as in Figure 23.20. They are among the 1,500 species called **brown algae** (Phaeophyta). Nearly all thrive in cool or temperate seawater, ranging from the intertidal zone to the open ocean. Masses of a floating brown alga, *Sargassum,* are the basis of a great floating ecosystem in the Sargasso Sea, which lies between the Azores and the Bahamas.

a

Different species appear olive-green, golden, or dark brown, depending on the pigments. Like chrysophytes, to which they may be related, the brown algae contain carotenoids such as fucoxanthin as well as chlorophylls $a$, $c_1$, and $c_2$. They range from microscopic, filamentous *Ectocarpus* to giant kelps, twenty to thirty meters long. Like red algae, they have diverse, complex life cycles that include sexual and asexual phases. In their life cycles, too, gametophytes alternate with sporophytes.

*Macrocystis, Laminaria,* and the other giant kelps are the largest, most complex protistans. Their multicelled sporophytes have stipes (stemlike parts), blades (leaflike parts), and holdfasts (anchoring structures). Hollow, gas-filled bladders impart buoyancy to the stipes and blades and keep them upright in the water. The stipes contain tubelike arrays of elongated cells. These tubes swiftly carry dissolved sugars and other products of photosynthesis to living cells through the kelp body. Flowering plants also transport sugars through similar kinds of tubes. This is a case of convergent evolution in two unrelated groups of large, multicelled organisms.

Giant kelp beds function as productive ecosystems. Think of them as underwater forests within which great numbers of diverse bacteria and protistans, as well as fishes and other animals, carry out their lives. The sizes of kelp forests are variable, depending on changes in ocean currents and in abundances of sea urchins (which feed on kelp debris) and sea otters (which feed on sea urchins). For example, in the late 1950s, a warm current displaced the cooler currents off the California coast. *Macrocystis* did not fare well with the temperature shift, and extensive beds off La Jolla and Palos Verdes almost disappeared. Without organic debris for the sea urchins to feed upon, these invertebrates fed directly on the kelps instead. Quantities of quicklime were dumped in the water to reduce the number of sea urchins, and the kelps made a comeback. So did fishes, lobster, abalone,

bladder

b

blade

stipe

holdfast

**Figure 23.20** (**a**) Closer view of *Postelsia,* the brown alga shown in Figure 23.1*a.* You will find it thriving along coasts exposed to heavy surf from Vancouver Island on down to central California. More than a hundred deeply grooved blades top a highly resilient stipe, which is fastened to rocky substrates by a mound of anchoring structures. Its spores never disperse far from the parent sporophyte. At low tide, they simply drip onto rocks from the grooved blades. (**b**) Diagram of *Macrocystis* and an underwater view of a kelp forest. A few species live in the coastal waters of North and South America, New Zealand, Tasmania, and most of the islands in subantarctic waters.

and other species that make the kelp beds their home.

*Macrocystis* and certain other brown algae are commercially harvested. Extracts from them are used in ice creams, puddings, jelly beans, salad dressings, beers, canned and frozen foods, cough syrups, toothpastes, cosmetics, floor polishes, and paper. Alginic acid from the cell walls of some species is used to make algins, which are added to various products as thickening, emulsifying, and suspension agents. In the Far East especially, people harvest kelps as sources of food and mineral salts, and as a fertilizer for crops.

---

The brown algae range in size from the microscopic to the largest of all of the multicelled protistans. Nearly all species live in marine habitats ranging from the intertidal zone to the surface waters of the open ocean.

---

Of all protistans, the **green algae** (Chlorophyta) are structurally and biochemically most like the plants and may be their closest relatives. All are photosynthetic. As in plants, their chlorophylls are the molecular types designated *a* and *b*, and they, too, store carbohydrates as starch grains inside their chloroplasts. The cell walls of some species are composed of cellulose, pectins, and other polysaccharides typical of plants.

Figure 23.21 is a sampling of the more than 7,000 known species. They include single-celled, sheetlike, tubular, filamentous, and colonial forms. You won't be able to see many without the aid of a microscope. Most, including the *Micrasterias* cell shown in Figure 23.1*b*, live in freshwater. *Micrasterias* is one of thousands of desmid species, which are important food producers in nutrient-poor ponds and peat bogs. Green algae also grow at the ocean surface, in marine sediments, just below the surface of soil, and on rocks, tree bark, other organisms, and snow. Some are symbionts with fungi, protozoans, and a few marine animals. A colonial form (*Volvox*) is a hollow, whirling sphere of 500 to 60,000 flagellated cells. Those white, powdery beaches in the tropics are largely the work of uncountable numbers of *Halimeda* cells, which formed calcified walls, then died and disintegrated (Figure 23.21*d*). Someday, green algae

**Figure 23.21**  Representative green algae. (**a**) One marine species (*Codium*) with a pronounced branching form. Although many green algae are microscopic, *C. magnum* is taller than you are. In 1956 still another species of *Codium*, from Washington, was accidentally introduced into the Connecticut River where it empties into the Atlantic. Puget Sound populations of this alga were kept in check by herbivores that evolved with them. In its new habitat, there were no native *Codium* species or herbivores that were adapted to grazing on it. The introduced alga spread along the coast as far north as Maine and as far south as North Carolina in a little more than two decades.

(**b**) Also from a marine habitat, *Acetabularia*, fancifully called the mermaid's wineglass. Each individual is a multinucleate cell mass with a rootlike structure, stalk, and cap in which gametes form.

(**c**) Sea lettuce (*Ulva*) grows in estuaries and attaches to kelps in the seas. Reproductive cells form within and are released from the margins of the sporophyte and gametophyte, both of which have the form shown in the diagram.

(**d**) *Halimeda incrassata.* Its large, branched cells form by repeated nuclear division. Cell walls form only during reproductive phases.

(**e**) *Udotea cyathiformis*, one of 100+ microscopic *Udotea* species in tropical and subtropical waters. Many are highly calcified.

(**f**) *Volvox*, a colony of interdependent cells that resemble free-living, flagellated cells of the genus *Chlamydomonas*.

(**g,h**) Above Utah's summer timberline, phycologist Ron Hoham's footprint in "red snow" reveals the presence of dormant cells of *Chlamydomonas nivalis*, a snow alga. Red accessory pigments protect the chlorophylls of these cells.

**h** Dormant cell

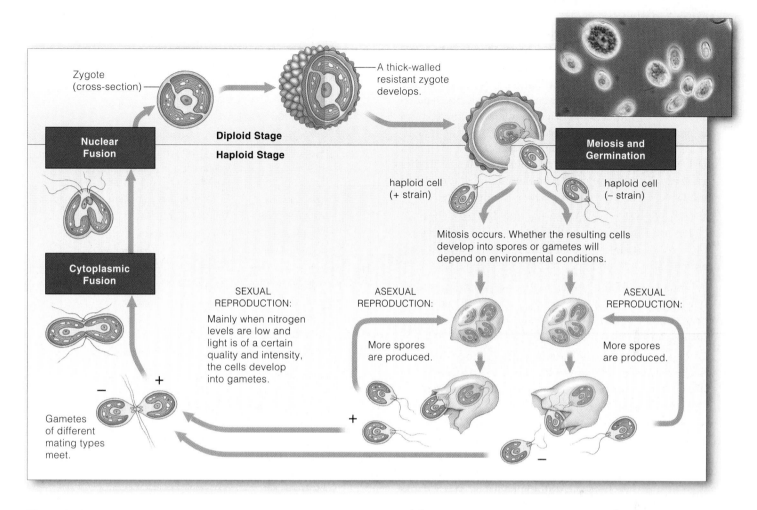

**Figure 23.22** Life cycle of a species of *Chlamydomonas*, one of the most common green algae of freshwater habitats. This single-celled species reproduces asexually most of the time and sexually under certain environmental conditions.

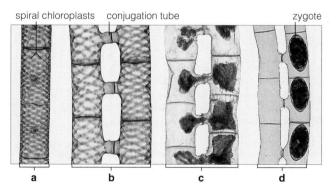

spiral chloroplasts   conjugation tube                    zygote

a          b          c          d

**Figure 23.23** One mode of sexual reproduction in *Spirogyra*, or watersilk. (**a**) The chloroplasts of this green alga are spiral and ribbonlike.

(**b**) Here, two different mating strains have made contact, and a conjugation tube has formed between cells of the adjacent haploid filaments. (**c,d**) The cellular contents of one strain pass through the tubes into cells of the other strain, where zygotes form. Zygotes develop thick walls. They undergo meiosis when they germinate and give rise to haploid filaments.

might even accompany astronauts on long missions in outer space. Green algae can grow in very small spaces, they release oxygen as a by-product of photosynthesis (the crew can't live without oxygen supplies), and they can take up carbon dioxide exhaled by the aerobically respiring crew.

Green algae employ diverse modes of reproduction. *Chlamydomonas* provides a classic example. This single-celled alga is twenty-five or so micrometers wide. It is able to reproduce sexually. Most of the time, however, it engages in asexual reproduction, with up to sixteen daughter cells forming by mitotic cell division within the confines of the parent cell wall. Daughter cells may live at home for a while, but sooner or later they leave by secreting enzymes that digest what's left of their parent. Figure 23.22 shows the life cycle of one species. To give a final example, Figure 23.23 shows *Spirogyra*, a filamentous green alga, reproducing sexually.

**The green algae show great diversity in size, morphology, life-styles, and habitats. The structure and biochemistry of some groups indicate they have close evolutionary ties to the plant kingdom.**

# SUMMARY

1. Recall that protistans and other eukaryotes differ from bacteria in key respects. At the least, eukaryotic cells have a double-membraned nucleus, mitochondria, endoplasmic reticulum, and larger ribosomes. These cells have microtubules, which are used as cytoskeletal elements, as spindles for chromosome movements, and in the 9 + 2 core of flagella and cilia. Eukaryotic cells have two or more chromosomes, each of which consists of DNA complexed with numerous histones and other proteins. Eukaryotic cells alone engage in mitosis and meiosis. Many have chloroplasts and other plastids.

2. Heterotrophic protistans include the water molds, chytrids, slime molds (cellular and plasmodial types), protozoans (true amoebas, animal-like flagellates, and ciliated species), and sporozoans. See Table 23.3.

3. The chytrids and water molds are decomposers or parasites. Like fungi, they secrete enzymes that digest organic matter, then absorb breakdown products. Some form a mycelium (a mesh of absorptive filaments).

4. Like animals, slime molds are predatory. For part of the life cycle, they crawl on substrates as free-living, phagocytic, amoebalike cells. Like fungi, they produce spore-producing structures.

5. Like animals, protozoans are predators, grazers, or parasites. The amoebas, foraminiferans, heliozoans, and radiolarians are amoeboid protozoans with or without skeletal elements. Animal-like protozoans are internal parasites or live freely in aquatic habitats. Many cause human diseases. Sporozoans are internal parasites that may form cysts during their life cycle. An example is *Plasmodium,* which causes malaria. Ciliated protozoans such as *Paramecium* typically use their cilia for motility.

6. The majority of euglenoids, chrysophytes (diatoms, golden algae, and yellow-green algae), dinoflagellates, and red, brown, and green algae are all photoautotrophic. Many are members of phytoplankton (Table 23.3).

7. The red, brown, and green algae differ from one another in their body plan, size, reproductive modes, and habitats. Most are multicelled photosynthesizers. Some species of brown algae are the largest protistans. Plants may be descended from early green algae.

8. It has taken us more than one chapter to consider features of protistans, the simplest eukaryotes, and to speculate on their evolutionary ties to other kingdoms. Table 23.4 summarizes key similarities and differences among the prokaryotes and eukaryotes.

---

## Review Questions

1. Outline some of the general characteristics of protistans. Then explain what is meant by this statement: Protistans are often classified by what they are *not.* CI

| Table 23.3 Summary of Major Groups of Protistans | |
|---|---|
| Group | Common Name |
| **HETEROTROPHIC** (decomposers, predators, grazers, parasites) | |
| Chytridiomycota | Chytrids |
| Oomycota | Water molds |
| Acrasiomycota | Cellular slime molds |
| Myxomycota | Plasmodial slime molds |
| Sarcodina | Amoeboid protozoans |
| | Rhizopods (naked amoebas, foraminiferans) |
| | Actinopods (radiolarians, heliozoans) |
| Mastigophora | Animal-like flagellated protozoans |
| Ciliophora | Ciliated protozoans |
| Apicomplexa | Major group of sporozoans |
| **PHOTOAUTOTROPHIC AND HETEROTROPHIC** | |
| Euglenophyta | Euglenoids |
| Pyrrhophyta | Dinoflagellates |
| **MOSTLY PHOTOAUTOTROPHIC** (photosynthesizers) | |
| Chrysophyta | Chrysophytes |
| | Golden algae |
| | Yellow-green algae |
| | Diatoms |
| | Coccolithophores |
| Rhodophyta | Red algae |
| Phaeophyta | Brown algae |
| Chlorophyta | Green algae |

2. Review Table 23.3, then cover the right column with a sheet of paper. Now list the common names for the major categories of protistans. *Table 23.3*

3. Correlate some structural features of a protistan from each of the groups listed with conditions in their environments:
   a. chytrids, water molds  *23.1*
   b. slime molds  *23.1*
   c. amoeboid protozoans  *23.3*
   d. ciliated protozoans  *23.4*
   e. animal-like flagellates, sporozoans  *23.5, 23.6*
   f. euglenoids, chrysophytes, dinoflagellates  *23.9, 23.10*
   g. red, brown, and green algae  *23.11–23.13*

4. Select three different protistan species, then briefly explain how they adversely affect our affairs, such as crop yields and human health.  *23.1, 23.3, 23.5–23.7*

5. Select and briefly explain how activities of photosynthetic protistan species have positive benefits for some communities of organisms, including human communities.  *23.10–23.13*

---

## Self-Quiz  (*Answers in Appendix III*)

1. Most chytrids and water molds are _____ decomposers in aquatic habitats.
   a. parasitic          c. autotrophic
   b. saprobic           d. chemosynthetic

## Table 23.4 Comparison of Prokaryotes With Eukaryotes

| | Prokaryotes | Eukaryotes |
|---|---|---|
| Organisms represented: | Bacteria only | Protistans, fungi, plants, and animals |
| Ancestry: | Two major, ancient lineages (archaebacteria, eubacteria) that evolved more than 3.5 billion years ago | Equally ancient prokaryotic ancestors gave rise to forerunners of eukaryotes which evolved more than 1.2 billion years ago. |
| Level of organization: | Single-celled | Protistans, single-celled or multicelled. Nearly all others are multicelled, with a division of labor among differentiated cells, tissues, and often organs. |
| Typical cell size: | Small (1–10 micrometers) | Large (10–100 micrometers) |
| Cell wall: | Most have distinctive walls. | Cellulose or chitin; none in animal cells |
| Membrane-bound organelles: | Very rarely | Typically profuse |
| Modes of metabolism: | Both anaerobic and aerobic | Aerobic modes predominate. |
| Genetic material: | Bacterial chromosome (and sometimes plasmids) | Complex chromosomes (DNA, many associated proteins) within a nucleus |
| Mode of cell division: | Prokaryotic fission, mostly; also budding | Nuclear division (mitosis, meiosis, or both), associated with one of various modes of cytoplasmic division |

2. _____ are free-living, amoebalike cells that crawl on rotting plant parts and engulf bacteria, spores, and organic compounds.
    a. water molds       c. sporozoans
    b. amoeboid protozoans   d. slime molds

3. In a _____ life cycle, amoeboid cells aggregate and migrate as a mass. Cells in the mass differentiate, forming reproductive structures and spores or gametes.
    a. slime mold   b. water mold   c. protozoan   d. chytrid

4. Amoebas, foraminiferans, and radiolarians are _____ .
    a. ciliated protozoans      c. amoeboid protozoans
    b. animal-like protozoans   d. sporozoans

5. Which disease is associated with trypanosomes, a type of parasitic, flagellated protozoan?
    a. Chagas disease       c. malaria
    b. toxoplasmosis      d. amoebic dysentery

6. Euglenoids and chrysophytes are mostly _____ .
    a. photoautotrophic     c. heterotrophic
    b. chemoautotrophic    d. omnivorous

7. Single-celled photosynthetic protistans, including most of the euglenoids, chrysophytes, and dinoflagellates, are members of the _____ , the "pastures" of most aquatic habitats.
    a. zooplankton      c. brown algae
    b. red algae       d. phytoplankton

8. Algin is used in ice cream, pudding, salad dressing, jelly beans, beer, cough syrup, toothpaste, cosmetics, and other products. Certain _____ are sources of algin.
    a. green algae      c. red algae
    b. brown algae     d. dinoflagellates

## Critical Thinking

1. Suppose you decide to vacation in a developing country where sanitation practices and standards of personal hygiene are poor. Having read about some of the parasitic protozoans lurking in water and damp soil, what would you consider safe to drink once you arrive there? Which kinds of foods might be best to avoid and what kinds of food preparations might make them safe to eat?

2. As you read in this chapter, red tides are associated with "algal blooms." Such blooms may follow the enrichment of aquatic habitats with water that drains into them from heavily fertilized croplands or from concentrated sources of raw sewage. After thinking about it, do you accept the resulting destruction of aquatic species, birds, and other forms of wildlife as an unfortunate but necessary side effect of human activities? If you do not accept it, how would you stop the water pollution?

And if you could stop it, what sorts of measures would you take to feed the enormous human population, which is now extremely dependent on high-yield (and very heavily fertilized) crops? How would you propose to dispose of or prevent the accumulation of fecal matter and other wastes of 6 billion people?

## Selected Key Terms

actinopod 23.3
algal bloom 23.10
amoeboid protozoan 23.3
animal-like flagellate 23.5
binary fission 23.2
brown alga 23.12
chrysophyte 23.10
chytrid 23.1
ciliated protozoan 23.4
coccolithophore 23.10
conjugation (protozoan) 23.4
contractile vacuole 23.6
cyst 23.2
diatom 23.10
dinoflagellate 23.10
euglenoid 23.9
golden alga 23.10
green alga 23.13
hypha 23.1
mycelium 23.1
pellicle 23.4
phytoplankton 23.8
plankton 23.3
protistan CI
protozoan 23.2
pseudopod 23.3
red alga 23.11
red tide 23.10
rhizopod 23.3
saprobe 23.1
slime mold 23.1
sporozoan 23.6
water mold 23.1
yellow-green alga 23.10

## Readings   See also www.infotrac-college.com

Bold, H., and M. Wynne. 1985. *Introduction to the Algae.* Second edition. Englewood Cliffs, New Jersey: Prentice-Hall. Includes descriptions of the economic importance of major algal groups.

Margulis, L. 1993. *Symbiosis in Cell Evolution.* Second edition. New York: Freeman. Paperback.

Margulis, L., and K. Schwartz. 1992. *Five Kingdoms.* Second edition. New York: Freeman. Paperback.

Satchell, M. 28 July 1997. "The Cell From Hell." *U.S. News and World Report,* 26–28. Among other cases, the article correlates hundreds of millions of gallons of raw sewage from North Carolina's factory-like hog farms with an algal bloom (by the dinoflagellate *Pfiesteria piscicada*) that killed 14 million fish.

# 24 FUNGI

## Ode to the Fungus Among Us

When push comes to shove, plants need certain fungi more than they need us. (In fact, they don't need us at all.) Fungi were there, as symbionts, when plants first invaded the land. **Symbiosis**, recall, refers to species that live closely together. In cases called **mutualism**, their interaction benefits both partners or does one of them no harm. Lichens and mycorrhizae are like this.

A **lichen** is a vegetative body in which a fungus has become intertwined with one or more photosynthetic organisms. From the Antarctic to the Arctic, you find lichens surviving in a variety of habitats that are just too hostile to support most organisms.

Lichens absorb mineral ions from substrates; some absorb nitrogen from the air. They make antibiotics against bacteria that can decompose lichens. They also make toxins against invertebrate larvae that graze on them. Coincidentally, their metabolic products help form new soils or enrich existing ones. This is what happens when lichens colonize new habitats, such as bare soil exposed by a retreating glacier. Conditions improve so much, other species move in and replace the pioneers. Actually, this may have happened when plants first invaded the land. Cyanobacteria-containing lichens even help maintain ecosystems. They put captured nitrogen into a form that plants use. *Lobaria* provides old-growth forests in the Pacific Northwest with up to 20 percent of their required nitrogen (Figure 24.1).

Lichens serve as early warnings of deteriorating environmental conditions. How? They absorb toxins but cannot get rid of them. From extensive studies in New York City and in England, we know that when lichens die around cities, air pollution is getting bad.

**a** SULFUR SHELF FUNGUS *Polyporus*

Fungi, too, enter into mutualistic interactions with young tree roots. Together they form a **mycorrhiza** (plural, mycorrhizae), which means "fungus-root." The fungus absorbs sugars from the plant, which absorbs minerals from its partner. The underground parts of a fungus thread through soil and afford a huge surface area for absorption. The fungus rapidly takes up many ions of phosphorus and other minerals when they are abundant and releases them to the plant when the ions are scarce. Many plants do not grow well at all in the absence of mycorrhizae.

Fungi help plants in still another way, because many are premier **decomposers**. Figure 24.2 shows just a few of these. Like all other decomposers, fungi break down organic compounds in their surroundings. However, few organisms besides fungi digest their dinner out on the table, so to speak. As fungi grow in or on organic matter, they secrete enzymes that digest it into bits that individual cells can absorb. Such a mode of nutrition is called **extracellular digestion and absorption**. Plants —the primary producers of nearly all ecosystems—also benefit because they can take up some of the released carbon and other nutrients.

Keep this global perspective in mind. Why? As you will see, the activities of some fungi do cause diseases in humans, pets and farm animals, ornamental plants,

**Figure 24.1** *Lobaria oregana*, one of the lichens.

**b** PURPLE CORAL FUNGUS *Clavaria*    **c** RUBBER CUP FUNGUS *Sarcosoma*

**d** BIG LAUGHING MUSHROOM *Gymnophilus*

**e** TRUMPET CHANTERELLE *Craterellus*    **f** SCARLET HOOD *Hygrophorus*

**Figure 24.2** Fungal species from southeastern Virginia. This sampling hints at the rich diversity within the kingdom Fungi.

and important crop plants. Some species are notorious spoilers of food supplies. Others help us manufacture substances ranging from antibiotics to cheeses. We tend to assign "value" to fungi and other organisms in terms of their direct effect on our lives. There is nothing wrong with battling dangerous species and admiring beneficial ones, as long as we do not lose sight of the greater roles of fungi or any other kind of organism in nature.

## KEY CONCEPTS

**1**. Fungi are heterotrophs. Together with heterotrophic bacteria, they are the biosphere's decomposers. Saprobic types obtain nutrients from nonliving organic matter. Parasitic types obtain them from tissues of living hosts.

**2**. Fungi secrete enzymes that digest food outside their body, then fungal cells absorb breakdown products. Their metabolic activities release carbon dioxide to the atmosphere and return many nutrients to the soil, where they become available to producer organisms.

**3**. Most fungi are multicelled. A mycelium, which is the food-absorbing portion of a fungal body, develops during the fungal life cycle. Each mycelium is a mesh of hyphae. The hyphae are elongated filaments that grow and develop by repeated mitotic cell divisions.

**4**. Commonly, a portion of the fungal hyphae becomes modified and weaves together to form a reproductive structure in or upon which fungal spores develop. A "mushroom" is such a structure. Germinating spores grow and develop into a new mycelium.

**5**. Many fungi are symbionts. Lichens consist of certain fungi that are partnered with algae and other organisms. Mycorrhizae consist of fungal species locked in mutually beneficial relationships with young roots of land plants. Metabolic activities of cells making up the fungal hyphae provide the plants with nutrients, and the plants provide the fungi with carbohydrates.

**6**. We tend to assign value to plants and fungi in terms of their direct effect on our lives. Our battles with the "bad" ones and reliance on the "good" ones should start from a solid understanding of their long-established roles in nature.

## Mode of Nutrition

**Fungi** are heterotrophs, meaning they require organic compounds that other organisms synthesize. Most are **saprobes**; they obtain nutrients from nonliving organic matter and so cause its decay. Others are **parasites**; they extract nutrients from tissues of a living host. When the cells of any species grow in or on organic matter, they secrete digestive enzymes and then absorb breakdown products. Again, their mode of extracellular digestion helps plants, which readily absorb some of the released nutrients and carbon dioxide by-products. Without the fungi and heterotrophic bacteria, communities would slowly become buried in their own garbage, nutrients would not be cycled, and life could not go on.

## Major Groups

For many of us, "fungi" are drab mushrooms sold in grocery stores. These are simply fungal body parts, and they are produced by a fungus with stunningly diverse relatives. The few species in Figure 24.2 don't do justice to the 56,000 fungal species we know about—and there may be at least a million more we don't know about!

We know, from the fossil record, that fungi evolved before 900 million years ago. Some accompanied plants onto the land 430 million years ago. About 100 million years later, three major lineages were well established. We call them the **zygomycetes** (Zygomycota), **sac fungi** (Ascomycota), and **club fungi** (Basidiomycota). Other, puzzling kinds known as "imperfect fungi" are lumped together but are not a formal taxonomic group. The vast majority of species in all these groups are multicelled.

## Key Features of Fungal Life Cycles

Fungi reproduce asexually quite often, but given the opportunity, they also reproduce sexually. They form great numbers of nonmotile spores. As in plants, their **spores** are reproductive cells or multicelled structures, often walled, that germinate after dispersal from the parent. In multicelled species, spores give rise to a mesh of branched filaments. The mesh, a **mycelium** (plural, mycelia), rapidly grows over or into organic matter and has a good surface-to-volume ratio for food absorption. Each filament in a mycelium is called a **hypha** (plural, hyphae). Hyphal cells commonly have chitin-reinforced walls. Their cytoplasm interconnects, so that nutrients flow unimpeded throughout the mycelium.

---

Fungi are major decomposers that engage in extracellular digestion and absorption of organic matter. Multicelled types form absorptive mycelia and spore-producing structures.

## A Sampling of Spectacular Diversity

Fungal life cycles and life-styles show dizzying variety. The most we can do here is to sample a few species, starting with the club fungi. The 25,000 or so club fungi include mushrooms, shelf fungi, coral fungi, puffballs, and stinkhorns. Figures 24.2 through 24.5 and 1.7d show examples. Some of the saprobic species are important decomposers of plant litter. As described later, other species are symbionts that live in close association with the young roots of trees in forests. The fungal rusts and smuts can destroy fields of wheat, corn, and other crop plants. Cultivation of the common mushroom (*Agaricus brunnescens*) is a multimillion-dollar business. It is the mushroom of grocery-store and pizza-topping fame. Yet some of its relatives produce toxins that can kill you or any other organism that nibbles on them.

Have you ever wondered which organisms are the oldest and the largest? *Armillaria bulbosa* is one of them. The mycelium of one individual, discovered in a forest in northern Michigan, extends through fifteen hectares of soil. (Each hectare is the equivalent of 10,000 square meters.) By some estimates, this fungus weighs more than 10,000 kilograms and has been spreading beneath the forest floor for more than 1,500 years!

**Figure 24.3** Two club fungi. (**a**) The light-red coral fungus *Ramaria*. (**b**) The shelf fungus *Polyporus*. With the exception of the rubber cup fungus, all of the fungal species shown in Figure 24.2 are club fungi.

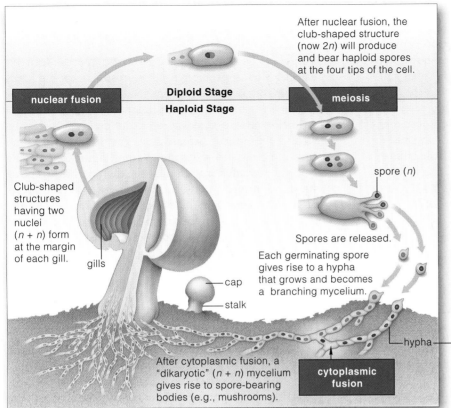

After nuclear fusion, the club-shaped structure (now 2n) will produce and bear haploid spores at the four tips of the cell.

**nuclear fusion**

**Diploid Stage**

**Haploid Stage**

**meiosis**

Club-shaped structures having two nuclei (n + n) form at the margin of each gill.

gills

cap

stalk

spore (n)

Spores are released.

Each germinating spore gives rise to a hypha that grows and becomes a branching mycelium.

hypha

After cytoplasmic fusion, a "dikaryotic" (n + n) mycelium gives rise to spore-bearing bodies (e.g., mushrooms).

**cytoplasmic fusion**

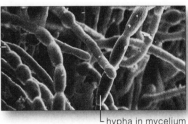

hypha in mycelium

**Figure 24.4** Generalized life cycle for many club fungi. When hyphal cells of two compatible mating strains grow together, their cytoplasm (but not nuclei) fuse. Cell divisions result in a dikaryotic mycelium (its cells each have two nuclei). Mushrooms form, with club-shaped, spore-bearing structures on their gill surface. The two nuclei in each structure fuse to form a diploid zygote, which starts the cycle anew.

When you see mushrooms or any other fungus growing outdoors, think twice before devouring them. Unlike cultivated species, such as the common mushroom, many are toxic, including the two species shown in Figure 24.5. *No one should eat any fungi that were gathered in the wild until they have been accurately identified as edible.* As the saying goes, there are old mushroom hunters and bold mushroom hunters, but no old, bold mushroom hunters.

a

b

**Figure 24.5** (**a**) Fly agaric mushroom (*Amanita muscaria*). It causes hallucinations when eaten. It was used to induce trances in ancient rituals in Central America, Russia, and India. (**b**) From California, *A. ocreata*. It can be fatal if eaten. A close relative, the death cap mushroom (*A. phalloides*), lives up to its name. Nibble as little as five milligrams of its toxin, and vomiting and diarrhea will begin eight to twenty-four hours later. Your liver and kidneys will degenerate; death may follow within a few days.

## *Example of a Fungal Life Cycle*

Probably you already know what a common mushroom (*A. brunnescens*) looks like, so let's use it as our example of a fungal life cycle. Like most of the other club fungi, this species produces short-lived reproductive bodies—**mushrooms**—that are merely its aboveground parts; its living mycelium is buried in soil or decaying wood. A mushroom has a stalk and a cap. Fine tissue sheets, or gills, line the cap's inner surface. On these gills, club-

shaped, spore-bearing structures form. The spores that form here are **basidiospores**. When a spore dispersed from a particular strain of mushroom lands on a suitable site, it germinates and gives rise to a haploid mycelium.

Suppose hyphae of two compatible mating strains make contact. They may undergo cytoplasmic fusion, but their nuclei do not fuse at once. The fused part may be the start of a *dikaryotic* mycelium, in which hyphal cells have one nucleus of each mating type (Figure 24.4). An extensive mycelium forms. And when conditions are favorable, mushrooms form. Each spore-producing structure of the mushroom is initially dikaryotic, but then its two nuclei fuse and form a short-lived zygote. The zygote undergoes meiosis, haploid spores develop outside on small stalks, then air currents disperse them.

Club fungi, the fungal group with the greatest diversity, have distinctive club-shaped, spore-bearing structures.

A fungus has a thing about spores. It produces sexual spores, asexual spores, or both, depending on contact with a suitable hypha, food availability, and how cool or damp conditions are. Its spores are usually small and dry, and air currents disperse them. Each spore that germinates can be the start of a hypha and a mycelium. Stalked reproductive structures may develop on many of the hyphae and produce asexual spores. After these spores germinate, each may be the start of still *another* extensive mycelium. In no time at all, that one fungus and staggering numbers of its descendants are busily decomposing organic stuff or pirating nutrients from a host! Look at what can happen to a slice of stale bread:

Each fungal class produces unique sexual spores. Club fungi form basidiospores, zygomycetes form spores by way of zygosporangia, and sac fungi form ascospores.

**Figure 24.6** Life cycle of the black bread mold *Rhizopus stolonifer*. Asexual phases are common. Different mating strains (+ and −) also reproduce sexually. Either way, haploid spores form and give rise to mycelia. Chemical attraction between a + hypha and a − hypha causes them to fuse. Two gametangia form, each with several haploid nuclei. Later their nuclei fuse to form a zygote. The zygote develops a thick wall, thus becoming a zygospore, and may remain dormant for several months. Meiosis occurs as the zygospore germinates, and asexual spores form.

## Producers of Zygosporangia

Consider the zygomycetes. Parasitic species feed on insects. Most saprobic types live in soil, decaying plant or animal material, and stored food. You just saw what *Rhizopus stolonifer*, the black bread mold, does to bread. A thick-walled sexual spore—a diploid zygote—forms when it reproduces sexually. A **zygosporangium**, a thin, clear covering, encloses the zygote (Figure 24.6*a*). The zygote proceeds through meiosis, then it gives rise to a specialized hypha that bears a spore sac. Some number of spores form inside the sporangium, and each may give rise to a new mycelium. Stalked hyphae grow out of such mycelia. Asexual spores form inside a spore sac perched on top of each stalk (Figure 24.6*b*).

## Producers of Ascospores

We know of more than 30,000 kinds of sac fungi. The vast majority are multicelled. Most form sexual spores called **ascospores** within sac-shaped cells. They alone form these cells, which are called asci (singular, ascus). Reproductive structures that consist of tightly interwoven hyphae enclose the asci of multicelled species. They resemble flasks, globes, and shallow cups. Figures 24.2*c* and 24.7 show some examples.

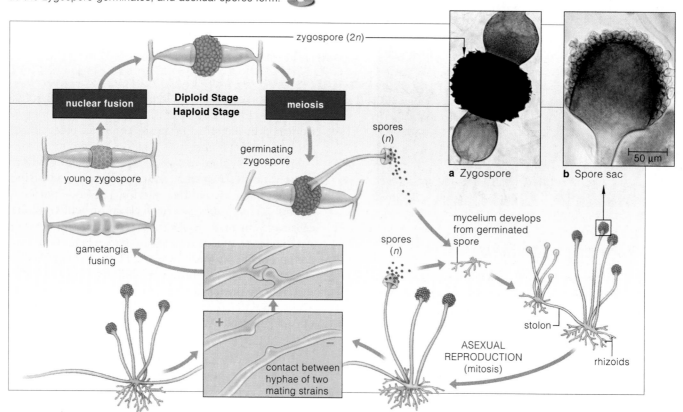

zygospore (2*n*)

| nuclear fusion | **Diploid Stage** | meiosis |

**Haploid Stage**

young zygospore

germinating zygospore

spores (*n*)

gametangia fusing

spores (*n*)

mycelium develops from germinated spore

a Zygospore    b Spore sac    50 μm

stolon

contact between hyphae of two mating strains

ASEXUAL REPRODUCTION (mitosis)

rhizoids

ascospore (sexual spore)

spore sac

b  ascocarp    c  ascocarp    d  conidia (chains of asexual spores)    e  budding yeast cell

**Figure 24.7**  Sac fungi. (**a**) Diagram and (**b**) photograph of *Sarcoscypha coccinia,* the scarlet cup fungus. Saclike structures on the cup's inner surface produce sexual spores (ascospores) by meiosis. (**c**) One of the morels (*Morchella esculenta*). This edible species has a poisonous relative. (**d**) From *Eupenicillium,* chains of asexual spores of a type called conidiospores. These drift away from the chains, like dust, after even the slightest jiggling. "Conidia" means dust. (**e**) Cells of *Candida albicans,* agent of "yeast" infections of the vagina, mouth, intestines, and skin.

spore-bearing hypha of this ascocarp

a

The vast majority of sac fungi are multicelled. They include high-priced truffles and morels (Figure 24.7*c*). Truffles are underground symbionts with roots of hazelnut and oak trees. Pigs and dogs are trained to snuffle out truffles in the woods. In France, truffles are now being cultivated on the roots of inoculated seedlings.

Other multicelled sac fungi include certain species of *Penicillium* that "flavor" Camembert and Roquefort cheeses and species that make penicillins, widely used as antibiotics. We use *Aspergillus* to make citric acid for candies and soft drinks, and to ferment soybeans for soy sauce. Most red, bluish-green, and brown fungal molds that spoil stored food also are multicelled. One species, the salmon-colored *Neurospora sitophila,* can run amok in bakeries and in research laboratories. It produces so many spores that it is extremely difficult to eradicate. One of its relatives, *N. crassa,* is an important organism in genetic research.

Sac fungi also include about 500 species of single-celled yeasts (although still other yeasts are classified as club fungi). Yeasts reproduce sexually when two cells fuse and become a spore-producing sac. Some types live in the nectar of flowers and on fruits and leaves. Bakers and vintners put fermenting by-products of vast populations of yeasts to use. For example, the carbon dioxide by-products of *Saccharomyces cerevisiae* leaven bread. The commercial production of wine and beer depends on its ethanol end product. Many yeast strains with desirable properties have been developed through artificial selection and genetic engineering. Then again, *Candida albicans,* a notorious relative of "good" yeasts, causes vexing infections in humans (Figure 24.7*e*).

roundworm    noose formed by hypha

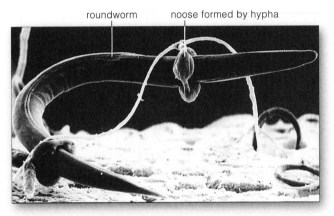

**Figure 24.8**  *Arthrobtrys dactyloides,* an imperfect fungus. This predatory species forms a nooselike ring that swells rapidly with turgor pressure when stimulated. The "hole" in the noose shrinks and captures a worm, into which hyphae will grow.

## Elusive Spores of the Imperfect Fungi

Imperfect fungi are set aside in a taxonomic holding station not because they are somehow defective but mainly because no one has yet discovered what kind of sexual spores they produce (if any). Figure 24.8 shows one of the species, a puzzling predatory fungus, that is awaiting formal classification. Investigators recently reunited the previously orphaned *Aspergillus, Candida,* and *Penicillium* with their kin—other sac fungi.

**Through their exuberant and rapid production of asexual and sexual spores, fungi take quick advantage of available organic matter, whether it has been discarded or is part of a living or dead organism. Their penchant for spore production is central to their success as decomposers and parasites.**

# THE SYMBIONTS REVISITED

Recall, from the introduction, that symbiosis refers to species that live together in close ecological association. Often one is a parasite's victim, not a partner. In cases of mutualism, interaction benefits both partners or does one of them no harm. Here are more detailed examples.

## Lichens

In the single vegetative body called a lichen, a fungus is intertwined with one or more photosynthetic species. The fungal part is the *myco*biont. The photosynthetic part is the *photo*biont. Of about 13,500 known types of lichens, nearly half incorporate sac fungi. Only 100 or so species serve as photobionts, and most often these are green algae and cyanobacteria.

A lichen forms after the tip of a fungal hypha binds with a suitable host cell. Both lose their wall, and their cytoplasm fuses or the hypha induces the host cell to cup around it. The mycobiont and the photobiont grow and multiply together. The lichen commonly has distinct layers. The overall pattern of growth may be leaflike, flattened, pendulous, or erect (Figures 24.1 and 24.9).

Lichens typically colonize sites that are hostile for most organisms, including sunbaked or frozen rocks, fence posts, gravestones, and plants, even the tops of

c

d

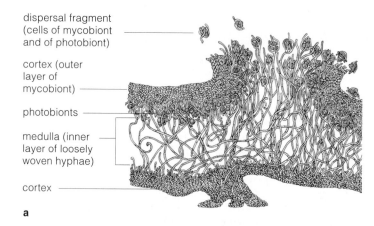

dispersal fragment (cells of mycobiont and of photobiont)

cortex (outer layer of mycobiont)

photobionts

medulla (inner layer of loosely woven hyphae)

cortex

a

b

**Figure 24.9** (**a**) Sketch of a stratified lichen, cross-section. (**b**) Encrusting lichens on a rock. (**c**) Leaflike lichen on a birch tree. (**d**) *Usnea*, a pendant lichen commonly called old man's beard. (**e**) *Cladonia rangiferina*, an erect, branching lichen.

e

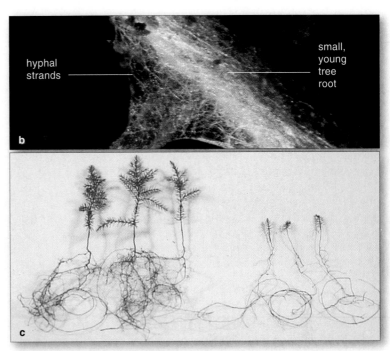

**Figure 24.10** (**a**) Lodgepole pine (*Pinus contorta*) seedling, longitudinal section. Notice the extent of the mycorrhiza compared to the shoot system, which is only about four centimeters tall. (**b**) The mycorrhiza of a hemlock tree. (**c**) Effects of the presence or absence of mycorrhizae on plant growth. The juniper seedlings at the left are six months old. They were grown in sterilized, phosphorus-poor soil with a mycorrhizal fungus. The seedlings at the right were grown without the fungus.

giant Douglas firs. Almost always, the fungus is the largest component. Cyanobacteria typically reside in a separate structure inside or outside the main body. The fungus gets a long-term source of nutrients, which it absorbs from photobiont cells. Nutrient withdrawals affect the photobiont's growth a bit, but the lichen may help shelter it. If more than one fungus is present in the lichen, it might be a mycobiont, a parasite, or even an opportunist that is using the lichen as a substrate.

### Mycorrhizae

Fungi, recall, also are mutualists with young tree roots, as mycorrhizae (Figure 24.10*a*). Without mycorrhizae, plants cannot grow as efficiently. In *ecto*mycorrhizae, hyphae form a dense net around living cells in roots but do not penetrate them (Figure 24.10*b*). Other hyphae form a velvety wrapping around roots as the mycelium radiates through soil. Ectomycorrhizae are common in temperate forests. They help the trees survive seasonal shifts in temperature and rainfall. About 5,000 fungal species, mostly club fungi, enter into such associations.

The more common *endo*mycorrhizae form in about 80 percent of all vascular plants. These fungal hyphae penetrate plant cells, as they do in lichens. Fewer than 200 species of zygomycetes serve as the fungal partner.

Their hyphae branch extensively, forming tree-shaped absorptive structures in plant cells. Hyphae also extend for several centimeters into the soil. Chapter 30 offers a closer look at these beneficial species.

### As Fungi Go, So Go the Forests

Since the early 1900s, collectors have recorded data on wild mushroom populations in European forests. As the records tell us, the number and kinds of fungi are declining at alarming rates. Mushroom gatherers can't be the cause, because inedible as well as edible species are vanishing. However, the decline does correlate with rising air pollution. Vehicle exhaust, smoke from coal burning, and emissions from nitrogen fertilizers pump ozone, nitrogen oxides, and sulfur oxides into the air. Normally as a tree ages, one species of mycorrhizal fungus gives way to another, in predictable patterns. When fungi die, trees lose this vital support system, and they become vulnerable to severe frost and drought. Are the North American forests at risk, also? Conditions there are deteriorating in comparable ways.

---

**Lichens and mycorrhizae are both symbiotic associations between fungi and other organisms, with mutual benefits.**

---

**24.5**

# A LOOK AT THE UNLOVED FEW

You know you are a serious student of biology when you view organisms objectively in terms of their place in nature, not in terms of their impact on humans in general and you in particular. As a student you salute saprobic fungi as vital decomposers and praise parasitic fungi that help keep populations of harmful insects and weeds in check. The true test is when you open the fridge to get a bowl of high-priced raspberries and discover a fungus beat you to them. The true test is when a fungus starts feeding on warm, damp tissues between your toes and turns skin scaly, reddened, and cracked (Figure 24.11*a*).

Which home gardeners wax poetic about black spot or powdery mildew on their roses? Which farmers happily give up millions of dollars a year to sac fungi that attack corn, wheat, peaches, and apples (Figure 24.11*b*)? Who cares that the sac fungus *Cryphonectria parasitica* blitzed most of the chestnut trees in eastern North America, leaving those once-magnificent trees to sprout as stubby versions of their former selves?

Who willingly inhales airborne spores of *Ajellomyces capsulatus*? After landing on soil, these dimorphic beasties form mycelia. When they alight in moist lung tissues, they form populations of yeastlike cells that cause a respiratory disease, *histoplasmosis*. Macrophages converge on infected tissues and engulf invading cells. Usually the defenders eliminate the threat, but their aggregations become calcified as the tissue heals. With heavy spore exposure, pneumonia develops. In rare cases, calcified masses form in the lymph nodes, liver, and other organs. Progressive histoplasmosis is usually lethal. The fungus thrives in regions where the nitrogen-rich droppings of many chickens, starlings, pigeons, and other birds pile up, as in spore-containing regions drained by the Ohio and Mississippi rivers. Kick up dust here and you put yourself at risk. Go spelunking in caves where roosting bats busily pile up droppings on the cave floor, and you invite the spores into your lungs.

| Table 24.1 | Some Pathogenic and Toxic Fungi* |
|---|---|
| **ZYGOMYCETES** | |
| *Rhizopus* | Food spoilage |
| **ASCOMYCETES** | |
| *Ajellomyces capsulatus* | Histoplasmosis |
| *Aspergillus* (some) | Aspergilloses (allergic reactions, sinus, ear, and lung infections; *A. flavus* toxin linked to cancers) |
| *Candida albicans* | Infection of mucous membranes |
| *Claviceps purpurea* | Ergot of rye, ergotism |
| *Coccidioides immitis* | Valley fever |
| *Cryphonectria parasitica* | Chestnut blight |
| *Microsporum, Trichophyton, Epidermophyton* | Various species cause ringworms, of scalp, body, nails, beard, athlete's foot (Figure 24.11*a*) |
| *Monilinia fructicola* | Brown rot of peaches, other stone fruits |
| *Ophiostoma ulmi* | Dutch elm disease |
| *Venturia inaequalis* | Apple scab (Figure 24.11*b*) |
| *Verticillium* | Plant wilt |
| **BASIDIOMYCETES** | |
| *Amanita* (some species) | Severe mushroom poisoning |
| *Puccinia graminis* | Black stem wheat rust |
| *Tilletia indica* | Smut of cereal grains |
| *Ustilago maydis* | Smut of corn |

* After C. Alexopoulos, C. Mims, and M. Blackwell. 1996.

Household molds cause asthma as well as sinus, ear, and lung infections, hearing loss, memory loss, dizziness, and bleeding from the lungs. They have boosted asthma rates 300 percent in the past twenty years and have been linked to nearly 100 percent of the chronic sinus infections. *Stachybotrys atra, Memnoliella, Cladosporium,* and some *Aspergillus* and *Penicillium* species are in this nasty group.

Certain fungi even have had impact on human history. Consider one notorious species, *Claviceps purpurea,* which parasitizes rye and other cereal grains. Give it credit; we use some of its by-products (alkaloids) to treat migraine headaches and, following childbirth, to shrink the uterus to prevent hemorrhaging. However, the alkaloids are toxic when ingested in large amounts. Eat a lot of bread made with tainted rye flour and you end up with *ergotism.* The symptoms include vomiting, diarrhea, hallucinations, hysteria, and convulsions. Untreated, the disease can turn limbs gangrenous, and it can end in death.

Ergotism epidemics were common in Europe during the Middle Ages, when rye was a major crop. Ergotism also thwarted Peter the Great, the Russian czar who was obsessed with conquering ports along the Black Sea for his nearly landlocked empire. Soldiers laying siege to the ports ate mostly rye bread and fed rye to their horses. The former went into convulsions and the latter into "blind staggers." Possibly, outbreaks of ergotism were used as an excuse to launch witch-hunts in early American colonies.

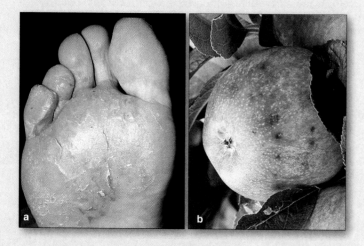

**Figure 24.11** Love those fungi! (**a**) Athlete's foot, courtesy of *Epidermophyton floccosum.* (**b**) Apple scab, trademark of *Venturia inaequalis.*

## SUMMARY

1. Fungi are heterotrophs and major decomposers. The saprobes feed on nonliving organic matter; parasites feed on the tissues of living organisms. Some species of fungi are symbiotic partners with other organisms. The cells of all species secrete digestive enzymes that break down food to small molecules, which the cells absorb.

2. Nearly all fungi are multicelled. The food-absorbing portion, the mycelium, consists of a mesh of filaments (hyphae). Aboveground reproductive structures, such as mushrooms, form from tightly interwoven hyphae.

3. The major groups of fungi are the zygomycetes, the ascomycetes (sac fungi), and the basidiomycetes (club fungi). Each is characterized by distinctive sexual and asexual spores. When a sexual phase cannot be detected or is absent from the life cycle, a fungus is assigned to an informal category called the imperfect fungi.

4. Lichens are mutualistic associations of fungi with photosynthetic partners (green algae and cyanobacteria, mostly). Mycorrhizae are mutualistic associations of a fungus and the young roots of plants. Fungal hyphae provide nutrients for their symbiont, which provides the fungus with carbohydrates.

### Review Questions

1. Describe the fungal mode of nutrition, and explain how the structure of mycelia facilitates this mode. *24.1*

2. How does a lichen differ from a mycorrhiza? *CI, 24.4*

### Self-Quiz (Answers in Appendix III)

1. New mycelia form after _____ germinate.
   a. hyphae   b. mycelia   c. spores   d. mushrooms

2. A "mushroom" is _____ .
   a. the food-absorbing part of a fungal body
   b. the part of the fungal body not constructed of hyphae
   c. a reproductive structure
   d. a nonessential part of the fungus

3. A mycorrhiza is a _____ .
   a. fungal disease of the foot     c. parasitic water mold
   b. fungus-plant relationship      d. fungus of barnyards

4. Parasitic fungi obtain nutrients from _____ .
   a. tissues of living hosts        c. only living animals
   b. nonliving organic matter       d. none of the above

5. Saprobic fungi derive nutrients from _____ .
   a. nonliving organic matter       c. root hairs
   b. living organisms               d. both b and c

6. Match the terms appropriately.
   _____ zygomycetes      a. mushrooms, shelf fungi
   _____ conidia          b. sac-shaped cells
   _____ hypha            c. chains of asexual spores
   _____ club fungi          in *Eupenicillium*
   _____ asci             d. each filament in a mycelium
   _____ sac fungi        e. black bread mold
                          f. truffles, morels, some yeasts

**Figure 24.12** Reproductive structures of *Pilobolus*, a name from a Greek word for "hat-thrower." The "hats" actually are spore sacs.

### Critical Thinking

1. *Pilobolus* is a type of fungus that commonly dines on horse dung. Each morning, stalked reproductive hyphae emerge from irregularly spaced piles of dung. By early afternoon, they have dispersed spores to sunlit grasses where horses feed. The spores pass through the horse gut unharmed and exit with their own pile of dung. At the tip of each stalked hypha is a dark-walled, spore-containing sac (Figure 24.12). Just below the sac, the stalk is differentiated into a vesicle, swollen with a fluid-filled central vacuole. At the base of the vesicle is a ring of light-sensitive, pigmented cytoplasm. The stalk bends as it grows until its wall is parallel with the sun's rays and light strikes all of the ring. When that happens, turgor pressure builds up inside the central vacuole until the vesicle ruptures. The forceful blast can propel spore sacs two meters away—which is amazing, considering that the stalk is less than ten millimeters tall. Reflect on the examples of fungi discussed in this chapter. Would you say *Pilobolus* is a zygomycete, club fungus, or sac fungus?

2. Diana sees in the laboratory that the fungus *Trichoderma* grows well in distilled water. It continues to do so even after she rigorously treats the water and glassware to remove all traces of organic carbon. This fungus is not a photoautotroph. Suggest a metabolic life-style that lets it grow under these conditions.

3. *Trichoderma* is being tested as a natural pest control agent. Laboratory experiments demonstrated that some strains of this fungus combat other fungi that cause plant diseases. Some even promote seed germination and plant growth. During one set of twenty trials, workers increased lettuce yields by 54 percent. What concerns must be addressed before *Trichoderma* can be released into the environment for commercial applications?

### Selected Key Terms

| | | |
|---|---|---|
| ascospore *24.3* | fungus *24.1* | sac fungus |
| basidiospore *24.2* | hypha *24.1* | (ascomycetes) *24.1* |
| club fungus | lichen *CI* | saprobe *24.1* |
| (basidiomycetes) *24.1* | mushroom *24.2* | spore (fungal) *24.1* |
| decomposer *CI* | mutualism *CI* | symbiosis *CI* |
| extracellular digestion | mycelium *24.1* | zygomycetes *24.1* |
| and absorption *CI* | mycorrhiza *CI* | zygosporangium *24.3* |
| | parasite *24.1* | |

### Readings   See also www.infotrac-college.com

Moore-Landecker, E. 1996. *Fundamentals of the Fungi.* Fourth edition. Englewood Cliffs, New Jersey: Prentice-Hall.

# 25 PLANTS

## *Pioneers In a New World*

Seven hundred million years ago, no shorebirds stirred and noisily announced the dawn of a new day. There were no crabs to clack their claws together and skitter off to burrows. The only sounds were the rhythmic muffled thuds of waves in the distance, at the outer limits of another low tide.

More than 3 billion years before, life had its beginning somewhere in the waters of the Earth. And now, quietly, the invasion of the land was under way.

Why did it happen? Astronomical numbers of photosynthetic cells had come and gone, and the oxygen-producing types had slowly changed the atmosphere. High above the Earth, the sun's energy had converted much of the oxygen into a dense ozone layer. That layer became a shield against lethal doses of ultraviolet radiation, which had kept early organisms beneath the water's surface.

Were cyanobacteria the first to adapt to intertidal zones, where mud dried out with each retreating tide? Were they the first to spread into shallow, freshwater streams flowing down to the coasts? Probably so. From fossil evidence, we know that later in time, green algae and fungi made the same journey together.

Every plant around you today is a descendant of ancient species of green algae that lived near the water's edge or made it onto the land. Diverse fungi still associate with nearly all of them. Together, plants and fungi became the basis of communities in coastal lowlands, near the snow line of mountains, and in just about all places in between (Figure 25.1).

We have a few tantalizing fossils of the first pioneers. We also are learning about them through comparative biochemistry and studies of existing species. Today, as in the late Precambrian, cyanobacteria and green algae grow in mats in nearshore waters and on the banks of freshwater streams (Figure 25.1a). After a volcano erupts or a glacier retreats, cyanobacteria are the first to colonize the barren rocks. Symbiotic associations between green algae and fungi follow. Gradually their organic products and remains accumulate and create pockets of soil. Then mosses and other species of plants

**Figure 25.1**  (**a**) Filaments of a green alga, massed in a shallow stream. More than 400 million years ago, green algal species that might have been ancestral to all plants, past and present, lived in similar streams that meandered down to the shores of early continents. (**b**) One land-dwelling descendant of those ancestral forms—a Ponderosa pine high above the floor of Yosemite Valley in the Sierra Nevada of California. (**c**) Flowers of one of the most highly prized flowering plants—orchids—growing on a branch of a living tree in a tropical rain forest. With this chapter, we turn to the beginning—and end of the line— of some ancient lineages.

can become established in the newly forming soil and further enrich it.

With this chapter we turn to the plant kingdom. Nearly all plants are multicelled photoautotrophs. They absorb energy from the sun, carbon dioxide from the air, and some minerals dissolved in water to synthesize organic compounds. These metabolic wizards also can split water molecules. In doing so, they get stupendous numbers of the electrons and hydrogen atoms required for growth into multicellular forms as tall as the giant redwoods, as extensive as an aspen forest that is one continuous clone.

We know of at least 295,000 kinds of existing plants. Be glad their ancient ancestors left the water. Without them, we humans and other land-dwelling animals never would have made it onto the evolutionary stage.

## KEY CONCEPTS

**1.** With very few exceptions, the plant kingdom consists of multicelled photoautotrophs. From earlier chapters, you know that plants use chlorophylls *a* and *b* as their main photosynthetic pigments. In this respect they are like green algae, which are their closest relatives.

**2.** Unlike their algal ancestors, which were adapted to aquatic habitats, nearly all existing plants live on land.

**3.** In general, plants are structurally adapted to intercept sunlight, absorb water and mineral ions, and conserve water. Their lignin-reinforced tissues permit upright growth. Root systems mine the soil for water and ions, and internal tissues conduct water and solutes to all living cells in belowground and aboveground parts.

**4.** Land plants are reproductively adapted to withstand dry periods. During the life cycle, a sporophyte develops roots, stems, and leaves. It holds on to its developing gametes and supplies them with water and food resources. And it disperses the new generation in ways that are responsive to the conditions of specific habitats.

**5.** Early divergences gave rise to the bryophytes, then seedless vascular plants, and then seed-bearing vascular plants. Of these categories, the seed producers were the most successful in radiating into drier environments.

**6.** The seed-bearing vascular plants called gymnosperms include the cycads, ginkgos, gnetophytes, and conifers. The angiosperms, another group of vascular plants, bear flowers as well as seeds. There are two classes of flowering plants, informally called the dicots and monocots.

### Overview of the Plant Kingdom

The plant kingdom includes at least 295,000 species of photoautotrophs and a few heterotrophs. Most kinds are **vascular plants**, defined in part by internal tissues that conduct water and solutes through roots, stems, and leaves. Fewer than 19,000 species are *non*vascular plants called **bryophytes**. Plants, like photoautotrophic bacteria and protistans, are producers for ecosystems.

Liverworts, hornworts, and mosses are bryophytes. The whisk ferns, lycophytes, horsetails, and ferns are *seedless* vascular plants. Cycads, ginkgos, gnetophytes, and conifers belong to a group of *seed-bearing* vascular plants called **gymnosperms**. The **angiosperms**, another group of vascular plants, bear flowers and seeds. Dicots and monocots are two classes of flowering plants.

The ancestors of plants evolved in the seas by 700 million years ago. About 265 million more years passed before simple stalked plants appeared along coasts and streams. Evolutionarily, the pace picked up after that. Within 60 million years, plants radiated through much of the land. Long-term changes in their structure and reproduction explain how the diversity came about.

### Evolution of Roots, Stems, and Leaves

Simple underground structures started to evolve when plants first colonized the land. In the lineages that led to vascular plants, they developed into root systems. Most **root systems** have many underground absorptive structures that collectively afford a large surface area. These rapidly take up soil water and dissolved mineral ions. In many species, root systems anchor the plant.

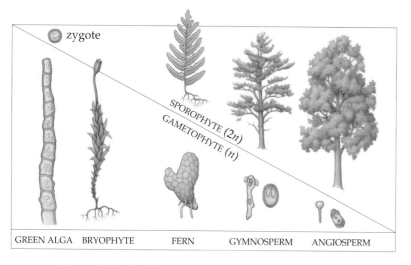

**Figure 25.2** Evolutionary trend among plants, from gametophyte (haploid) dominance to sporophyte (diploid) dominance in the life cycle. These representative existing species range from a green alga (*Ulothrix*) to a flowering plant. The trend occurred when early plants were colonizing habitats on land. See also Section 10.5.

Aboveground, **shoot systems** evolved. These have stems and leaves, which efficiently absorb energy from the sun and carbon dioxide from the air. Stems grew and branched extensively only after plants developed a biochemical capacity to synthesize and deposit **lignin**, an organic compound, in cell walls. Collectively, cells with lignified walls are strong enough to structurally support stems, which grow in patterns that increase the total light-intercepting surface of leaves.

Cellular pipelines for water and solutes evolved in many plants. The pipelines were major factors in the evolution of roots, stems, and leaves. They developed as components of xylem and phloem, which are two vascular tissues. **Xylem** distributes water and dissolved ions to all of the plant's living cells. **Phloem** distributes dissolved sugars and other photosynthetic products.

Life on land also depended on water conservation, which had not been a problem in most aquatic habitats. Shoots became protected by a **cuticle**, a waxy coat that helps conserve water on hot, dry days. Also, **stomata** (singular, stoma), tiny openings across the surfaces of leaves and some stems, helped control the absorption of carbon dioxide and restrict evaporative water loss. Later chapters describe these tissue specializations.

### From Haploid to Diploid Dominance

As early plants radiated into higher, drier places, their life cycles changed. Think about the gametes of algae, which can get together only in the presence of liquid water. As earlier chapters showed, a *haploid* (*n*) phase in the form of **gametophytes** (gamete-producing bodies) dominates their life cycles. The diploid (2*n*) phase is the zygote, which forms when gametes fuse at fertilization.

Now look at Figure 25.2. *In most plant life cycles, the diploid phase dominates.* After a diploid zygote forms at fertilization, mitotic cell divisions and cell enlargements transform it into a multicelled diploid body, of a type called a **sporophyte**. Pine trees are an example. In time, some cells of the sporophyte undergo meiosis and give rise to haploid cells of a type called **spores** (sporophyte means spore-producing body). Later, the spores divide by way of mitosis and give rise to the gametophytes.

The shift to diploid dominance was an adaptation to land habitats, most of which show seasonal changes in the availability of free water and dissolved nutrients. Long ago in those challenging habitats, natural selection must have favored sporophytes with well-developed root systems. Young roots of such systems interact with fungal symbionts (Section 24.4). The association, called a mycorrhiza, enhances the plant's uptake of water and scarce minerals, even during dry seasons.

Unlike algae and bryophytes, vascular plants have a sporophyte that is larger and structurally far more

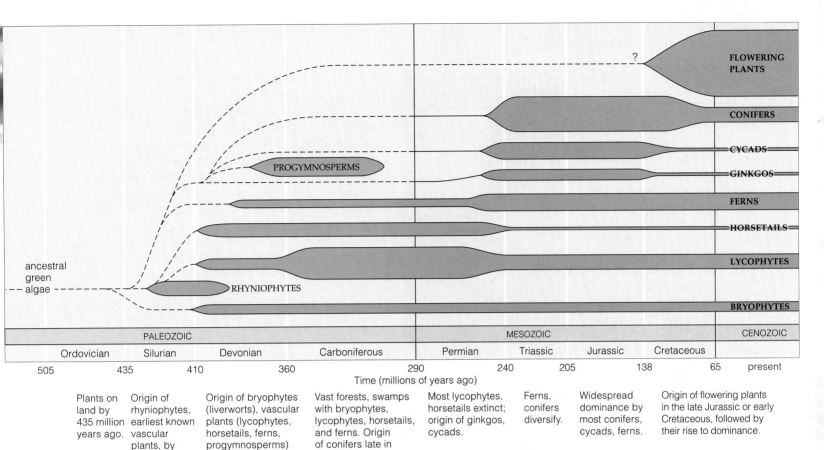

| | PALEOZOIC | | | | MESOZOIC | | | | CENOZOIC |
|---|---|---|---|---|---|---|---|---|---|
| | Ordovician | Silurian | Devonian | Carboniferous | Permian | Triassic | Jurassic | Cretaceous | |

| 505 | 435 | 410 | 360 | 290 | 240 | 205 | 138 | 65 | present |

Time (millions of years ago)

| Plants on land by 435 million years ago. | Origin of rhyniophytes, earliest known vascular plants, by mid-Silurian. | Origin of bryophytes (liverworts), vascular plants (lycophytes, horsetails, ferns, progymnosperms) by end of Devonian. | Vast forests, swamps with bryophytes, lycophytes, horsetails, and ferns. Origin of conifers late in Carboniferous. | Most lycophytes, horsetails extinct; origin of ginkgos, cycads. | Ferns, conifers diversify. | Widespread dominance by most conifers, cycads, ferns. | Origin of flowering plants in the late Jurassic or early Cretaceous, followed by their rise to dominance. |

**Figure 25.3** Milestones in plant evolution. Where a dashed line turns solid, this indicates when the lineage originated. The width of each lineage indicates variations in the range of diversity over time.

complex than the gametophyte. As you will see shortly, the gametophytes of seedless vascular plants develop independently of the sporophyte that produces them. But they protect their gametes, and after fertilization they nourish the embryo sporophytes. The sporophyte became most dominant among the gymnosperms and, later, the angiosperms. It retains, nourishes, and protects developing gametophytes as well as young sporophytes. *And it does so until environmental conditions are suitable for fertilization and for dispersal of the new generation.*

### Evolution of Pollen and Seeds

Like some seedless species, seed-bearing plants produce not one but two types of spores. We call this condition *hetero*spory, as opposed to *homo*spory (only one type). In both gymnosperms and angiosperms, one type of spore develops into female gametophytes, where eggs form and become fertilized. The other spore type gives rise to **pollen grains**, which develop into the mature, sperm-bearing male gametophytes. Pollen grains hitch rides on air currents, insects, birds, and so on; they do not require free-standing water to reach the eggs. In this respect they differ greatly from algae. The evolution of pollen grains contributed to the successful radiation of seed-bearing plants into high and dry habitats.

Seed production also was adaptive in drier habitats. Female gametophytes (and eggs) of seed-bearing plants form inside nutritive tissues and a jacket of cell layers. Each **seed** consists of an embryo sporophyte, nutritive tissues, and a protective coat, which develops from the jacket. Seeds can withstand hostile conditions. It was no coincidence that seed plants rose to dominance during Permian times, when shifts in climate were extreme.

Before turning to the spectrum of diversity among plants, take a look at Figure 25.3. You may wish to use it as a map of the branching evolutionary roads.

The plant kingdom includes multicelled, photosynthetic species called bryophytes, seedless vascular plants, and seed-bearing vascular plants. Most of these live on land.

In most lineages, structural adaptations to life on land included root and shoot systems, waxy cuticles, stomata, vascular tissues, and lignin-reinforced tissues.

Sporophytes with well-developed roots, stems, and leaves came to dominate the life cycle of most land plants. Parts of these complex sporophytes nourish and protect fertilized eggs and embryos until conditions favor their growth.

Some plants started to produce two types of spores, not one. This led to the evolution of male gametes well adapted for dispersal without liquid water and to the evolution of seeds.

## THE BRYOPHYTES

Today, the bryophyte lineage consists of about 18,600 species called **mosses**, **liverworts**, and **hornworts**. These nonvascular plants are mostly well adapted to grow in fully or seasonally moist habitats. However, you will find some mosses growing in deserts and on windswept plateaus of Antarctica. Mosses particularly are sensitive to air pollution. Where air quality is poor, mosses are few or absent.

All known bryophytes are less than twenty centimeters, or eight inches, tall. They do have leaflike, stemlike, and rootlike parts, but these do not contain xylem or phloem. Like lichens and some algae, bryophytes can dry out and then revive after absorbing moisture. Most have rhizoids. Rhizoids are elongated

cells or threadlike structures that attach gametophytes to the soil and serve as absorptive structures.

Bryophytes are the simplest plants to display three features that evolved early in land plants. *First*, a cuticle prevents water loss from aboveground parts. *Second*, a cellular jacket around the parts that produce sperm and eggs holds in moisture. *Third*, of all plants, bryophytes alone have large gametophytes that do not depend on sporophytes for nutrition. Instead, embryo sporophytes start to develop inside gametophyte tissues—and even at maturity, they are *attached to* the gamete-producing body and still gain some nutritional support from it.

With 10,000 species, mosses are the most common bryophytes. The gametophytes of some species grow in clusters and form low, cushiony mounds (Figure 25.4a). Those of others commonly grow in branched, feathery

**Figure 25.4** (**a**) Moss-covered rocks near a small cascading stream. (**b**,**c**) Photograph and life cycle of a moss (*Polytrichum*), one of the common bryophytes. The moss sporophyte remains attached to the gametophyte and is dependent upon it. The gametophyte provides it with nutrients and water.

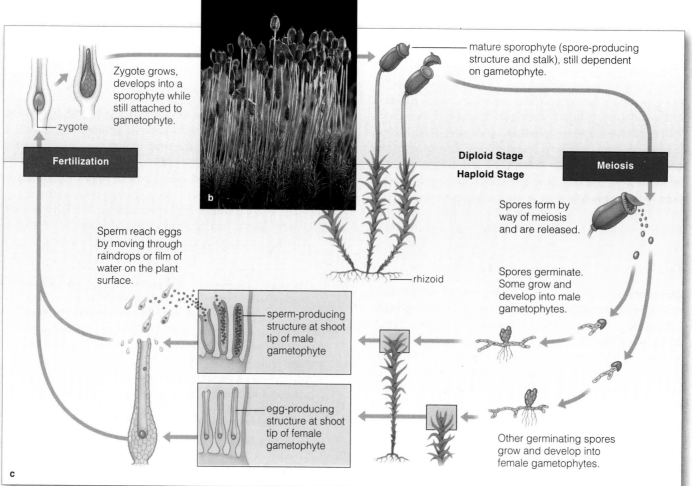

Zygote grows, develops into a sporophyte while still attached to gametophyte.

zygote

**Fertilization**

Sperm reach eggs by moving through raindrops or film of water on the plant surface.

sperm-producing structure at shoot tip of male gametophyte

egg-producing structure at shoot tip of female gametophyte

mature sporophyte (spore-producing structure and stalk), still dependent on gametophyte.

**Diploid Stage**

**Haploid Stage**

**Meiosis**

Spores form by way of meiosis and are released.

Spores germinate. Some grow and develop into male gametophytes.

Other germinating spores grow and develop into female gametophytes.

rhizoid

a

b

**Figure 25.5** (**a**) Peat bog in Ireland. This family is cutting blocks of peat and stacking them to dry as a fuel source for their home. Peat also is harvested on a scale that is large enough to generate electricity in peat-burning power plants.

(**b**) Gametophyte of a peat moss (*Sphagnum*). A few sporophytes are attached to them. Brown, jacketed structures on the white stalks are the sporophytes. Because of its fine antiseptic properties and high absorbency, peat moss was used as an emergency poultice on wounds of soldiers during World War I.

male gametophyte     female gametophyte     gemmae

a        b        c

**Figure 25.6** *Marchantia*, one of the liverworts. Like other liverworts, this nonvascular plant reproduces sexually. Unlike the other types, it forms (**a**) male reproductive parts and (**b**) female reproductive parts on different plants. (**c**) *Marchantia* also reproduces asexually by way of gemmae, multicelled vegetative bodies that develop in tiny cups on the plant body. Gemmae grow and develop into individual plants after splashing raindrops transport them to suitable sites.

patterns on tree trunks and branches in humid climates. Eggs and sperm develop in tiny, jacketed vessels at the shoot tips of gametophytes. Sperm reach the eggs by swimming through a film of water on plant parts. After fertilization, the zygotes give rise to sporophytes, each composed of a stalk and a jacketed structure in which spores will develop.

Figure 25.5 shows one of 350 kinds of peat mosses (*Sphagnum*). Whereas most bryophytes grow slowly, the peat mosses can grow fast enough to yield twelve metric tons of organic matter per hectare, which is about twice as much as corn plants yield. They soak up five times as much water as cotton does, owing to large, dead cells in their leaflike parts. They also produce acids that inhibit the growth of bacterial and fungal decomposers. Their remains accumulate into compressed, excessively moist mats called **peat bogs**. In cold and temperate regions, peat bogs cover an area equal to one-half of the United States. Only the most acid-tolerant plants, such as larch,

cranberries, blueberries, and Venus flytraps, can grow in the bogs, which can be as acidic as vinegar.

Nearly all the peat harvested and dried in Ireland and elsewhere is burned to generate electricity in power plants. Compared to coal burning, peat fires generate fewer pollutants. Every so often, peat harvesters come across exceptionally well-preserved bodies of humans who lived 2,000 to 3,000 years ago. Some ancient bogs apparently were sites of ceremonial human sacrifices.

So as not to dwell on the macabre, let us leave this section with the liverworts and their interesting ways of reproducing, as described in Figure 25.6.

Bryophytes are nonvascular plants with flagellated sperm that require liquid water to reach and fertilize the eggs.

A sporophyte of these plants develops within gametophyte tissues. It remains attached to the gametophyte and receives some nutritional support from it.

# EXISTING SEEDLESS VASCULAR PLANTS

Figure 25.7*a* shows one of the early seedless vascular plants. Descendants of certain lineages are still with us; we call them **whisk ferns**, **lycophytes**, **horsetails**, and **ferns**. Like their ancestors, they differ from bryophytes in three key respects. The sporophyte does not remain attached to a gametophyte, it has true vascular tissues, and it is the larger, longer lived phase of the life cycle.

Most seedless vascular plants live in wet, humid places, and their gametophytes lack vascular tissues. Water droplets clinging to the plants are the only means by which flagellated sperm can reach the eggs. The few species in dry habitats reproduce sexually during brief, seasonal pulses of heavy rains. In a sense, whisk ferns, lycophytes, horsetails, and ferns are the "amphibians" of the plant kingdom. They have not fully escaped the aquatic habitats of their ancestors.

## Whisk Ferns

Whisk ferns (Psilophyta), which are not ferns, resemble a whisk broom. Florist suppliers commonly cultivate them in Hawaii, Texas, Louisiana, Florida, Puerto Rico, and other tropical or subtropical regions. One genus,

*Psilotum*, is a unique vascular plant, for its sporophytes have no roots or leaves. The photosynthetic, branched stems have scalelike projections and, internally, xylem and phloem (Figure 25.7*b*). Belowground are **rhizomes**, branching, short, mostly horizontal absorptive stems.

## Lycophytes

About 350 million years ago, lycophytes (Lycophyta) included tree-sized members of swamp forests. About 1,100 far tinier species exist today. The most familiar are club mosses, members of communities in the Arctic, the tropics, and regions in between. Many form mats on forest floors. One type, called the resurrection plant, is common in Texas, New Mexico, and Mexico.

Sporophytes of most club mosses have leaves and a branching rhizome that gives rise to vascularized roots and stems. Some have nonphotosynthetic, cone-shaped leaf clusters with spore-producing structures (Figure 25.7*c*). Each cluster is a **strobilus** (plural, strobili). After spores disperse, they germinate and develop into small, free-living gametophytes. *Selaginella* is heterosporous; two kinds of spores develop in the same strobilus.

## Horsetails

Tree-sized sphenophytes (Sphenophyta) flourished in ancient swamp forests. Twenty-five or so smaller species of one genus, *Equisetum*, made it to the present. These are the horsetails. Their body plan changed very little over the past 300 million years.

Horsetails thrive in streambank muds, vacant lots, roadsides, and other disrupted habitats. Figure 25.7*d–f* shows the vegetative, photosynthetic stems and fertile stems of one species. Its spores give rise to free-living

**Figure 25.7** (**a**) *Cooksonia*, one of the earliest known vascular plants, no more than a few centimeters tall. It probably grew in mud flats. Its upright, branching stems had a cuticle. Its spores formed in structures at stem tips. Compare Figure 18.2*c*. (**b**) Sporophytes of a whisk fern (*Psilotum*), a seedless vascular plant. Pumpkin-shaped, spore-producing structures form at the ends of stubby branchlets. (**c**) Sporophyte of one of the lycophytes (*Lycopodium*). (**d**) Vegetative stem of *Equisetum*, which resembles a horsetail. (**e**) Fertile, nonphotosynthetic stems of *Equisetum*. (**f**) Closer look at the spore-producing structure of a fertile stem.

a

strobilus, an aggregation of spore-producing structures, at tip of a vegetative shoot of the horsetail sporophyte

**f** Each petal-shaped structure of a strobilus contains many spores that formed by way of meiotic cell divisions.

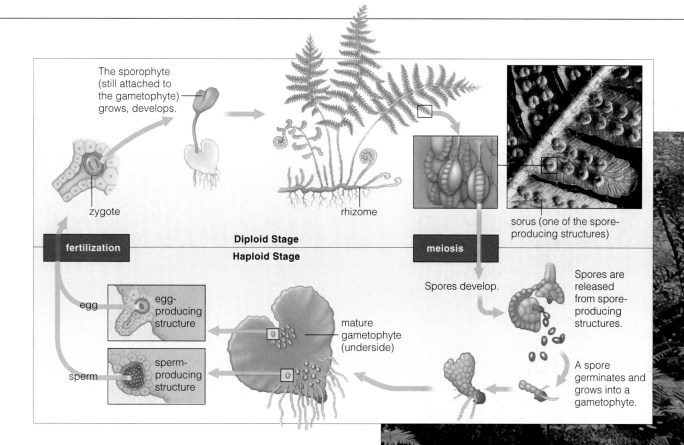

The sporophyte (still attached to the gametophyte) grows, develops.

zygote

sorus (one of the spore-producing structures)

**fertilization**

**Diploid Stage**

**Haploid Stage**

**meiosis**

egg

egg-producing structure

sperm

sperm-producing structure

mature gametophyte (underside)

Spores develop.

Spores are released from spore-producing structures.

A spore germinates and grows into a gametophyte.

rhizome

gametophytes 1 millimeter to 1 centimeter across. The sporophytes of most horsetails have rhizomes, hollow photosynthetic stems, and scale-shaped leaves. Strands of xylem and phloem are arrayed in a ring in the stems. Silica-reinforced ribs structurally support these stems and give them a texture like sandpaper. Pioneers of the American West, who did not have many places to store and wash towels, gathered horsetails on the westward journeys and used them as pot scrubbers.

## Ferns

With 12,000 or so species, the ferns (Pterophyta) are the largest and most diverse group of seedless vascular plants. All but about 380 are native to the tropics, but they are popular houseplants all over the world. Their size range is stunning. Some floating species are less than 1 centimeter across. Some tropical tree ferns are 25 meters (82 feet) tall. One climbing fern has a modified leaf stalk about 30 meters long.

Most ferns have vascularized rhizomes that give rise to roots and leaves. Exceptions include tropical tree ferns and epiphytes. (*Epiphyte* refers to any aerial plant that grows attached to tree trunks or branches.) While they develop, young fern leaves are coiled, rather like a fiddlehead. At maturity, the leaves (fronds) commonly are divided into leaflets.

You may have noticed rust-colored patches on the lower surface of many fern fronds. Each patch, a cluster

**Figure 25.8** Life cycle of a fern. The photograph shows ferns growing in a moist habitat in Indiana. Ferns with finely divided fronds are in the foreground.

of spore-producing structures, is a sorus (plural, sori). At dispersal time, the spore-producing structures snap open with such force that the released spores catapult through the air. Each spore that germinates develops into a small gametophyte. You can see an example, a green, heart-shaped gametophyte, in Figure 25.8.

Seedless vascular plants (whisk ferns, lycophytes, horsetails, and ferns) have sporophytes adapted to conditions on land. Yet they have not entirely escaped their aquatic ancestry. When they reproduce sexually, their flagellated sperm cannot reach the eggs unless liquid water is clinging to the plant.

## 25.4 ANCIENT CARBON TREASURES

Three hundred million years ago, about halfway through the Carboniferous, mild climates prevailed and swamp forests carpeted the wet lowlands of the continents. The absence of pronounced seasonal swings in temperature favored plant growth through much of the year. The plants having lignin-reinforced tissues and well-developed root and shoot systems had the competitive edge under these growth conditions, and some of them evolved into giants. Massive-stemmed lycophyte trees—the giant club mosses—topped out at nearly forty meters (Figure 25.9). Each of their strobili produced as many as 8 billion microspores or hundreds of megaspores (Section 25.5). Being so high above the forest floor, dispersal of the new generations was a cinch. Giant horsetails, including species of *Calamites*, were close to twenty meters tall. Often their aboveground stems, which grew from rapidly spreading underground rhizomes, formed dense thickets.

As it happened, the sea level rose and fell fifty times during the Carboniferous. Each time the sea receded, the swamp forests flourished. When the sea moved back in, forest trees became submerged and buried in sediments that protected them from decay. Gradually the sediments compressed the saturated, undecayed remains into peat. Each time more sediments accumulated, they subjected the peat to increased heat and pressure that made it even more compact. In this way, compressed organic remains were transformed into great seams of **coal** (Figure 25.9).

With its high percentage of carbon, coal is energy rich and is one of our premier "fossil fuels." It took a fantastic amount of photosynthesis, burial, and compaction to form each major seam of coal in the Earth. It has taken us only a few centuries to deplete much of the world's known coal deposits. Often you will hear about annual "production rates" for coal or some other fossil fuel. But how much do we really produce each year? None. We simply *extract* it from the Earth. Coal is a nonrenewable source of energy.

*Lepidodendron*

stem of a giant lycophyte (*Lepidodendron*)

seed fern (*Medullosa*); probably related to the progymnosperms, which may have been among the earliest seed-bearing plants

stem of a giant horsetail (*Calamites*)

**Figure 25.9** Reconstruction of a Carboniferous forest. The boxed inset shows part of a seam of coal.

Further reading: Student Guide to InfoTrac on web site →

# RISE OF THE SEED-BEARING PLANTS

About 360 million years ago, as the Devonian gave way to the Carboniferous, the first seed-bearing plants arose. In terms of diversity, numbers, and distribution, they would become the most successful groups of the plant kingdom. Seed ferns, gymnosperms, and (much later) angiosperms were the dominant groups. They differed from seedless vascular plants in three crucial respects.

First, seed-bearing plants produce pollen grains, the sperm-bearing male gametophytes. Remember, these plants produce two types of spores. Their **microspores** give rise to pollen grains. Unlike the spores of seedless vascular plants, they do not have a "tetrad scar," which marks the cleavage planes between four spores that form during meiotic cell division (Figure 25.10).

Like a suitcase, a pollen grain is a means of getting its contents (the sperm) to the eggs, even during times of prolonged drought. Seedless vascular plants do not have such an advantage; without predictable rains and moisture, their sperm simply cannot reach the eggs, and this has adverse effects on reproductive success. By contrast, pollen grains of gymnosperms simply drift with air currents. Those of angiosperms also are loaded onto insects, birds, bats, and other animals that truck them to the eggs. **Pollination** is the name for the arrival of pollen grains on the female reproductive structures. By this process, seed-bearing plants escaped dependence on free water for fertilization.

Second, besides microspores, seed-bearing plants also produce **megaspores**. These develop inside **ovules**, the female reproductive structures which, at maturity, are seeds (Figure 25.11). Each ovule consists of a female gametophyte (with its egg cell), nutrient-rich tissue, and a jacket of cell layers which, recall, develops into the seed coat. One zygote will form inside the ovule when a sperm reaches and fertilizes the egg. An embryo sporophyte will develop, and when the time comes to say good-bye to the parent plant, the coat around the seed will afford protection for the journey.

Third, compared to seedless vascular plants, the gymnosperms had water-conserving traits, including thicker cuticles and stomata recessed below the leaf surface. The traits gave gymnosperms competitive advantages in drier, cooler environments. Such environments were ushered in when Carboniferous gave way to the Permian. Before then, **seed ferns** of the type shown in Figure 25.9 rose to dominance, and they prevailed for about 70 million years. The seed ferns probably have evolutionary links with **progymnosperms**, which might have been one of the earliest plants to produce seeds. (At the least, they bore seedlike structures.) When the global climate did become cooler and drier, the swamplands disappeared. So did seed ferns. New kinds of seed-bearers—cycads, conifers, and other gymnosperms—would replace them.

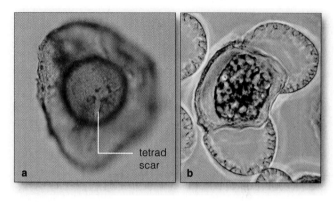

**Figure 25.10** (**a**) From the Devonian-Carboniferous boundary, a fossilized spore of a lycophyte. Its tetrad scar is typical of the spores of seedless plants. (**b**) A pine pollen grain (*Pinus*). LIke the pollen of other seed-bearing plants, it has no tetrad scar.

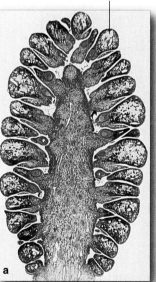 

ovule, which may mature into a seed

seed (mature ovule); the surrounding ovary matured to form the fleshy fruit

**Figure 25.11**
(**a**) Longitudinal section through a seed-bearing cone of a pine. These seeds are not protected by sporophyte tissues; they are exposed on scales of a cone. (**b**) A mature ovule, or seed, of a peach (*Prunus*), one of the angiosperms. It is protected by a seed coat, which is enclosed within the fleshy, edible tissue of the fruit.

In time, gymnosperms were the primary producers that sustained the dinosaurs. What some folks call the Age of Dinosaurs, botanists call the Age of Cycads.

---

**Seed-bearing plants rely on pollen grains, ovules that mature into seeds, and tissue adaptations to dry conditions.**

---

With a bit of history behind us, we turn now to a survey of some of the existing gymnosperms. Unlike the seeds of flowering plants, which are enclosed in a chamber called an ovary, gymnosperm seeds are perched, in an exposed way, on a spore-producing structure. (*Gymnos* means naked; *sperma* is taken to mean seed.)

## Conifers

**Conifers** (Coniferophyta) are woody trees and shrubs that have needlelike or scalelike leaves and that bear seeds exposed on cone scales. Conifer **cones** are clusters of modified leaves that surround the spore-producing structures. Figure 25.12 gives examples.

Most conifers shed some leaves all year long yet stay leafy, or *evergreen*. A few are *deciduous*, meaning that they shed all of their leaves in the fall. Among the conifers are the most abundant trees of the Northern Hemisphere (pines), the tallest (the coast redwoods), as well as the oldest (the bristlecone pine; one 4,725-year-old tree sprouted when Egyptians were building the Great Sphinx). Also in this group are firs, yews, spruces, junipers, larches, cypresses, the bald cypress, podocarps, and the dawn redwood.

## Lesser Known Gymnosperms

CYCADS  About 100 species of **cycads** (Cycadophyta) made it to the present day. Figure 25.13 shows examples of their pollen-bearing and seed-bearing cones, which form on separate plants. Insects and air currents transfer pollen from "male" plants to "female" plants. At first glance, you might mistake cycad leaves for those of a palm tree, but palms are flowering plants.

These plants mainly inhabit tropical and subtropical areas. Two species (*Zamia*) grow wild in Florida and are planted as ornamentals. Elsewhere, cycad seeds and cycad trunks that are ground into a flour are made edible by removing their toxic alkaloids. Many cycad species are vulnerable to extinction.

**Figure 25.12**  (**a**) Bristlecone pine (*Pinus longaeva*) growing near the timberline in the Sierra Nevada. (**b**) Male pine cones releasing pollen. (**c**) Appearance of a female pine cone at the time of pollination. (**d**) From a juniper (*Juniperus*), cones with a berrylike appearance. These cones are made of fused-together, fleshy scales.

**Figure 25.13**
(**a**) Pollen-bearing cone of a "male" cycad (*Zamia*).
(**b**) Seed-bearing cone of a "female" cycad. Of all the existing species of gymnosperms, the cycads produce the largest seed-bearing cones. Some of these grow as long as one meter and weigh more than fifteen kilograms.

**Figure 25.14** (**a**) Ginkgo tree. (**b**) A fossilized ginkgo leaf compared with a leaf from its existing descendant. The fossil formed at the Cretaceous–Tertiary boundary. Even though 65 million years have passed since then, the leaf structure has not changed much, if at all. (**c**) Pollen-bearing cones and (**d**) fleshy-coated seeds of this type of gymnosperm.

**Figure 25.15** (**a**) Sporophyte of *Ephedra* and (**b**) its pollen-bearing cones and (**c**) a seed-bearing cone. (**d**) Sporophyte of *Welwitschia mirabilis* and (**e**) its seed-bearing cones.

GINKGOS The **ginkgos** (Ginkgophyta) were a diverse group in dinosaur times. The only surviving species is the maidenhair tree, *Ginkgo biloba*. Like a few species of larch and some other gymnosperms, these plants are deciduous. Several thousand years ago, ginkgo trees were widely planted around temples in China. Then the natural populations nearly became extinct, even though ginkgos seem hardier than many other trees. Perhaps they became targets for firewood. Today, male ginkgo trees are again widely planted. They have attractive, fan-shaped leaves and are resistant to insects, disease, and air pollutants. Female trees are not favored. Their thick, fleshy seeds, which are the size of small plums, give off an awful stench (Figure 25.14).

GNETOPHYTES At present, there are three genera of woody plants known as the **gnetophytes** (Gnetophyta). Trees and leathery leafed vines of *Gnetum* thrive in the humid tropics. The shrubby *Ephedra* lives in California deserts and some other arid regions (Figure 25.15a–c). Photosynthesis proceeds in its green stems. *Welwitschia mirabilis* grows in hot deserts of south and west Africa. Its sporophyte is mainly a deep-reaching taproot. Its exposed part, a woody disk-shaped stem, has cones and one or two strap-shaped leaves that split lengthwise repeatedly as the plant ages (Figure 25.15d, e).

Conifers, cycads, ginkgos, and gnetophytes are groups of existing gymnosperms. Like their ancestors, they bear seeds on the exposed surfaces of cones and other spore-producing structures.

# A CLOSER LOOK AT THE CONIFERS

Before we leave the gymnosperms, let's use conifers as an example of reproductive strategies. Depending on the species, a conifer's life cycle lasts a year or more.

Who among us hasn't noticed the woody, shelflike scales of "a pine cone"? The scales are actually parts of a mature female cone in which megaspores formed and developed into female gametophytes. Pine trees also produce male cones, in which microspores form and develop into pollen grains (Figure 25.16). Each spring, millions of pollen grains drift away from the male cones.

Pollination is completed when some land on ovules of female cones. After each germinates, a tubular structure forms from it. This germinating pollen grain, the sperm-bearing male gametophyte, grows toward the egg in a female gametophyte. For species of pines, fertilization occurs months or a year after pollination.

As in other gymnosperms, seed formation begins at the ovule (Figure 25.16). An embryo sporophyte starts developing from the fertilized egg. The outer layers of the jacket around the female gametophyte and embryo mature into a hard coat. The seed coat will protect the embryo sporophyte after it is dispersed from the parent plant. The nutrients will help it through the critical time

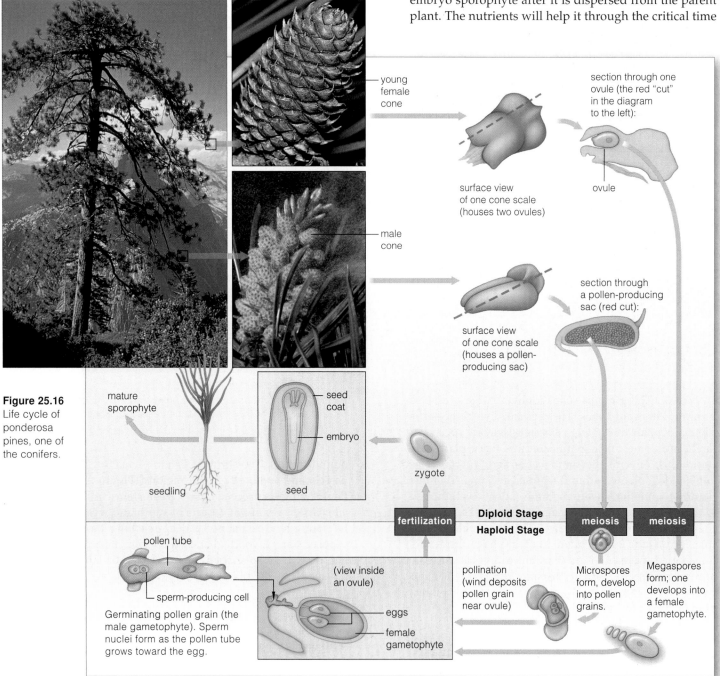

**Figure 25.16** Life cycle of ponderosa pines, one of the conifers.

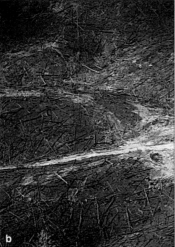

**Figure 25.17** Sampling of the rampant deforestation under way around the world. Only deforested tracts in North America are shown; Section 51.4 focuses on the tropical rain forests.

From America's heartland, logged-over acreage in Arkansas (**a**). From the eastern seaboard, one of the denuded bits of North Carolina (**b**). Clear-cut peaks in Washington (**c**) and Alaska (**d**). These are not isolated examples. In the early 1980s, 400 million board feet of timber were being cut in Washington's Olympic Peninsula every year. In Arkansas, about one-third of the Ouachita National Forest was clear-cut. Its once-diverse forest communities have been replaced by "tree farms" of a single species of pine. Throughout the world, huge tracts of land that were deforested years ago still show no signs of recovery.

of germination, before its roots and shoots become fully functional.

Conifers dominated many land habitats during the Mesozoic, but their slow reproductive pace put them at a competitive disadvantage when the flowering plants began their stunning adaptive radiation (Section 21.6). Coniferous forests still predominate in the far north, at higher elevations, and in some parts of the Southern Hemisphere. However, existing conifers face more than competition with flowering plants for resources. Now

they are vulnerable to **deforestation**: the removal of all trees from large tracts, as by clear-cutting (Figure 25.17). Conifers just have the bad luck to be premier sources of lumber, paper, and other wood products required in human societies. We return to this topic in Chapter 51.

Where flowering plants flourish, conifers are at a competitive disadvantage, partly because they take so long to reproduce. Rampant deforestation isn't helping them one bit, either.

# ANGIOSPERMS—THE FLOWERING, SEED-BEARING PLANTS

Only angiosperms produce specialized reproductive structures called **flowers** (Figure 25.18). *Angeion,* which means vessel, refers to the female reproductive parts at the center of a flower. The enlarged base of the "vessel" is the floral ovary, where ovules and seeds develop.

Most flowering plants coevolved with **pollinators**—insects, bats, birds, and other animals that withdraw nectar or pollen from a flower and, in so doing, transfer pollen to its female reproductive parts. The recruitment of animals as assistants in reproduction probably has contributed to the success of flowering plants, which have dominated the land for 100 million years.

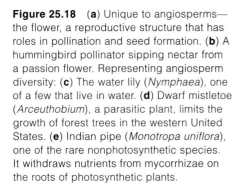

After fertilization in the reproductive structures called flowers, seeds develop from ovules and some tissues of the parent sporophyte.

seed
coat
endosperm
(nutritive tissue)
embryo
sporophyte

SEED

ovule — ovary

**a**

FLOWER

**Figure 25.18** (**a**) Unique to angiosperms—the flower, a reproductive structure that has roles in pollination and seed formation. (**b**) A hummingbird pollinator sipping nectar from a passion flower. Representing angiosperm diversity: (**c**) The water lily (*Nymphaea*), one of a few that live in water. (**d**) Dwarf mistletoe (*Arceuthobium*), a parasitic plant, limits the growth of forest trees in the western United States. (**e**) Indian pipe (*Monotropa uniflora*), one of the rare nonphotosynthetic species. It withdraws nutrients from mycorrhizae on the roots of photosynthetic plants.

**b**

**c**  **d**  **e**

At least 260,000 species thrive in a variety of habitats. They range in size from tiny duckweeds (a millimeter or so long) to towering *Eucalyptus* trees (some are more than 100 meters tall). A few species, including mistletoes and Indian pipe, aren't even photosynthetic. They withdraw nutrients directly from other plants or from mycorrhizae.

There are two classes of flowering plants, called the **dicots** and **monocots** (more formally, the Dicotyledonae and Monocotyledonae). Among the 180,000 dicots are most herbaceous (nonwoody) plants, such as cabbages and daisies; most flowering shrubs and trees, such as oaks and apple trees; water lilies; and cacti. Among 80,000 or so species of monocots are the orchids, palms, lilies, and grasses, including rye, sugarcane, corn, rice, and wheat, as well as many other highly valued crop plants (Appendix I).

The next unit deals with the structure and function of flowering plants. For now, simply start thinking about how a large sporophyte dominates the life cycles. It retains and nourishes gametophytes; its sperm are dispersed within pollen grains. Endosperm, a nutritive tissue, surrounds embryo sporophytes inside the seeds of flowering plants. As the seeds develop, the ovaries (along with other structures) mature into fruits. Fruits protect and help disperse embryos. Figure 25.19, in the next section, is an overview of these life cycle events.

---

**Angiosperms are the most successful plants in terms of their diversity, numbers, and distribution. They alone produce flowers. Most species coevolved with animal pollinators.**

---

# VISUAL OVERVIEW OF FLOWERING PLANT LIFE CYCLES

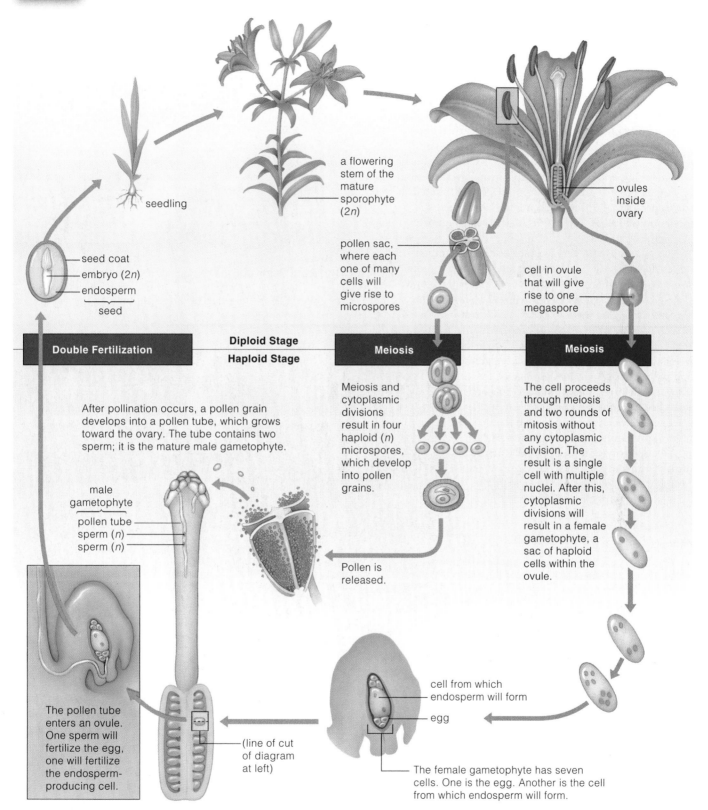

seedling

a flowering stem of the mature sporophyte (2*n*)

ovules inside ovary

seed coat
embryo (2*n*)
endosperm
seed

pollen sac, where each one of many cells will give rise to microspores

cell in ovule that will give rise to one megaspore

**Double Fertilization**

**Diploid Stage**
**Haploid Stage**

**Meiosis**

**Meiosis**

After pollination occurs, a pollen grain develops into a pollen tube, which grows toward the ovary. The tube contains two sperm; it is the mature male gametophyte.

Meiosis and cytoplasmic divisions result in four haploid (*n*) microspores, which develop into pollen grains.

The cell proceeds through meiosis and two rounds of mitosis without any cytoplasmic division. The result is a single cell with multiple nuclei. After this, cytoplasmic divisions will result in a female gametophyte, a sac of haploid cells within the ovule.

male gametophyte
pollen tube
sperm (*n*)
sperm (*n*)

Pollen is released.

The pollen tube enters an ovule. One sperm will fertilize the egg, one will fertilize the endosperm-producing cell.

(line of cut of diagram at left)

cell from which endosperm will form

egg

The female gametophyte has seven cells. One is the egg. Another is the cell from which endosperm will form.

**Figure 25.19** Representative flowering plant life cycle. This example is for a lily (*Lilium*), one of the monocots. "Double" fertilization is a distinctive feature of flowering plant life cycles. A male gametophyte delivers two sperm to an ovule. One sperm fertilizes the egg, and the other fertilizes a cell that gives rise to endosperm, a tissue that will nourish the forthcoming embryo. Section 31.3 provides a closer look at flowering plant life cycles, using a dicot as the example.

# 25.10 SEED PLANTS AND PEOPLE

Which plants provide taste thrills and which kill? Starting with trials and errors of the earliest human species, we started collecting intimate knowledge of plants. Certainly by 300,000 years ago, *Homo erectus* clans in China were stashing pine nuts, walnuts, hazelnuts, and rose hips in caves and roasting seeds. At least by then, grown-ups were teaching children which plants are edible and which are toxic. Through language, youngsters learned about the plants that had evolved in their parts of the world.

About 11,000 years ago, people started to domesticate wheat, barley, and other plants, this being a way to count on having reliable quantities of food. Of an estimated 3,000 species that different populations recognized as food, only about 200 became *the* major crops (Figure 25.20).

Plant lore still threads through our lives in other ways. We learned to use flowers to grace our homes and enrich our customs (Figure 25.21). We learned to use fast-growing, soft-wooded conifers for lumber, fuel, and paper—and slow-growing, hard-wooded flowering plants, such as cherry, maple, and mahogany, for fine furniture. We make twine and rope from leaves of century plants (*Agave*), cords and textiles from leaf fibers of Manila hemp, and thatched roofs from grasses and palm fronds. Insecticides derived from Mexican cockroach plants kill cockroaches, fleas, lice, and flies. Extracts of neem tree leaves kill nematodes, insects, and mites but not the natural predators of those common pests.

And what about those flowers! Their oils impart scents to perfumes. Eucalyptus and camphor oils have medical uses. Digitalin extracted from foxglove (*Digitalis purpurea*)

**Figure 25.21**   A bride tossing her bouquet to single women who attended her wedding, a ritual that supposedly reveals who will be married next. (A sixty-eight-year-old optimist caught this one.)

**Figure 25.20**   A few of the prized flowering plants.

(**a**) Triticale, a popular hybrid grain; parental stocks are wheat and rye (*Secale*). It combines the high yield of wheat with rye's tolerance of truly harsh environmental conditions. (**b**) From the American Midwest, mechanized harvesting under way in a field of common bread wheat, *Triticum*.

(**c**) Fruits of *Theobroma cacao*. Each fruit holds as many as forty seeds, which are processed to make cocoa butter or chocolate essences. Americans who on average buy 8–10 pounds of chocolate per year might not be happy that a pathogenic fungus (*Monilia*) may be driving *T. cacao* to extinction.

(**d**) Indonesians gathering tender shoots of tea plants, evergreen shrubs related to camellias. Plants growing on hillsides in moist, cool regions yield leaves with the best flavors. Only the terminal bud and two or three of the youngest leaves are picked for the finest teas

(**e**) From Hawaii, a field of sugarcane (*Saccharum officinarum*). Wild stock of this cultivated species may have evolved in New Guinea. Sap extracted from its cut stems is boiled down to make sucrose crystals (table sugar) and syrups.

Further reading: Student Guide to InfoTrac on web site →

*DIGITALIS PURPUREA*

*NICOTIANA RUSTICUM*

*HYOSCYAMUS*

*CANNABIS SATIVA*

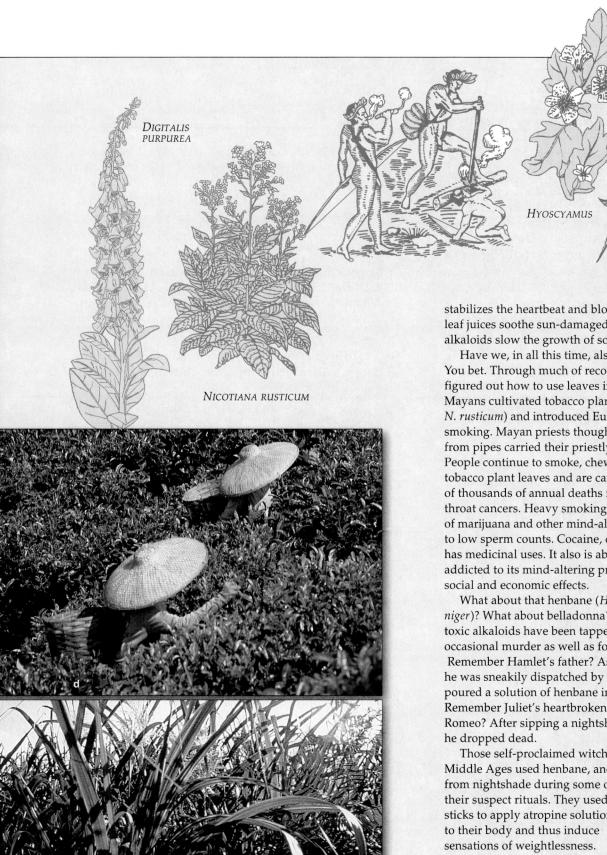

stabilizes the heartbeat and blood circulation. *Aloe vera* leaf juices soothe sun-damaged skin. Periwinkle leaf alkaloids slow the growth of some cancer cells.

Have we, in all this time, also learned to abuse plants? You bet. Through much of recorded history, people also figured out how to use leaves in harmful ways. Ancient Mayans cultivated tobacco plants (*Nicotiana tabacum* and *N. rusticum*) and introduced European explorers to tobacco smoking. Mayan priests thought that the smoke rising from pipes carried their priestly thoughts to the gods. People continue to smoke, chew, and tuck in their mouth tobacco plant leaves and are candidates for the hundreds of thousands of annual deaths from lung, mouth, and throat cancers. Heavy smoking of *Cannabis sativa*, source of marijuana and other mind-altering substances, is linked to low sperm counts. Cocaine, derived from coca leaves, has medicinal uses. It also is abused by millions who are addicted to its mind-altering properties, with devastating social and economic effects.

What about that henbane (*Hyoscyamus niger*)? What about belladonna? Their toxic alkaloids have been tapped for the occasional murder as well as for medicine. Remember Hamlet's father? As he slept, he was sneakily dispatched by someone who poured a solution of henbane into his ear. Remember Juliet's heartbroken, suicidal Romeo? After sipping a nightshade potion, he dropped dead.

Those self-proclaimed witches of the Middle Ages used henbane, and atropine from nightshade during some of their suspect rituals. They used sticks to apply atropine solutions to their body and thus induce sensations of weightlessness. During their atrophine-induced sprees, they assumed that they were flying off to rendezvous with demons. Hence those Halloween cartoons of witches flying hither and yon on their broomsticks.

*ATROPA BELLADONNA*

## SUMMARY

1. Green algae probably gave rise to plants, which had invaded the land by 435 million years ago. Nearly all species of plants are multicelled photoautotrophs. Table 25.1 summarizes and compares the major phyla.

2. Several trends in plant evolution may be identified by comparing different lineages (see also Table 25.2):

a. Structural adaptations to dry conditions, such as vascular tissues (xylem and phloem).

b. A shift from haploid to diploid dominance during the life cycle. Complex sporophytes evolved; they hold on to, nourish, and protect spores and gametophytes.

c. A shift from one to two spore types (homospory to heterospory) that, among gymnosperms and flowering plants, led to the evolution of pollen grains and seeds.

3. Mosses, liverworts, and hornworts are bryophytes, nonvascular plants that have no well-developed xylem or phloem and that require free water for fertilization.

### Table 25.1 Comparison of Major Existing Plant Groups

*Nonvascular land plants. Fertilization requires free water. Haploid dominance. Cuticle, stomata present in some.*

| | |
|---|---|
| **BRYOPHYTES** | 18,600 species. Moist, humid habitats. |

*Seedless vascular plants. Fertilization requires free water. Diploid dominance. Cuticle, stomata present.*

| | |
|---|---|
| **WHISK FERNS** | 7 species, sporophytes with no obvious roots or leaves. *Psilotum*. |
| **LYCOPHYTES** | 1,100 species with simple leaves. Mostly wet or shady habitats. |
| **HORSETAILS** | 25 species of single genus. Swamps, disturbed habitats. |
| **FERNS** | 12,000 species. Wet, humid habitats in mostly tropical, temperate regions. |

*Gymnosperms—vascular plants with "naked seeds." Free water not required for fertilization. Diploid dominance. Cuticle, stomata present.*

| | |
|---|---|
| **CONIFERS** | 550 species, mostly evergreen, woody trees and shrubs having pollen- and seed-bearing cones. Widespread distribution. |
| **CYCADS** | 185 slow-growing tropical, subtropical species. |
| **GINKGO** | 1 species, a tree with fleshy-coated seeds. |
| **GNETOPHYTES** | 70 species. Limited to some deserts, tropics. |

*Angiosperms—vascular plants with flowers and protected seeds. Free water not required for fertilization. Diploid dominance. Cuticle, stomata present.*

**FLOWERING PLANTS**

| | |
|---|---|
| Monocots | 80,000 species. Floral parts often arranged in threes or in multiples of three; one seed leaf; parallel leaf veins common. |
| Dicots | At least 180,000 species. Floral parts often arranged in fours, fives, or multiples of these; two seed leaves; net-veined leaves common. |

### Table 25.2 Evolutionary Trends Among Plants

| Bryophytes | Ferns | Gymnosperms | Angiosperms |
|---|---|---|---|

Nonvascular ⟶ Vascular ————————————⟶

Haploid ⟶ Diploid ————————————⟶
dominance dominance

Spores of ⟶ Spores of ————————————⟶
one type two types

Motile gametes ——————————⟶ Nonmotile ⟶
gametes*

Seedless ————————⟶ Seeds ——————⟶

\* Require pollination by wind, insects, animals, etc.

4. Nearly all vascular plants are adapted to land. Their cuticle and stomata conserve water. Root systems mine soil for nutrients. Upright and branched growth patterns of shoot systems intercept sunlight and carbon dioxide. Tissues enclose and protect spores and gametes.

5. The seedless vascular plants include the whisk ferns, lycophytes, horsetails, and ferns. Their flagellated sperm require ample water to swim to the eggs.

6. Gymnosperms and flowering plants (angiosperms) are vascular plants. Both produce pollen grains (mature microspores that develop into male gametophytes) and megaspores (which give rise to female gametophytes).

a. Megaspores develop within ovules: reproductive structures that contain (1) the egg-producing female gametophytes, (2) the precursor of nutritive tissue, and (3) a jacket of cell layers, the outer portion of which will develop into the seed coat. A mature ovule is a seed.

b. The evolution of pollen grains freed these plants from dependence on water for fertilization. Their seeds are efficient means of dispersing new generations, even during hostile conditions. Pollen grains and seeds were key adaptations in the move to high and dry habitats.

7. Only angiosperms produce flowers. Most coevolved with pollinators, such as insects, which enhance the transfer of pollen grains to female reproductive parts. Their seeds contain a nutritive tissue (endosperm) and are usually surrounded by fruit, which aids in dispersal.

### Review Questions

1. Identify a few structural and reproductive modifications that helped plants invade and diversify in habitats on land. *25.1*

2. Does the haploid phase or diploid phase dominate the life cycles of most plants? *25.1*

3. Name representatives of the following groups of plants and then compare their main characteristics: (*also refer to Table 25.1*)

a. Bryophytes and seedless vascular plants *25.2–25.4*
b. Gymnosperms and angiosperms *25.6–25.8*

4. Distinguish between:

a. Root system and shoot system *25.1*
b. Xylem and phloem *25.1*
c. Sporophyte and gametophyte *25.1*
d. Ovule and seed *25.1, 25.5*
e. Microspore and megaspore *25.5*

**Figure 25.22** Where many conifers end up.

## Self-Quiz (*Answers in Appendix III*)

1. Which of the following statements is *not* true?
   a. Monocots and dicots are two classes of angiosperms.
   b. Bryophytes are nonvascular plants.
   c. Lycophytes and angiosperms are both vascular plants.
   d. Gymnosperms are the simplest vascular plants.

2. Of all land plants, bryophytes alone have independent
   _____ and attached, dependent _____ .
   a. sporophytes; gametophytes    c. rhizoids; zygotes
   b. gametophytes; sporophytes    d. rhizoids; stalked
                                                           sporangia

3. Whisk ferns, lycophytes, horsetails, and ferns are classified
   as _____ plants.
   a. multicelled aquatic    c. seedless vascular
   b. nonvascular seed       d. seed-bearing vascular

4. Which does *not* apply to gymnosperms and angiosperms?
   a. vascular tissues
   b. diploid dominance
   c. single spore type
   d. all of the above

5. A seed is _____ .
   a. a female gametophyte    c. a mature pollen tube
   b. a mature ovule          d. an immature embryo

6. Match the terms appropriately.
   _____ gymnosperm      a. gamete-producing body
   _____ sporophyte      b. help control water loss
   _____ lycophyte       c. "naked" seeds
   _____ ovary           d. protects, nourishes, disperses
   _____ bryophyte          embryo sporophyte
   _____ gametophyte     e. spore-producing body
   _____ stomata         f. nonvascular land plant
   _____ angiosperm seed g. seedless vascular plant
                                      h. usually a fruit at maturity

## Critical Thinking

1. Figure 25.22 shows a forest in the Nahmint Valley of British Columbia, before and after logging. It also shows wood frames of homes that are in the process of being built. Reflect on these photographs and Figure 25.17. To stop the loggers, would you chain yourself to a tree in an old-growth forest scheduled for clear-cutting? If your answer is yes, would you also give up the chance of owning a wood-frame home (as most homes are in developed countries)? What about forest products, including newspapers, toilet tissue, and fireplace wood?

2. With respect to question 1, multiply each of your answers by 6 billion (there are almost that many people in the world) and describe what might happen when, inevitably, we run out of trees. Also describe what you might consider to be some of the pros and cons of tree farms of, say, a single species of pine.

3. Elliot Meyerowitz of the California Institute of Technology has studied the genetic basis of flower formation in *Arabidopsis thaliana*. By inducing mutations in seeds of this small weed, he discovered three genes (call them *A*, *B*, and *C*) that interact in different parts of a developing flower. Gene interactions lead to the formation of different structures—sepals, petals, stamens, and carpels—from the same mass of undifferentiated tissue. From what you know of gene regulation, suggest ways in which the *A*, *B*, and *C* genes might be controlling flower development.

4. Genes nearly identical to the *A*, *B*, and *C* genes of *A. thaliana* also have been isolated from snapdragons and other flowering plants. These plants had evolved by 150 million years ago. They quickly rose to dominance in nearly all land habitats.
   Review the general introduction to adaptive radiation in Section 21.6. Then speculate on how the spectacular and rapid adaptive radiation of flowering plants came about.

## Selected Key Terms

| | | |
|---|---|---|
| angiosperm *25.1* | hornwort *25.2* | pollinator *25.8* |
| bryophyte *25.1* | horsetail *25.3* | progymnosperm *25.5* |
| coal *25.4* | lignin *25.1* | rhizome *25.3* |
| conifer *25.6* | liverwort *25.2* | root system *25.1* |
| cuticle *25.1* | lycophyte *25.3* | seed *25.1* |
| cycad *25.6* | megaspore *25.5* | seed fern *25.5* |
| deforestation *25.7* | microspore *25.5* | shoot system *25.1* |
| dicot *25.8* | monocot *25.8* | spore *25.1* |
| fern *25.4* | moss *25.2* | sporophyte *25.1* |
| flower *25.8* | ovule *25.5* | stoma (stomata) *25.1* |
| gametophyte *25.1* | peat bog *25.2* | strobilus *25.3* |
| ginkgo *25.6* | phloem *25.1* | vascular plant *25.1* |
| gnetophyte *25.6* | pollen grain *25.1* | whisk fern *25.3* |
| gymnosperm *25.1* | pollination *25.5* | xylem *25.1* |

## Readings *See also www.infotrac-college.com*

Gray, J., and W. Shear. September–October 1992. "Early Life on Land." *American Scientist* 80:444–456.

Moore, R., W. D. Clark, and K. Stern. 1995. *Botany*. Dubuque, Iowa: W. C. Brown. Beautifully illustrated.

# ANIMALS: THE INVERTEBRATES

## *Madeleine's Limbs*

In August of 1994, about 900 million years after the first animals appeared on Earth, Madeleine made *her* entrance. As they are wont to do, her grandmothers and aunts made a quick count on the sly—arms, legs, ears, and eyes, two of each; fully formed mouth and nose—just to be sure these were present and accounted for.

One grandmother, having been too long in the company of biologists, experienced an epiphany as she witnessed Madeleine's birth. In that profound instant she sensed ancestral connections emerging from the distant past and through her, into the future.

Madeleine's body plan did not emerge out of thin air. Thirty-five thousand years ago, people just like us were having children just like Madeleine. And if we are interpreting the fossil record correctly, then five million years ago the offspring of individuals on the road to modern humans resembled her in some respects but not others. Sixty million years ago, primate ancestors of those individuals were giving birth precariously, up in the trees. Two hundred and fifty million years ago, mammalian ancestors of those primates were giving birth—and so on back in time to the very first animals, which had no limbs or eyes or noses at all.

We have very few clues to what those first animals looked like, but one thing is clear. By the dawn of the Cambrian, they had given rise to all major groups of invertebrates—animals without backbones—even to Madeleine's backboned but limbless ancestors.

And what stories those Cambrian animals tell! One bunch flourished 530 million years ago, in a submerged basin that had formed between a reef and the coast of an early continent. Protected from ocean currents, the sediments had piled up against the steep reef. About 500 feet below the surface, the water was oxygenated and clear. Small, well-developed animals lived in, on, and above the dimly lit, muddy sediments (Figure 26.1).

Like castles built from wet sand along a seashore, their living quarters were unstable. Part of the bank above the community slumped abruptly and obliterated it. The sediments from that underwater avalanche kept scavengers from reaching and removing all traces of the dead. Gradually through time, muddy silt rained down on the tomb. Increased pressure and chemical changes transformed the sediments into finely stratified shale, and the soft parts of the flattened animals became shimmering, mineralized films.

Sixty-five million years ago, part of the seafloor was plowing under the North American plate, and western Canada's mountain ranges were slowly rising. By 1909, the fossils had traveled high into the eastern mountains of British Columbia. In that year a fossil hunter tripped over a chunk of shale, which split apart into thin, fine layers—and so the Burgess Shale story came to light.

In this chapter and the next, you will be comparing body plans of different groups of animals. Such comparisons give insight into evolutionary relatedness and help us to construct family trees, such as the one in Figure 26.2. Don't assume that structurally simple animals of the most ancient lineages are somehow primitive or evolutionarily stunted. As you will see, they, too, are exquisitely adapted to their environment.

**Figure 26.1** Reconstruction of a few Cambrian animals known from fossils of the Burgess Shale, British Columbia.

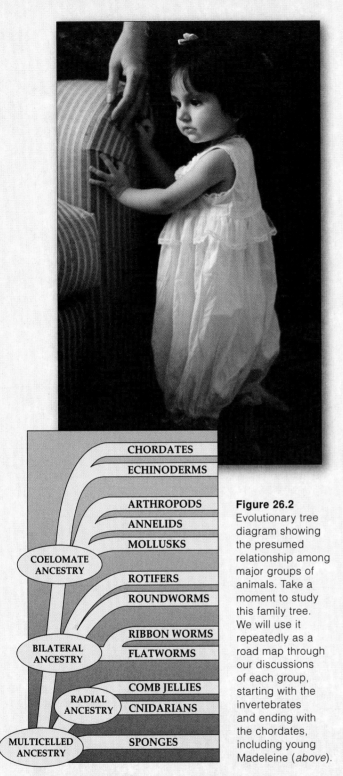

CHORDATES

ECHINODERMS

ARTHROPODS

ANNELIDS

MOLLUSKS

COELOMATE
ANCESTRY

ROTIFERS

ROUNDWORMS

RIBBON WORMS

BILATERAL
ANCESTRY

FLATWORMS

COMB JELLIES

RADIAL
ANCESTRY

CNIDARIANS

MULTICELLED
ANCESTRY

SPONGES

SINGLE-CELLED, PROTISTAN-LIKE ANCESTORS

**Figure 26.2**
Evolutionary tree diagram showing the presumed relationship among major groups of animals. Take a moment to study this family tree. We will use it repeatedly as a road map through our discussions of each group, starting with the invertebrates and ending with the chordates, including young Madeleine (*above*).

## KEY CONCEPTS

**1.** All animals are multicelled, aerobic heterotrophs that ingest or parasitize other organisms. Nearly all kinds have tissues, organs, and organ systems, and most are motile during at least part of their life cycle. Animals reproduce sexually and often asexually, and their embryos develop through a series of continuous stages.

**2.** Animals originated in the late Precambrian. More than 2 million existing animal species have been identified. Of these, more than 1,950,000 are invertebrates (animals with no backbone). Fewer than 50,000 species are vertebrates (animals with a backbone).

**3.** Comparisons of the body plans of existing animals, in conjunction with the fossil record, reveal that there were several trends in the evolution of certain lineages. The most revealing aspects of an animal's body plan are its type of symmetry, gut, and cavity (if any) between the gut and body wall; whether it has a distinct head end; and whether it is divided into a series of segments.

**4.** The placozoans and sponges are structurally simple animals with no body symmetry. Both are at the cellular level of body construction. Cnidarians and comb jellies show radial symmetry, and they are at the tissue level of body construction.

**5.** Flatworms, roundworms, rotifers, and nearly all other animals that are more complex than the cnidarians show bilateral symmetry. They consist of tissues, organs, and organ systems.

**6.** Not long after the flatworms evolved, divergences gave rise to two major lineages. One evolutionary branching gave rise to the mollusks, annelids, and arthropods. The other gave rise to the echinoderms and chordates.

**7.** By biological measures, including diversity, sheer numbers, and distribution, the arthropods—especially insects—have been the most successful animal group.

As you poke through the branches of the animal family tree, keep the greater evolutionary story in mind. At each branch point, microevolutionary processes gave rise to workable changes in body plans. Madeleine's uniquely human traits, and yours, emerged through modification of certain traits that had evolved earlier in countless generations of vertebrates and, before them, in ancient invertebrate forms.

## General Characteristics of Animals

What, exactly, are **animals**? We can only define them by a list of characteristics, not with a sentence or two. *First*, animals are multicelled. In most cases their body cells form tissues that become arranged as organs and organ systems. The body cells of nearly all species have a diploid chromosome number. *Second*, all animals are heterotrophs that get carbon and energy by ingesting other organisms or by absorbing nutrients from them. *Third*, animals require oxygen for aerobic respiration. *Fourth*, animals reproduce sexually and, in many cases, asexually. *Fifth*, most animals are motile during at least part of the life cycle. *Sixth*, the life cycle includes stages of embryonic development. Briefly, mitotic cell divisions transform the animal zygote into a multicelled embryo. The embryonic cells give rise to primary tissue layers, **ectoderm**, **endoderm**, and usually **mesoderm**. These layers in turn give rise to all tissues and organs of the adult, as described in Sections 33.6 and 44.2.

## Diversity in Body Plans

Mammals, birds, reptiles, amphibians, and fishes are the most familiar animals. All are **vertebrates**, the only animals with a "backbone." And yet, of probably more than 2 million species of animals, fewer than 50,000 are vertebrates! What we call the **invertebrates** are animals with plenty of diverse features, but not a backbone.

We group the animals into more than thirty phyla. Table 26.1 lists the groups described in this book. The characteristics they share with one another arose early in time, before divergences from a common ancestor gave rise to separate lineages. Later, as morphological differences accumulated among them, the lineages took off in amazingly diverse directions. How might we get a conceptual handle on their modern-day descendants—on animals as different as flatworms, hummingbirds, spiders, toads, humans, and giraffes? We can compare their similarities and differences with respect to five basic features. These are body symmetry, cephalization, type of gut, type of body cavity, and segmentation.

BODY SYMMETRY AND CEPHALIZATION    With very few exceptions, animals are radial or bilateral. Those with **radial symmetry** have body parts arranged regularly around a central axis, like spokes of a bike wheel. Thus a cut down the center of a hydra (Figure 26.3*a*) divides it into equal halves; another cut at right angles to the first divides it into equal quarters. Radial animals live in water. Their body plan is adapted to intercepting food that is coming toward them from any direction.

Animals with **bilateral symmetry** have a right half and left half that are mirror images of each other. Most

### Table 26.1  Animal Phyla Described in This Book

| Phylum | Some Representatives | Existing Species |
|---|---|---|
| PLACOZOA (*Trichoplax*) | Simplest animal; like a tiny plate, but with layers of cells | 1 |
| PORIFERA (poriferans) | Sponges | 8,000 |
| CNIDARIA (cnidarians) | Hydrozoans, jellyfishes, corals, sea anemones | 11,000 |
| PLATYHELMINTHES (flatworms) | Turbellarians, flukes, tapeworms | 15,000 |
| CTENOPHORA (comb jellies) | Venus's girdle | 100 |
| NEMERTEA (ribbon worms) | Proboscis-equipped worms closely related to flatworms | 800 |
| NEMATODA (roundworms) | Pinworms, hookworms | 20,000 |
| ROTIFERA (rotifers) | Tiny body with crown of cilia, great internal complexity | 2,000 |
| MOLLUSCA (mollusks) | Snails, slugs, clams, squids, octopuses | 110,000 |
| ANNELIDA (segmented worms) | Leeches, earthworms, polychaetes | 15,000 |
| ARTHROPODA (arthropods) | Crustaceans, spiders, insects | 1,000,000+ |
| ECHINODERMATA (echinoderms) | Sea stars, sea urchins | 6,000 |
| CHORDATA (chordates) | Invertebrate chordates: Tunicates, lancelets | 2,100 |
| | Vertebrates: Fishes | 21,000 |
| | Amphibians | 4,900 |
| | Reptiles | 7,000 |
| | Birds | 8,600 |
| | Mammals | 4,500 |

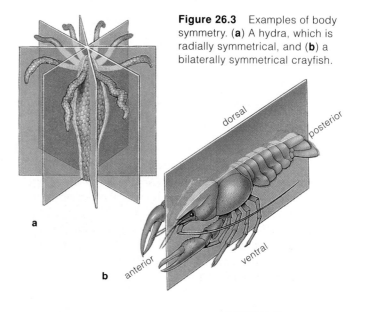

**Figure 26.3** Examples of body symmetry. (**a**) A hydra, which is radially symmetrical, and (**b**) a bilaterally symmetrical crayfish.

dorsal

posterior

anterior

ventral

a

b

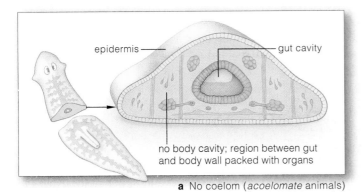

**a** No coelom (*acoelomate* animals)

epidermis — gut cavity

no body cavity; region between gut and body wall packed with organs

**b** Pseudocoel (*pseudocoelomate* animals)

epidermis — gut cavity

unlined body cavity (pseudocoel) around gut

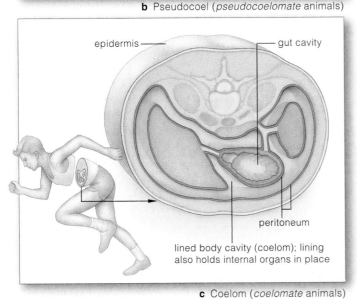

epidermis — gut cavity

peritoneum

lined body cavity (coelom); lining also holds internal organs in place

**c** Coelom (*coelomate* animals)

**Figure 26.4** Type of body cavity (if any) in animals.

have an *anterior* end (head) and an opposite, *posterior* end. They have a *dorsal* surface (a back) and an opposite, *ventral* surface (Figure 26.3b). As fossils show, this body plan evolved among the first forward-creeping species. Their forward end would have encountered food and other stimuli first, so there must have been selection for **cephalization**. With this evolutionary process, sensory structures and nerve cells became concentrated in the head. The joint evolution of bilateral body plans and cephalization resulted in pairs of muscles and pairs of sensory structures, nerves, and brain regions.

**TYPE OF GUT**  The **gut** is a tubular or saclike region in the body in which food is digested, then absorbed into the internal environment. Saclike guts have one opening (a mouth) for taking in food and expelling residues. Other guts are part of a tubelike, "complete" digestive system with an opening at two ends (mouth and anus). Different parts of the system have specialized functions, such as preparing, digesting, and storing material. As more efficient digestive systems evolved, this helped pave the way for increases in body size and activity.

**BODY CAVITIES**  In between the gut and the body wall of most bilateral animals is a body cavity (Figure 26.4). One type of cavity, a **coelom**, has a unique tissue lining called a peritoneum. This lining also encloses organs in the coelom and helps hold them in place. For example, your body has a coelom, and a sheetlike muscle called a diaphragm divides it into two smaller cavities. Your heart and lungs are positioned in the upper (thoracic) cavity, and your stomach, intestines, and other organs occupy the lower (abdominal) cavity.

Some invertebrates don't have a body cavity; tissues fill the region between their gut and body wall. Others have a pseudocoel ("false coelom"), a body cavity with no peritoneum. The coelom was a key innovation by which larger and more complex animals evolved from ancestral forms. It favored increases in size and activity by cushioning and protecting internal organs.

**SEGMENTATION**  Segmented animals have a repeating series of body units that may or may not be similar to one another. The many segments of earthworms have a similar outward appearance. Insect segments are fused into three units (head, thorax, and abdomen) and differ greatly from one another. Especially among the insects, diverse head parts, legs, wings, and other appendages evolved from less specialized segments.

Animals are multicelled, and most have tissues, organs, and organ systems. Most have a diploid chromosome number.

Animals are aerobically respiring heterotrophs that ingest other organisms or absorb nutrients from them.

Animals reproduce sexually and, in many cases, asexually. They go through a period of embryonic development, and most are motile during at least part of the life cycle.

Body plans of animals differ with respect to five features: body symmetry, cephalization, type of gut, type of body cavity, and segmentation.

Judging from recent genetic evidence and radiometric dating of fossilized tracks, burrows, and microscopic embryos, animals originated between 1.2 billion and 670 million years ago, during the Precambrian. *Where did they come from?* They probably evolved from protistan lineages, but we don't know which ones (Section 23.3).

By one hypothesis, the forerunners of animals were ciliates, like *Paramecium*, and had multiple nuclei in a one-celled body. As they evolved, each nucleus became compartmentalized in individual cells of a multicelled body. However, we don't know of any existing animal that develops by compartmentalization.

By another hypothesis, multicelled animals arose from spherical colonies of a number of flagellated cells, maybe like the *Volvox* colonies shown in Section 23.13. In time, as a result of mutation, some cells in the colony became modified in ways that enhanced reproduction and other specialized tasks. And so began the division of labor that characterizes multicellularity.

Suppose such colonies became flattened and started creeping around on the seafloor. A creeping life-style could have favored the evolution of layers of cells, such as those of *Trichoplax adhaerens*. This is the only known **placozoan** (Placozoa, after *plax*, meaning plate, and *zoon*, which means animal). *Trichoplax* is a soft-bodied marine animal, shaped a bit like a tiny pita bread. It has no symmetry and no mouth. Its several thousand cells are arranged into two distinctly different layers. It briefly humps up when its body glides over food, as in Figure 26.5. Gland cells in the lower layer secrete digestive enzymes onto the food, then individual cells absorb the breakdown products. Reproduction might be asexual (by budding or fission) or sexual, by mechanisms not yet understood. In sum, structurally and functionally, *Trichoplax* is as simple as animals get.

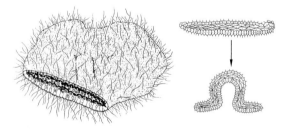

**Figure 26.5** Cutaway views of *Trichoplax adhaerens*, an animal with a two-layer body measuring about three millimeters across.

Possibly the question of origins requires more than one answer. It may be that different lineages descended from more than one group of protistan-like ancestors.

**Multicelled animals arose from protistan-like ancestors that may have resembled ciliates, colonial flagellates, or both.**

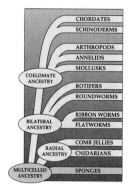

**Sponges** (Porifera) are animals with no symmetry, tissues, or organs. Yet they are one of nature's success stories. They have been abundant in the seas ever since the Precambrian, especially in the waters off coasts and along coral reefs. Of about 8,000 known species, only a hundred or so live in freshwater habitats. Many small worms, shrimps, and other animals make their home in or on a sponge body. Some sponges are large enough to sit in. Others are as small as a fingernail. Figures 26.6 and 26.7 show only a few of the flattened, sprawling, compact, lobed, tubular, cuplike, and vaselike shapes.

Regardless of its shape, the body of a sponge is not symmetrical. Flattened cells do line the outer surface and inner cavities. But these linings are not much more specialized than the cell layers of *Trichoplax*, and they differ from the tissues of other animals. Amoeboid cells live in a gelatin-like substance between the two linings (Figure 26.7b). Spicules, tough fibers, or both stiffen the sponge body. The fibers are made of spongin, a protein, and the sharp, glasslike spicules are made of calcium carbonate or silica.

The skeletal elements may be a reason why sponges as a group have endured so long. Cleveland Hickman put it this way: Most potential predators discover that sampling a sponge is about as pleasant as eating a mouthful of glass splinters embedded in fibrous gelatin. Besides, chemically speaking, many sponges stink.

Water flows into the sponge body through many microscopic pores and chambers, then out through one or more large openings. It does so when thousands or millions of **collar cells** beat their flagella. The cells are components of the body's inner lining. Their "collars" are food-trapping structures called microvilli (Figure 26.7c). Bacteria and other food in the water get trapped in the collars, then are engulfed by phagocytosis. Some

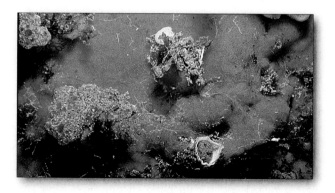

**Figure 26.6** A sprawling, red-orange sponge, one of many types that encrust underwater ledges in temperate seas.

**Figure 26.7** (**a**,**b**) Body plan of a simple sponge. The outer lining consists of flattened cells. It also has some contractile cells, most of which are arranged around the large opening at the top. These cells contract slowly, independently of the other cell types, and so influence water flow through the body. Amoeboid cells inside the gelatin-like matrix between the inner and outer linings secrete materials from which the body's spicules and fibers are constructed. Other amoeboid cells digest and transport food. They do not lose the capacity to divide, and their cellular descendants can differentiate into any other type of sponge cell. They also have roles in asexual processes, such as gemmule formation.

(**c**) A great many flagellated cells line inner canals and chambers of the sponge body. Each of these phagocytic cells has a collar of food-trapping structures called microvilli. Fine filaments connect the microvilli to each other; they form a "sieve" that strains food particles from the water. At the base of the collar, the cell engulfs the trapped food.

(**d**) An example of skeletal elements—Venus's flower basket (*Euplectella*). This marine sponge has six-rayed spicules of silica fused in a rigid, elaborate network. A thin layer of cells stretches over all interconnecting spicules. At the base of the body is an anchoring tuft of spicules.

(**e**) Basket sponge releasing a cloud of sperm into the water.

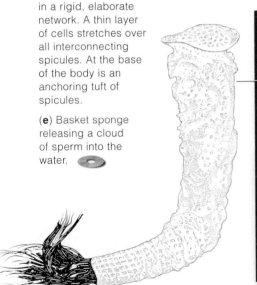

of the food also is transferred to the amoebalike cells for further breakdown, storage, and distribution.

Sponges reproduce sexually. Most release sperm into the water (Figure 26.7*e*). But sponges retain eggs until after fertilization and new embryos have started their development. A young sponge proceeds through a microscopic, swimming larval stage. A **larva** (plural, larvae) is a sexually immature stage that grows and develops into an adult, the sexually mature form of the species. As you will see, the life cycle of many animal species includes larval stages.

Some kinds of sponges also reproduce asexually by fragmentation; small fragments break away from the parent and grow into new sponges. Most freshwater species also reproduce asexually by way of gemmules. These are clusters of sponge cells, some of which form a hard covering around others. The clusters inside are protected from extreme cold or drying out. Later, when favorable conditions return, the gemmules germinate and establish a new colony of sponges.

Sponges have no symmetry, tissues, or organs; they are at the cellular level of construction. Yet they have successfully endured through time, possibly because most predators find their spicule-rich and often stinky bodies unappetizing.

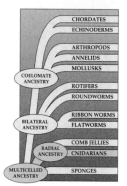

Look at the two jellyfishes in Figure 26.8. They are scyphozoans which, along with anthozoans (such as sea anemones) and hydrozoans (such as *Hydra*) are tentacled, radial animals known as **cnidarians**. Most live in the seas. Of 11,000 known species of the phylum Cnidaria, fewer than 50 are adapted to freshwater habitats.

### Regarding the Nematocysts

Of all animals, cnidarians alone produce **nematocysts**: capsules that house dischargeable, tubular threads. The threads of some types have prey-piercing barbs and an open tip that delivers toxins (Figure 26.9). Other nematocysts discharge threads that ooze a sticky substance from their tip or long threads that entangle prey. Many swimmers have learned that the toxin-tipped threads of some species can sting. Hence the name of the phylum, Cnidaria, after the Greek word for nettle.

### Cnidarian Body Plans

The **medusa** (plural, medusae) and **polyp** are the most common cnidarian body forms. Both have a saclike gut (Figure 26.8a,b). Medusae float. Some look like bells and others like upside-down saucers. The mouth, centered under the bell, may have extensions that assist in prey

capture and feeding. Polyps have a tubelike body with a tentacle-fringed mouth at one end. Usually the other end is attached to a substrate. Unlike *Trichoplax* (which is rather like a gut on the run when draped over food), a cnidarian has a permanent food-processing chamber. Its gut has gastrodermis, a sheetlike lining that incorporates glandular cells. The cells secrete digestive enzymes. An epidermis lines the rest of the body's surfaces (Figure 26.10). Each lining is an **epithelium** (plural,

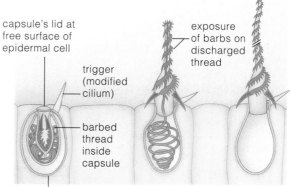

**Figure 26.9** One type of nematocyst before and after a prey organism (not shown) touched its trigger. The contact made the capsule more "leaky" to water. Water diffused inward, turgor pressure built up within the capsule, and the thread was forced to turn inside out. The thread's tip pierced the prey's body.

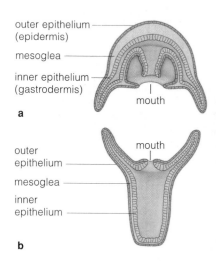

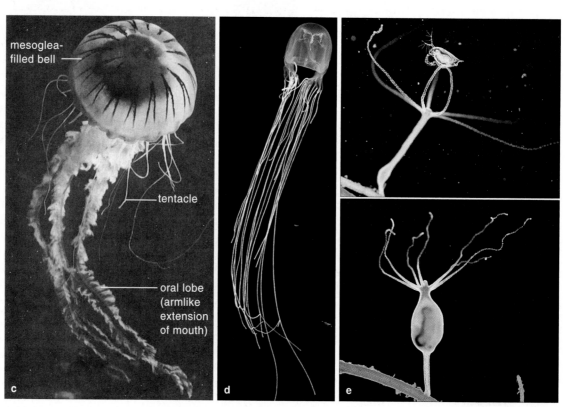

**Figure 26.8** Cnidarian body plans. (**a**,**b**) Medusa and polyp, midsection. Two jellyfish: (**c**) sea nettle (*Chrysaora*) and (**d**) sea wasp (*Chironix*), which has tentacles up to fifteen meters long. Its toxin can kill a human in minutes. (**e**) A hydrozoan polyp (*Hydra*) attached to a substrate, capturing and then digesting its prey.

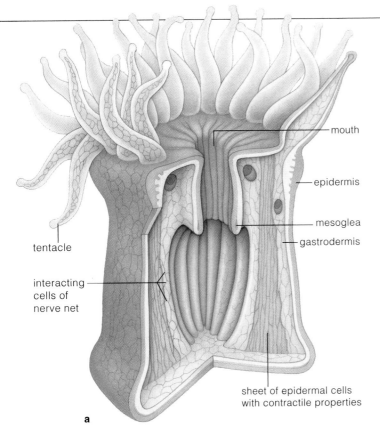

**Figure 26.10** (**a**) Tissue organization of sea anemones. (**b**) Tentacles fringing the mouth of a sea anemone, which eats many fishes but not clownfishes. A clownfish swims out, captures food, and returns to the tentacles, which protect it from predators. The sea anemone eats food scraps falling from the fish mouth. This is a case of mutualism, a two-way flow of benefits between species. (**c**) A sea anemone using its hydrostatic skeleton to escape from a sea star. It closes its mouth, and the force generated by contractile cells in epithelial tissues acts against water in the gut. The body changes shape and the anemone thrashes about, generally in a direction away from the predator.

epithelia), a tissue having a free surface that faces the environment or some type of fluid inside the body. Any animal that is structurally more complex than sponges incorporates epithelia. Cnidarian epithelia house **nerve cells**. Such cells receive signals from receptors that can detect changes in the surroundings. They send signals to **contractile cells**, which carry out suitable responses. (When stimulated, contractile cells *shorten*, then return to their original length when stimulation stops.) The nerve cells interact as a "nerve net," a simple nervous tissue, to control movement and changes in shape.

Between the epidermis and gastrodermis is a layer of gelatinous secreted material, the mesoglea ("middle jelly"). Jellyfishes contain enough mesoglea to impart buoyancy and serve as a firm yet deformable skeleton against which contractile cells act. Imagine many cells contracting in coordinated ways in a jellyfish bell. Their action narrows the bell and forces water to jet out from underneath it, and the jet propels the jellyfish forward. The bell returns to its original position, cells contract

again, and the animal is propelled forward. Although the swimming movements are not Olympian, they work well enough for an animal that secures food by dragging its tentacles through the water for small prey.

Any fluid-filled cavity or cell mass against which contractile cells can act is a **hydrostatic skeleton**. With coordinated contractions, the cavity's volume or mass does not change but rather is shunted about, so that the shape of the body changes. The contractile cells of most polyps, which have little mesoglea, act against water in their gut. As you will see, nearly all animals have some form of skeletal–muscular system of movement.

---

Cnidarians are radial animals with tentacles, a saclike gut, epithelia, a nerve net, and a hydrostatic skeleton. They alone produce nematocysts.

Cnidarians are at the tissue level of construction; compared with sponges, they have layers of cells interacting in more coordinated fashion in the performance of specific tasks.

---

## VARIATIONS ON THE CNIDARIAN BODY PLAN

### Various Stages in Cnidarian Life Cycles

From the preceding section, you may have concluded that the body form of a particular cnidarian is a medusa or a polyp. This is indeed the case for many species. However, the life cycle of *Obelia, Physalia,* and some other cnidarians includes both forms. By using the life cycle of *Obelia* as an example, you can get a general sense of how these forms grow and develop (Figure 26.11).

The medusa is the sexual stage of the cnidarian life cycle. It has simple **gonads**, which are primary (gamete-producing) reproductive organs. Either the epidermis or gastrodermis houses the gonads, which release gametes by rupturing. Most of the zygotes formed at fertilization develop into **planulas**—a kind of swimming or creeping larva, usually with ciliated epidermal cells. In time, a mouth opens at one end, the larva is transformed into a polyp or medusa, and the cycle begins anew.

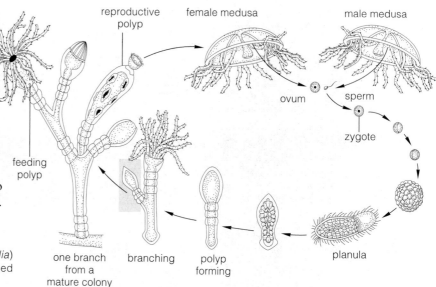

**Figure 26.11** *Right:* Life cycle of a hydrozoan (*Obelia*) that includes medusa and polyp stages. An established colony may contain thousands of feeding polyps.

### A Sampling of Colonial Types

The reef-forming corals and other colonial anthozoans are a fine example of variations on the basic cnidarian body plan. The reef formers thrive in clear, warm water that is at least 20°C (68°F). As Figure 26.12 shows, the colonies consist of polyps that have secreted calcium-

**Figure 26.12** (**a**) One of the reef-building, colonial corals. External skeletons of their polyps interconnect with one another. Dinoflagellate mutualists live in the polyp tissues. (**b**) Aerial view of a barrier reef, mainly an accumulation of the compacted skeletons of certain coral species. (**c**) Not all corals are reef builders; cup corals such as *Tubastrea* are solitary forms. With tentacles extended, their polyps look like tiny sea anemones.

reinforced external skeletons, which interconnect with one another. Over time, the skeletons accumulate and so become the main building material for reefs. Today the most impressive accumulations parallel the eastern coast of Australia for about 1,600 kilometers. We call their remains the Great Barrier Reef.

Reef-building corals receive nutrient inputs from the changing tides and from dinoflagellate mutualists living in their tissues. Masses of these photosynthetic protistans supply corals with oxygen, recycle their mineral wastes, and adjust the pH of the surrounding water in a way that enhances the rate of calcium deposition for their host's skeletons. The host corals reciprocate by giving the dinoflagellates a relatively safe, sunlit habitat that has plenty of dissolved carbon dioxide and mineral ions. In the sunlit parts of a reef, nutrients are cycled quickly, directly, and efficiently.

As a final example of cnidarian diversity, consider *Physalia*, informally called the Portuguese man-of-war. The toxin in the nematocysts of this infamous colonial hydrozoan poses a danger to bathers and fishermen as

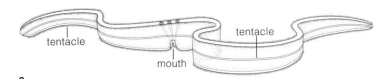

If you have ever been diving in coastal waters anywhere from the tropics to the poles, you may have observed cnidarians that look a bit like jellyfishes with combs running down the sides. These are **comb jellies**, of phylum Ctenophora (meaning "comb-bearing"). All are weak-swimming predators in planktonic communities, and all show modified radial symmetry. Slice them in two equal halves, then slice them into quarters, and two of those quarters will be mirror images of the other two.

A comb jelly has eight rows of comblike structures made of thick, fused cilia (Figure 26.14). All the combs in a row beat in waves and propel the animal forward, usually (and suitably) mouth first. Some species have two long, muscular tentacles with branches that are equipped with sticky cells. Comb jellies do not produce nematocysts, but sometimes they opportunistically save and use the ones from jellyfish they have eaten. Others use their sticky lips to capture prey.

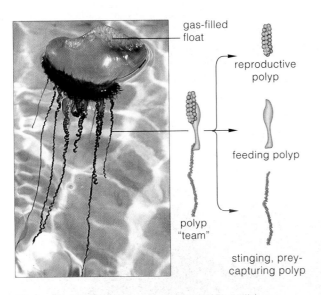

Figure 26.13 Portuguese man-of-war (*Physalia*).

tentacle    tentacle

mouth

**a**

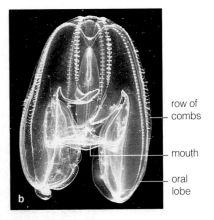

**Figure 26.14** Examples of comb jellies. (**a**) Sketch of Venus's girdle (*Cestum veneris*), a transparent animal about 1.5 meters (more than 4 feet) long.

(**b**) *Mnemiopsis*. Its oral lobes alternately expand and contract to waft prey into the mouth. It is about 15 centimeters (6 inches) long.

row of combs

mouth

oral lobe

**b**

well as to prey organisms (fish). Although *Physalia* lives mainly in warm waters, currents sometimes move it up to the Atlantic coasts of North America and Europe. A blue, gas-filled float that develops from the planula keeps the colony near the water's surface, where winds move it about (Figure 26.13). Under the float, groups of polyps and medusae interact as "teams" in feeding, reproduction, defense, and other specialized tasks.

---

Cnidarians, including the colonial forms, show notable variation in their body plans and life-styles.

Evolutionarily, comb jellies are interesting because they have cells with multiple cilia. This trait evolved in many of the complex animals. In addition, comb jellies are the simplest animals having embryonic tissues that resemble mesoderm. Mesoderm is the embryonic source of muscles and organs of the circulatory, excretory, and reproductive systems of most complex animals.

---

Comb jellies have a modified radial body and comblike structures of modified cilia. They have an embryonic tissue that resembles the mesoderm of more complex animals.

# ACOELOMATE ANIMALS—AND THE SIMPLEST ORGAN SYSTEMS

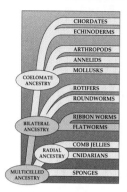

When we move beyond the cnidarians in our survey, we find animals that range from flatworms to humans. All of these animals have simple or complex organs. An **organ** is an association of one or more kinds of tissues, arranged in particular proportions and patterns. Most often, one organ interacts with others to carry out an activity that helps the body function. By definition, two or more organs that are interacting efficiently in the performance of some task represent an **organ system**.

We turn now to the simplest animals at the *organ-system* level of construction. They are not what you would call breathtakingly complex but, as you will see shortly, a few parasitic types can make our lives miserable.

Parasites, recall, reside in or on living hosts and feed on their tissues. Most do not kill their hosts, at least not until after they reproduce. A *definitive* host harbors the mature stage of the parasite's life cycle. One or more *intermediate* hosts harbor immature stages.

## Flatworms

Among the 15,000 or so known species of **flatworms** (phylum Platyhelminthes) are turbellarians, flukes, and tapeworms. Most of these bilateral, cephalized animals have simple organ systems in a flattened body, as in Figure 26.15. Their digestive system, for example, has a pharynx (a muscular tube, which flatworms use for feeding) as well as a saclike, often branching gut. They differ in their reproductive systems, but most species are **hermaphrodites**: individuals have female *and* male gonads. Each has a penis (sperm-delivery structure), so two flatworms can reproduce sexually by exchanging sperm. Secretions from certain flatworm glands form a protective capsule around the fertilized eggs.

**TURBELLARIANS** Most turbellarians (class Turbellaria) live in the seas; only planarians and a few others live in freshwater. Some eat tiny animals or suck tissues from dead or wounded ones. A planarian can divide in half at its midsection, then each half can regenerate missing parts. Thus it reproduces asexually by *transverse* fission. Like you, a planarian can adjust the composition and volume of its body fluids. Its water-regulating system has one or more branched tubes called protonephridia (singular, protonephridium). These tubes extend from pores at the body surface to bulb-shaped flame cells in tissues. When excess water diffuses into the flame cells, a tuft of cilia "flickering" in the bulb drives the water through the tubes to the outside (Figure 26.15*b*).

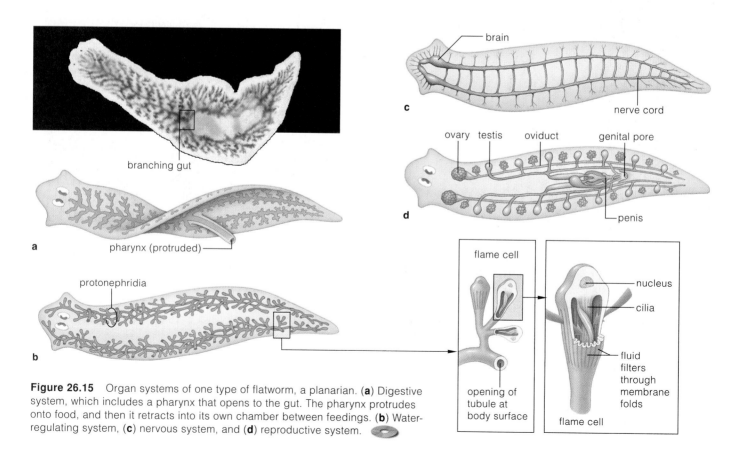

**Figure 26.15** Organ systems of one type of flatworm, a planarian. (**a**) Digestive system, which includes a pharynx that opens to the gut. The pharynx protrudes onto food, and then it retracts into its own chamber between feedings. (**b**) Water-regulating system, (**c**) nervous system, and (**d**) reproductive system.

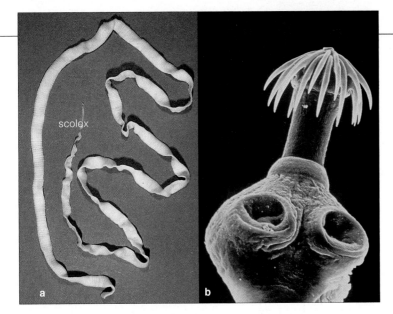

scolex

a          b

**Figure 26.16** (**a**) Tapeworm scolex. This
one attaches to the gut lining of a shorebird,
its primary host. (**b**) Sheep tapeworm.

They may have done so by increased cephalization and
the development of tissues derived from mesoderm.

## Ribbon Worms

**Ribbon worms** (phylum Nemertea) are bilateral, soft-
bodied, elongated predators that swallow or suck tissue
fluids from small worms, mollusks, and crustaceans.
Most crawl, burrow, or lurk in shallow marine habitats,
as in Figure 26.17, but some live in freshwater or humid
tropical habitats. Like flatworms, which may be their
close relatives, ribbon worms have a ciliated surface,
and they secrete mucus and then move by beating their
cilia through it. Ribbon worms also resemble flatworms
in their tissue organization. However, they differ from
flatworms in having a circulatory system, a complete
gut, and separation of sexes. Ribbon worms also have a
proboscis, which in this case is a tubular, prey-piercing,

**FLUKES** Flukes (class Trematoda) are parasitic worms.
Their complicated life cycle has sexual phases, many
asexual phases, and one to four hosts. Almost always, a
snail or a clam is the initial host for larval or juvenile
stages, and a vertebrate is the definitive host. Section
29.5 takes a closer look a
blood fluke (*Schistosoma*).
Schistosomes parasitize as
many as 200 million people.

**TAPEWORMS** Tapeworms
(class Cestoda) parasitize
intestines of vertebrates. It
seems probable that the
ancestral tapeworms had a
gut but later lost it during
their evolution in animal
intestines, which happen
to be habitats that are rich
in predigested food. Their
existing descendants attach
to the intestinal wall with
a scolex, a structure that
has suckers, hooks, or both
(Figure 26.16). **Proglottids**
bud just behind the scolex;
each is a new unit of the tapeworm body. These units
are hermaphroditic; they mate and transfer sperm with
one another. The older proglottids (those farthest from
the scolex) store fertilized eggs. They break off from the
younger ones, then they leave the body in feces. Later,
the eggs may enter an intermediate host. Section 26.9
describes how proglottids form during the life cycle of
a representative tapeworm.

In some respects, the simplest turbellarians, larval
flukes, and larval tapeworms resemble the planulas of
cnidarians. The resemblance inspires speculation that
bilateral animals evolved from planula-like ancestors.

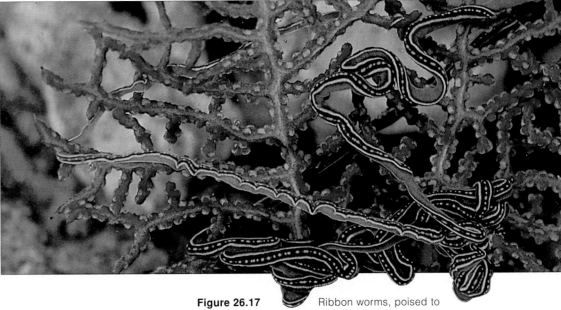

**Figure 26.17** Ribbon worms, poised to
ambush small prey that pass by their lair in an orange-
colored gorgonian (horny coral) colony.

venom-delivering device tucked inside the head end.
Muscle contractions force the proboscis inside out and
into prey, which the venom paralyzes.

---

Flatworms are among the simplest bilateral, cephalized
animals with organ systems. Ribbon worms may be related
to them, but they have a complete gut, a circulatory system,
and other traits that are notable departures from flatworms.

---

# ROUNDWORMS

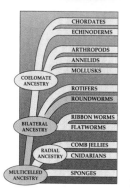

The **Roundworms** (phylum Nematoda) are pseudocoelomate worms. They thrive in nearly all environments, and they may be the most abundant animals alive. In most sediments beneath shallow fresh water or saltwater, there might be up to a million roundworms in each square meter. Many thousands may occupy a handful of rich topsoil in which scavenging types make fast work of dead earthworms or rotting plants. There are 20,000 known species, but there may be a hundred times more. Figure 26.18 shows a typical roundworm's body plan.

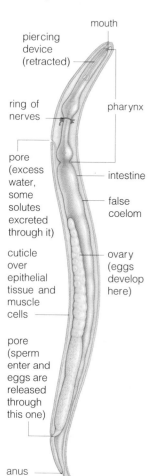

mouth

piercing device (retracted)

ring of nerves

pharynx

pore (excess water, some solutes excreted through it)

intestine

false coelom

cuticle over epithelial tissue and muscle cells

ovary (eggs develop here)

pore (sperm enter and eggs are released through this one)

anus

**Figure 26.18** Body plan of a roundworm (*Paratylenchus*). This one parasitizes the roots of plants.

A roundworm is bilateral. But its body is cylindrical, typically tapered at both ends, and covered by a protective cuticle. Animal **cuticles** are tough, often flexible body coverings. A roundworm is the simplest animal equipped with a complete digestive system. Between the gut and body wall is a false coelom, typically jam-packed with reproductive organs. The cells in every tissue absorb nutrients from the coelomic fluid and give up wastes to it.

Parasitic species can severely damage their hosts, which include humans, cats, dogs, cows, sheep, soybeans, potatoes, and other crop plants. Although thin, they can grow long; the roundworms in female sperm whales may be nine meters long, stretched out. Elephantiasis, one of the world's most rapidly spreading diseases, is the work of a few species. (See also Sections 26.9 and 48.6.)

Most roundworms are free-living types that are harmless or beneficial, as when they cycle nutrients for communities. One species, *Caenorhabditis elegans*, is used in research into inheritance, development, and aging. The *C. elegans* genome was the first of any multicelled organism to be fully sequenced (Section 44.7).

Roundworms are cylindrical, bilateral, cephalized animals with a false coelom and a complete digestive system. Most cycle nutrients in communities; the parasites are notorious.

---

*Focus on Health*

# A ROGUE'S GALLERY OF WORMS

A number of parasitic flatworms and roundworms call the human body home. In any year, for instance, 200 million people house blood flukes responsible for *schistosomiasis*. One of these, the Southeast Asian blood fluke *Schistosoma japonicum*, requires a human definitive host, standing water in which its larvae can swim, and an aquatic snail as an intermediate host. The flukes grow, become sexually mature, and mate inside a human host (Figure 26.19*a*). After being fertilized, a female's eggs (*b*) leave the human body in feces and hatch into ciliated, swimming larvae (*c*) that burrow into a snail (*d*) and multiply asexually. In time, many fork-tailed larvae develop (*e*). These leave the snail and swim about until they contact human skin (*f*). They bore in and migrate to thin-walled intestinal veins, then the cycle begins anew. In infected humans, white blood cells that defend the body attack the masses of fluke eggs, and masses of grainy debris form in tissues. In time, the liver, spleen, bladder, and kidneys deteriorate.

Some tapeworms parasitize humans. Different species use pigs, freshwater fish, or cattle as intermediate hosts. Humans become infected when they eat pork, fish, or beef that is raw, improperly pickled, or insufficiently cooked—and contaminated with tapeworm larvae (Figure 26.20).

Or consider a parasitic roundworm that causes thin, serpentlike ridges in human skin. For several thousand years, healers have been extracting the "serpents" by winding them out slowly, painfully, around a stick. The roundworms called pinworms and hookworms cause other problems. *Enterobius vermicularis*, a pinworm of temperate regions, parasitizes humans. It lives in the large intestine, but at night the centimeter-long females migrate to the anal region of the host and lay eggs. Their presence causes itching, and scratchings made in response transfer

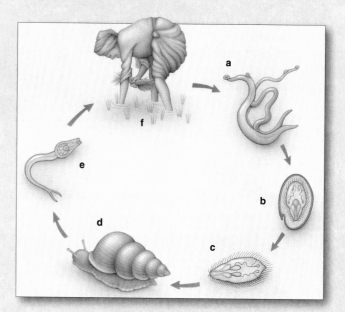

**Figure 26.19** Life cycle of a dangerous blood fluke, *Schistosoma japonicum*.

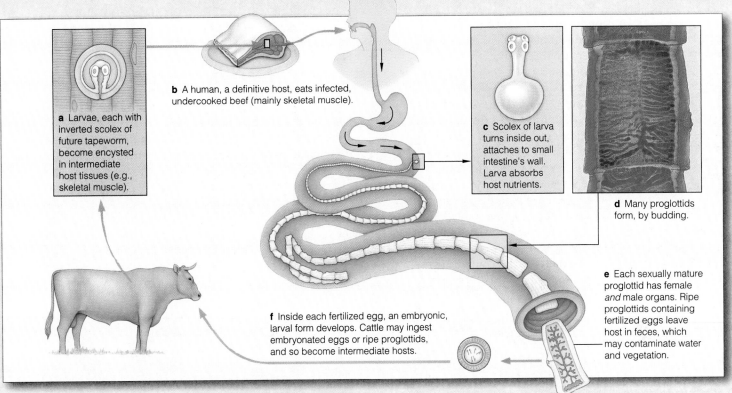

**a** Larvae, each with inverted scolex of future tapeworm, become encysted in intermediate host tissues (e.g., skeletal muscle).

**b** A human, a definitive host, eats infected, undercooked beef (mainly skeletal muscle).

**c** Scolex of larva turns inside out, attaches to small intestine's wall. Larva absorbs host nutrients.

**d** Many proglottids form, by budding.

**e** Each sexually mature proglottid has female *and* male organs. Ripe proglottids containing fertilized eggs leave host in feces, which may contaminate water and vegetation.

**f** Inside each fertilized egg, an embryonic, larval form develops. Cattle may ingest embryonated eggs or ripe proglottids, and so become intermediate hosts.

**Figure 26.20** Life cycle of a beef tapeworm, *Taenia saginata*.

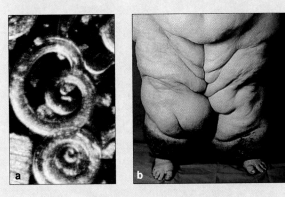

**Figure 26.21** (**a**) Juveniles of a roundworm, *Trichinella spiralis*, inside the muscle tissue of a host animal. (**b**) Legs of a woman parasitized by the roundworm *Wuchereria bancrofti*.

some eggs to hands, then to other objects. Newly laid eggs contain embryonic pinworms. Within a few hours, they develop into juveniles and are ready to hatch if another human inadvertently ingests them.

The hookworms are especially vexing in impoverished areas of the tropics and subtropics. Adult hookworms live in a host's small intestine. Using their toothlike devices or sharp ridges around the mouth, they cut into the intestinal wall, feed on blood and other tissues, and so compete with their host for nutrients. Adult females, about a centimeter long, can release a thousand eggs daily. These leave the body in feces, then hatch into juveniles. When it meets up with bare skin of a host, a juvenile hookworms cuts its

way inside. The parasite travels the bloodstream to the lungs. It works its way into the air spaces and moves up the windpipe. When the host swallows, the parasite is transported to the stomach and then the small intestine, where it may mature and live for several years.

Another roundworm, *Trichinella spiralis*, causes painful, sometimes fatal symptoms. Adults live in the lining of the small intestine. Females release juveniles (Figure 26.21*a*), which work their way into blood vessels and travel to muscles. There they become encysted; they secrete a covering around themselves and enter a resting stage. Humans become infected mainly by eating insufficiently cooked meat from pigs or some game animals. It is not easy to detect the encysted juveniles when fresh meat is being examined, even in a slaughterhouse.

Figure 26.21*b* shows the results of prolonged, repeated infections by *Wuchereria bancrofti*, another roundworm. Adult worms become lodged in lymph nodes, organs that filter lymph (excess tissue fluid) that normally flows into the bloodstream. The worms obstruct lymph flow. When an obstruction causes fluid to back up and accumulate in tissues, legs and other body regions may enlarge grossly. We call this condition *elephantiasis*.

A mosquito is *Wuchereria*'s intermediate host. Females of this parasite produce active young that travel about at night in the bloodstream. When a mosquito sucks blood from an infected person, the juveniles enter the insect's tissues. In time they move close to the insect's sucking device and enter a new host when it draws blood again.

## ROTIFERS

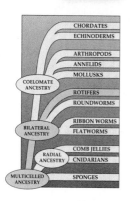

Like roundworms, the **rotifers** (Rotifera) are bilateral, cephalized animals with a false coelom. All but about 5 percent live in freshwater, such as lakes, ponds, and even films of water on mosses and other plants. We typically find between 40 and 500 rotifers in a liter of pondwater; 5,000 were recorded on a few occasions. They eat bacteria and microscopic algae. Most types are not even a millimeter long, yet rarely have so many organs been packed in so little space. As Figure 26.22 shows, rotifers have a pharynx, an esophagus, digestive glands, a stomach, protonephridia, and usually an intestine and anus. Nerve cell bodies clustered in the head end integrate body activities. Certain rotifers contain "eyes" (clusterings of absorptive pigments). Two "toes" exude

**Figure 26.22**
A rotifer (*Philodina roseola*). Males are unknown in this species and many others. Females produce diploid eggs that become diploid females. The females of other species do the same, but they also can produce haploid eggs that develop directly into haploid males. If a haploid egg happens to be fertilized by a male, it develops into a female. The males show up only occasionally. They are dwarfed, and they are short-lived.

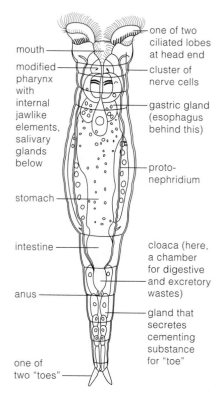

substances that attach free-living species to substrates at feeding time. A feature unique to rotifers is a crown of cilia at the head end that assists in swimming and in wafting food toward the mouth. Its rhythmic motions reminded early microscopists of a turning wheel; hence the name of the phylum ("rotifer" means wheel-bearer).

---

**The rotifers are bilateral, cephalized animals with ciliated lobes at their head end and a false coelom that is packed with diverse organs.**

---

## TWO MAJOR DIVERGENCES

Bilateral animals not much more complex than modern flatworms apparently evolved in the late precambrian. Soon afterward, some species gave rise to two lineages of coelomate animals (Figure 26.2). We call these great lineages the **protostomes** and **deuterostomes**. Mollusks, annelids, and arthropods are protostomes. Echinoderms and chordates are deuterostomes.

As a result of mutations, embryos of the animals in each lineage develop differently from the fertilized egg. For instance, mitotic cell divisions cut the egg cytoplasm repeatedly along prescribed planes to form a tiny ball of cells—the early embryo (Section 44.3). Protostomes undergo *spiral* cleavage. In this developmental pattern, the earliest cuts are made at oblique angles relative to the main body axis. But deuterostomes undergo *radial* cleavage, a pattern in which the earliest cuts are made parallel with and perpendicular to the axis:

Early protostome embryo. Its four cells are undergoing cleavages *oblique to* the original body axis:

Early deuterostome embryo. Its four cells are undergoing cleavages *parallel with* and *perpendicular to* the original body axis:

Also, the very first opening to form at the surface of a protostome embryo becomes the mouth; an anus forms elsewhere. In a deuterostome embryo, the first opening becomes the anus; the second becomes the mouth. As another example, the coelom of a protostome arises from spaces in the mesoderm, but a deuterostome coelom forms from outpouchings of the gut wall:

*How a coelom forms in a protostome embryo:* pouch will form mesoderm around coelom developing gut

*How a coelom forms in a deuterostome embryo:* solid mass of mesoderm developing gut — coelom

Such modifications to the embryonic stages of the two kinds of animals led to major differences in body plans.

---

**Soon after the coelomate animals evolved in Cambrian times, two great lineages––the protostomes and deuterostomes— evolved through mutations that affected how their embryos develop, and this led to major differences in body plans.**

---

# A SAMPLING OF MOLLUSCAN DIVERSITY

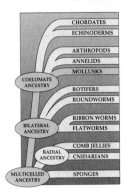

CHORDATES
ECHINODERMS
ARTHROPODS
ANNELIDS
MOLLUSKS
COELOMATE ANCESTRY
ROTIFERS
ROUNDWORMS
RIBBON WORMS
BILATERAL ANCESTRY
FLATWORMS
COMB JELLIES
RADIAL ANCESTRY
CNIDARIANS
MULTICELLED ANCESTRY
SPONGES

From children's books and explorations in gardens, few of us would have trouble recognizing a land snail when we see one (Figure 26.23*a*). Yet few of us know much about its 110,000 relatives in one of the largest of all animal groups, the phylum Mollusca. As the name implies, **mollusks** have fleshy soft bodies (*molluscus*, a Latin word, means soft). These bilateral animals have a small coelom. *Most* have a shell, or a reduced version of one, made of calcium carbonate and protein. These components are secreted from cells of a tissue that drapes like a skirt over the body mass. This tissue, the **mantle**, is unique to mollusks. The respiratory organs, gills of a type called ctenidia, contain thin-walled leaflets for gas exchange. *Most* mollusks have a fleshy foot. *Many* have a radula,

because their foot spreads out as they crawl. Many species have spirally coiled or conical shells. Coiling compacts the organs into a mass that can be balanced above the body, rather like a backpack. Other species have a reduced shell or none at all (Figure 26.23*a,b*).

Chitons are slow-moving or sedentary grazers with a dorsal shell divided into eight plates (Figure 26.23*c*). The bivalves—animals having a "two-valved shell"—include clams, scallops, oysters, and mussels (Figure 26.23*d*). Some bivalves are only a millimeter across. A few giant clams are over a meter across and weigh 225 kilograms (close to 500 pounds). Humans have eaten one type of bivalve or another since prehistoric times.

Cephalopods are highly active predators of the seas. This class includes the swiftest invertebrates (squids), the largest known invertebrate (giant squid), and the smartest (octopuses, Figure 26.23*e*). For instance, show

**Figure 26.23** A few mollusks. (**a**) Land snail, a gastropod. (**b**) Two sea slugs, a type of gastropod called nudibranchs. Different gastropods creep, swim, or float as they graze upon, prey upon, or parasitize other organisms. (**c**) Chiton from the intertidal zone of California's Monterey Bay. With its broad foot, it creeps over and clings to substrates. (**d**) Scallop, one of the bivalves. Light-sensitive "eyes" (small dark dots) fringe two halves of its shell. Many bivalves, including a few pearl producers, have shells lined with iridescent mother-of-pearl. (**e**) Octopus. Its eyes resemble yours, although they form in a different way. Like other cephalopods, it has a well-developed nervous system.

a tonguelike, toothed organ that shreds food destined for the gut. Mollusks with a well-developed head have eyes and tentacles, but not all have a head. Beyond the generalizations, there are no "typical" mollusks. They range from tiny snails in treetops to huge predators of the seas. In this section and the next, we sample four classes: chitons, gastropods, bivalves, and cephalopods.

Gastropods ("belly foots") make up the largest class (90,000 species of snails and slugs). They are so named

an octopus an object with a distinctive shape and then give it a mild electric shock, and it will thereafter avoid that object. With respect to memory and the capacity to learn, octopuses and some of the squids are the world's most complex invertebrates.

---

**Mollusks are bilateral, soft-bodied, coelomate animals that differ tremendously in body details, size, and life-styles.**

---

Maybe it was their fleshy, soft bodies—so forgiving of chance evolutionary changes in morphology—that gave the ancestors of mollusks the potential to diversify in so many ways and to radiate into so many habitats. Let's explore this idea by using a few characteristics of our representative mollusks. As a point of departure, start by studying the body plan shown in Figure 26.24.

## Twisting and Detwisting of Soft Bodies

Notice, in Figure 26.24*a,* how evolution put an unusual twist in the soft snail body. Its anus dumps wastes near the mouth! As a gastropod embryo develops, a cavity between its mantle and shell twists counterclockwise by 180°. So does nearly all of the visceral mass (the gut, heart, gills, and other internal organs). This process, called **torsion**, occurs only in gastropods:

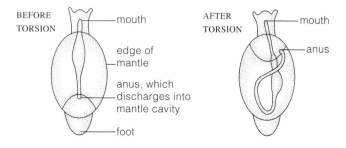

Such a drastic rearrangement of body parts could have come about through mutations that affected retractor muscles, which attach a gastropod embryo or larva to its shell. The muscles on the body's right side develop before those on the left. As they grow toward the front of the body, they drag the mantle cavity and organs with them. It takes them a few hours at most to do this.

By one hypothesis, torsion proved to be adaptive, for the head could withdraw into the mantle cavity in times of danger. However, in some 1985 experiments by J. Pennington and F. Chia, predators ate just as many torsioned larvae as "pre-torsioned" ones. Was torsion a bad evolutionary experiment? By putting the gills, anus, and kidneys above the mouth, it certainly created a potentially awful sanitation problem. At the very least, the uptake of discharged wastes in ancestral torsioned species must have been distasteful.

In fact, we find evolutionary compensations for this state of affairs. Most gastropods now have enough cilia in this region to create currents that sweep the wastes away. Also, torsion is not as pronounced as it once was in some lineages. Nudibranchs have even undergone an apparent detorsion; the soft larval body twists, but then it untwists. At some point in their evolution, they also ended up losing most of their mantle cavity and all of their ctenidia. Most species have other outgrowths that function in gas exchange (Figures 26.23*b* and 26.24*c*).

## Hiding Out, One Way or Another

If you were small, edible, and soft of body, an external shell would be a distinct advantage, as it is for chitons and clams. When a chiton is disturbed by predators or surf or exposed by a receding tide, it hunkers under its shell. Muscles in its foot pull the body mass down, and the mantle's edge around the shell's rim presses like a suction cup against a rock. That eight-plated shell is flexible. Pluck a chiton from a rock, and it rolls up in a ball until it can unroll and become reattached elsewhere.

Besides having a shell, protection also can be had by hiding in sediments and other substances. A bivalve's head is not much to speak of, but its foot is usually large and specialized for burrowing. Bivalves that dig

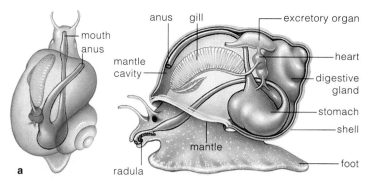

**Figure 26.24** (**a**) Body plan of an aquatic snail, a gastropod. (**b**) Close-up of a radula, a feeding device of some mollusks, including snails. As a radula rhythmically protracts and retracts, it rasps food and draws it toward the gut on the retraction stroke. (**c**) Sea slug *Aplysia,* also called a sea hare. Two flaps above its dorsal surface are foot extensions that undulate and ventilate the mantle cavity. Like many other gastropods, *Aplysia* is hermaphroditic. It can function as a male, a female, or both.

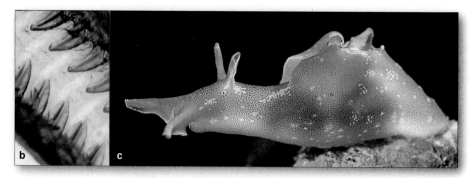

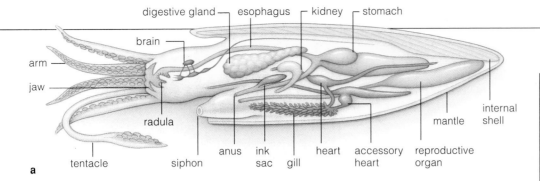

digestive gland — esophagus — kidney — stomach
brain
arm
jaw
radula
internal shell
mantle
tentacle — siphon — anus — ink sac — gill — heart — accessory heart — reproductive organ

**a**

in sand or mud have paired siphons: extensions of the mantle edges, fused into tubes (Figure 26.25a). Water is drawn into the mantle cavity through one siphon and leaves through the other, carrying wastes. One wonders what preyed on the ancestors of geoducks of the Pacific Northwest, which have siphons more than a meter long.

### On the Cephalopod Need for Speed

Some 500 million years ago the cephalopods, with their buoyant, chambered shells, were the supreme predators in Ordovician seas (Section 21.5). And yet, of a lineage having more than 7,000 ancestral species, the shell of all but one of the existing descendant species is reduced or gone (Figure 26.26). What happened? This evolutionary trend coincided with an adaptive radiation of the bony fishes—which preyed on cephalopods or were strong competitors for the same prey. During what may have been a long-term race for speed and wits, cephalopods lost their thick external shell and became streamlined and highly active. Of all mollusks, they now have the largest brain relative to body size and display the most

**Figure 26.26** (**a**) Body plan (generalized ) of a cuttlefish, a cephalopod. Its tentacles, thinner than the arms, are specialized for capturing prey. (**b**) A squid (*Dosidiscus*) and a diver inspecting each other. (**c**) A chambered nautilus, the only existing cephalopod that has an external shell. Being so active, cephalopods have great demands for oxygen. They are the only mollusks having a closed circulatory system. Blood is pumped from a main heart to two gills, each with an accessory (booster) heart at its base that speeds blood flow, oxygen uptake (for muscle cells especially), and carbon dioxide removal.

**b**

**c**

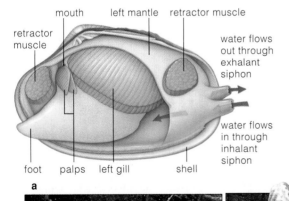

mouth — left mantle — retractor muscle
retractor muscle
water flows out through exhalant siphon
water flows in through inhalant siphon
foot — palps — left gill — shell

**a**

**b**

**Figure 26.25** (**a**) Body plan of a clam, with half of its shell removed. In nearly all bivalves, gills serve in collecting food and in respiration. As water moves through the mantle cavity, mucus on the gills traps food. Cilia move the mucus and food to palps, where final sorting takes place before suitable bits are driven to the mouth. (**b**) A scallop escaping from a sea star by clapping its valves and producing a propulsive water jet.

complex behavior. Nerves connect their brain to muscles that respond swiftly to food or danger. They have highly efficient blood circulation and respiration (Figure 26.26). Except for the chambered nautilus, they can discharge dark fluid from their ink sac, maybe to confuse predators.

*Jet propulsion* became the name of the game. Cephalopods force a jet of water from the mantle cavity and a funnel-shaped siphon. As mantle muscles relax, water is drawn into the cavity. As they contract, a water jet is squeezed out. When the mantle's free edge closes down on the head at the same time, a jet shoots through the siphon. The brain controls the siphon's activity and the direction of escape or pursuit.

**Lively stories emerge when evolutionary theory is used to interpret the fossil record and the range of existing species diversity, as we have done for the mollusks.**

# ANNELIDS—SEGMENTS GALORE

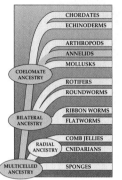

Maybe you've noticed earthworms after a downpour, when they wriggle out of their burrows to avoid drowning. Earthworms are among 15,000 or so species of bilateral, segmented animals known as the **annelids** (Annelida). Their relatives are a few kinds of leeches and the polychaetes, which are far more diverse but not nearly as well known (Figure 26.27). The phylum name means "ringed forms." But the rings are actually a series of repeating body units, and the segmentation is pronounced. Also, except for leeches, nearly all segments have pairs or clusters of chitin-reinforced bristles on each side of the body. The bristles are also called setae or chaetae, but these are just formal names for bristles. When pushed into soil, the bristles provide the traction required for crawling or burrowing. They have become broadened paddles in some swimming species. Earthworms, one of the oligochaetes, have few setae per body segment, and marine polychaete worms typically have many of them (*oligo–*, few; *poly–*, many).

## Advantages of Segmentation

A segmented body has great evolutionary potential, for individual parts can undergo modification and become highly adapted for specialized tasks. Although most of an earthworm's segments are similar, the leeches have suckers at both ends, and polychaetes have an elaborate head and fleshy-lobed appendages known as parapods ("closely resembling feet"). By analyzing the existing species, we catch glimpses of developments that led to increases in size and to more complex internal organs.

## Annelid Adaptations—A Case Study

The earthworms are examples of familiar annelids. As Figure 26.28 shows, partitions divide their body into a series of coelomic chambers. In most of the chambers we find repeats of muscles, blood vessels, branching nerves, and other

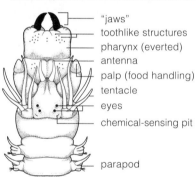

"jaws"
toothlike structures
pharynx (everted)
antenna
palp (food handling)
tentacle
eyes
chemical-sensing pit

parapod

c

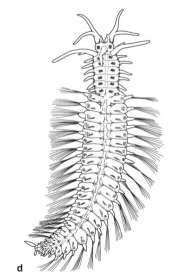

d

**Figure 26.28** (**a**) A familiar annelid—one of the earthworms. (**b–d**) Polychaetes. These give a sense of the dizzying variety of modifications that have evolved in this group, starting with a segmented, coelomic body plan. The type shown in (**b**) is one of the tube-dwellers. Featherlike structures at its head end are coated with mucus. After the mucus has trapped bacteria and other bits of food, coordinated beating of cilia sweeps them to the mouth. Most polychaetes live in marine habitats. They actually are one of the most common types of animals along coasts. Many are predators or scavengers; others dine on algae.

before feeding

after feeding

**Figure 26.27** Leeches. Most leeches have sharp jaws and a blood-sucking device. This one is shown before and after gorging on human blood. For at least 2,000 years, *Hirudo medicinalis*, a freshwater leech, has been employed as a blood-letting tool to "cure" problems ranging from nosebleeds to obesity.

Leeches are still used, but more selectively, as when they draw off pooled blood after surgeons reattach a severed ear, lip, or fingertip. A patient's body cannot do this on its own until the severed blood circulation routes are reestablished.

**Figure 26.29** Earthworm body plan. (**a**) Midbody, transverse section. (**b**) A nephridium, one of many functional units that help maintain the volume and the composition of body fluids. (**c**) Portion of the closed circulatory system. The system is linked in its functioning with nephridia. (**d**) Part of the digestive system, near the worm's head end. (**e**) Part of the nervous system.

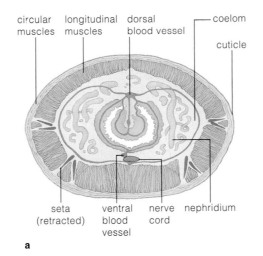

circular muscles — longitudinal muscles — dorsal blood vessel — coelom — cuticle — seta (retracted) — ventral blood vessel — nerve cord — nephridium

**a**

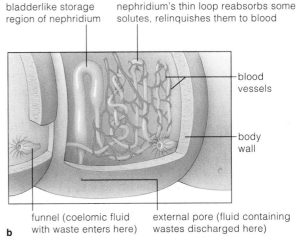

bladderlike storage region of nephridium — nephridium's thin loop reabsorbs some solutes, relinquishes them to blood — blood vessels — body wall — funnel (coelomic fluid with waste enters here) — external pore (fluid containing wastes discharged here)

**b**

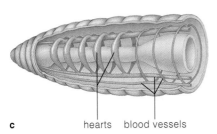

**c** — hearts — blood vessels

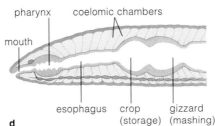

pharynx — coelomic chambers — mouth — esophagus — crop (storage) — gizzard (mashing)

**d**

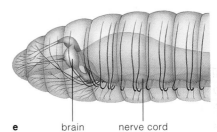

**e** — brain — nerve cord

organs. The gut extends through all the chambers, from mouth to anus. Like all annelids, an earthworm has a cuticle of secreted material that encapsulates the body surface. It bends easily, and it is permeable to water as well as to gases, which is one reason why annelids are restricted to aquatic habitats or moist habitats on land.

Earthworms are scavengers. They ingest moistened soil and mud that contains decomposing plant material and other organic matter. Each worm can eat its own weight every twenty-four hours. As numerous worms burrow and feed, they collectively aerate soil and lift nutrients to the surface, to the benefit of many plants.

As in other annelids, the fluid-cushioned coelomic chambers serve as a hydrostatic skeleton against which muscles act. Each segment's wall incorporates a layer of circular muscles (Figure 26.29a). When longitudinal muscles that bridge several segments contract, circular muscles relax, so the segments shorten and fatten. When the pattern reverses, the segments lengthen. While this is going on, bristles on different segments protract and retract. When the first few segments lengthen and their setae are not touching the ground, the front part of the body is extended. Bristles of the segments behind them plunge into the ground and anchor the worm. Next, the first few segments plunge *their* bristles into the ground, and the segments toward the posterior end of the body retract their bristles and are pulled forward. The whole worm moves forward when alternating contractions and elongations proceed along the body's length.

Figure 26.29*b* shows part of a system of **nephridia** (singular, nephridium), units that regulate the volume and composition of body fluids. In many annelids, cells of the units are similar to flame cells, which implies an evolutionary link between the flatworms and annelids. More often, a nephridium starts out as a funnel that collects excess fluid from one coelomic chamber. The funnel connects with a tubular part of the nephridium, which delivers fluid to a pore at the surface in the body wall of the next coelomic chamber.

The worm's head end has a rudimentary **brain**, an aggregation of nerve cell bodies that integrate sensory input as well as commands for muscle responses for the whole body. Paired **nerve cords**, each a bundle of long extensions of nerve cell bodies, lead away from the brain. They are pathways for rapid communication. In each body segment, the paired nerve cords broaden into a ganglion (plural, ganglia), a cluster of nerve cell bodies that controls local activity.

Finally, as is the case for most annelids, earthworms have a closed circulatory system, with blood confined in hearts and muscularized blood vessels. Contractions keep blood circulating in one direction. Smaller blood vessels service the gut, nerve cord, and body wall.

**Annelids are bilateral, coelomate, segmented worms that have complex organ systems. Some species show the degree of specialization possible with a segmented body plan.**

## Arthropod Diversity

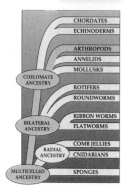

CHORDATES
ECHINODERMS
ARTHROPODS
ANNELIDS
MOLLUSKS
COELOMATE ANCESTRY
ROTIFERS
ROUNDWORMS
RIBBON WORMS
BILATERAL ANCESTRY
FLATWORMS
COMB JELLIES
RADIAL ANCESTRY
CNIDARIANS
MULTICELLED ANCESTRY
SPONGES

Evolutionarily speaking, "success" means having the greatest number of species, producing the most offspring, occupying the most habitats, effectively fending off threats and competition, and having the capacity to exploit the greatest amounts and kinds of food. These are the features that come to mind when we attempt to characterize the **arthropods** (Arthropoda). We have already identified over a million species—mostly insects—and researchers discover new ones weekly, mainly in tropical forests and the seas.

Of four major lineages, trilobites are extinct (Section 21.5). The other three are chelicerates (spiders and their relatives), crustaceans (including barnacles and crabs), and uniramians (centipedes, millipedes, and insects).

## Adaptations of Insects and Other Arthropods

Six important adaptations contributed to the success of arthropods in general and the insects in particular: a hardened exoskeleton, jointed appendages, fused and modified segments, specialized respiratory structures, efficient nervous system and sensory organs, and often a division of labor in the life cycle.

**HARDENED EXOSKELETONS** Arthropods have a cuticle of chitin, proteins, and surface waxes that may even be impregnated with calcium carbonate deposits. It acts as a rigid, protective external skeleton—an **exoskeleton**. Such cuticles might have evolved as defenses against predation. They took on added functions when some arthropods first invaded the land. They support a body deprived of water's buoyancy, and their waxy surface restricts evaporative water loss. Hard cuticles stop size increases, but arthropods grow in spurts by **molting**. At certain stages of their life cycle, they secrete a soft new cuticle under the old one, which they shed (Figure 26.30). The body mass increases first by uptake of air or water, then by repeated, rapid cell division before the new cuticle can harden.

**JOINTED APPENDAGES** If an arthropod had a cuticle that was uniformly hardened, it wouldn't move much. But the arthropod cuticle is thinner at *joints,* where

**Figure 26.30** Molting, demonstrated by a red-orange centipede wriggling out of its old exoskeleton.

two body parts abut. Muscles associated with the joints make the thinner cuticle bend in specific directions and move body parts. This jointed exoskeleton was a key innovation that led to appendages as diverse as wings, antennae, and legs (*arthropod* means "jointed foot").

**FUSED AND MODIFIED SEGMENTS** The first arthropods were segmented, like the annelid stock that presumably gave rise to them. In most existing descendants, serial repeats of the body wall and organs are masked; many fused-together segments are modified to perform more specialized functions. For example, in the ancestors of insects, different segments fused to form three regions—head, thorax, and abdomen—which morphologically diverged from one another in astonishing ways.

**RESPIRATORY STRUCTURES** Many aquatic arthropods depend on gills for gas exchange. Air-conducting tubes evolved among insects and other land-dwellers. Insect tracheas begin as pores on the body surface and branch into tubes that deliver oxygen directly to tissues. They support energy-consuming activities, such as flight.

**SPECIALIZED SENSORY STRUCTURES** Intricate eyes and other sensory organs contributed to arthropod success. Numerous species have a wide angle of vision and can process visual information from many directions.

**DIVISION OF LABOR** Moths, butterflies, beetles, flies, and many other species divide the job of surviving and reproducing among different stages of development. For many species, the new individual is a *juvenile*—a miniaturized version of the adult that simply changes in size and proportion until reaching sexual maturity. Other species show **metamorphosis**, meaning the body form changes from embryo to adult. Under hormonal commands, their size increases, tissues reorganize, and body parts are remodeled. Sections 26.19, 37.8, and 44.8 have examples of this transitional time. Immature stages such as caterpillars typically specialize in feeding and growing in size. The adult stage specializes primarily in dispersal and reproduction. For such species, then, the life cycle turns on a *division of labor*: the different stages of development become specialized in ways that are adaptive to environmental conditions, such as seasonal variation in food resources and water supplies.

---

As a group, the arthropods are exceptionally abundant and widespread, and they have enormously different life-styles.

Their success arises largely from their hardened, jointed exoskeletons; fused, modified body segments; specialized appendages; specialized respiratory, nervous, and sensory organs; and often a division of labor in the life cycle.

---

Further reading: Student Guide to InfoTrac on web site →

The chelicerates originated in shallow seas early in the Paleozoic. The only surviving marine species are a few mites, sea spiders, and horseshoe crabs (Figure 26.31). The familiar chelicerates on land—spiders, scorpions, ticks, and chigger mites—are all classified as arachnids. We might say this about the whole group: Never have so many been loved by so few.

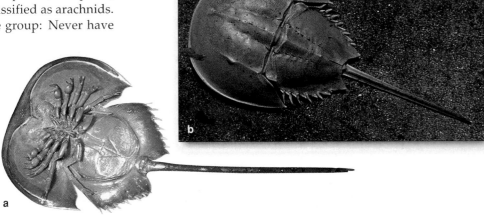

**Figure 26.31**    Horseshoe crab, (**a**) ventral and (**b**) dorsal views. The *five* pairs of legs of this chelicerate are one of its defining features. They are hidden beneath the hard, shieldlike cover.

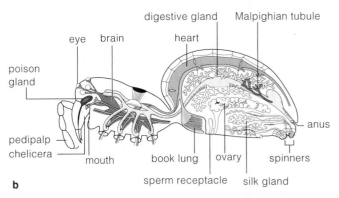

**Figure 26.32**    (**a**) Wolf spider. Like most spiders, it helps keep populations of insects in check. Its bite is harmless to humans. Section 26.20 describes two of its dangerous relatives. (**b**) A spider's internal organization.

Spiders and scorpions are efficient predators; they sting or bite and may subdue prey with venom. When most kinds sting or bite people, the reaction might be painful but only rarely does it lead to serious problems. Collectively, the spiders especially are beneficial in that they prey upon great numbers of pestiferous insects.

Bites of the blood-sucking ticks that parasitize deer, mice, and other vertebrates cause maddening itches and often serious diseases. For example, bites of some ticks transmit the bacterial agents of Rocky Mountain spotted fever or Lyme disease to humans (Sections 22.6 and 26.20). Most mites are free-living scavengers.

Arachnids have segments fused into a forebody and hindbody. The forebody's jointed appendages include four pairs of legs, a pair of pedipalps with primarily sensory functions, and a pair of chelicerae that inflict wounds and discharge venom. The appendages of the hindbody spin out silk threads for webs and for egg cases. Most webs are netlike. One type of spider spins a vertical thread with a ball of sticky material at the end. It uses a leg to swing the ball at insects passing by!

Inside the body is an *open* circulatory system, with a heart that pumps blood into body tissues, then receives blood through small openings in its wall. Some of the slowly circulating blood travels through moist folds of book lungs. These respiratory organs, which resemble loose pages of a book, greatly increase the surface area that is available for gas exchange with the air. Figure 26.32*b* shows the arrangement of book lungs and other major organs inside the spider body.

**The spiders, scorpions, and their relatives have a variety of appendages specialized for predatory or parasitic life-styles.**

## A LOOK AT THE CRUSTACEANS

Shrimps, lobsters, crabs, barnacles, pillbugs, and other crustaceans got their name because they have a hard yet flexible "crust" (an external skeleton)—but so do nearly all arthropods. Only some of the 35,000 species live in fresh water or on land. The vast majority live in marine habitats, where they are so abundant they have been dubbed the insects of the seas. Lobsters and crabs are "giants" of this subphylum; most crustaceans are less than a few centimeters long. All have major roles in food webs, and humans harvest many edible types.

By having many pairs of similar appendages along most of their length, the simplest crustaceans might resemble their annelid ancestors. In the more complex

lineages, unspecialized appendages evolved into diverse structures of the sort shown in Figure 26.33. The strong claws of lobsters and crabs, for example, are used to collect food, intimidate other animals, and sometimes dig burrows. Feathery appendages of barnacles comb microscopic bits of food from the water.

Many crustaceans have sixteen to twenty segments; some have more than sixty. In crabs, lobsters, and some other crustaceans, the dorsal cuticle extends back from the head as a shieldlike cover over some or all of the segments. The head has two pairs of antennae, one pair of jawlike appendages (mandibles), and two pairs of appendages for handling food. Crayfish, crabs, lobsters,

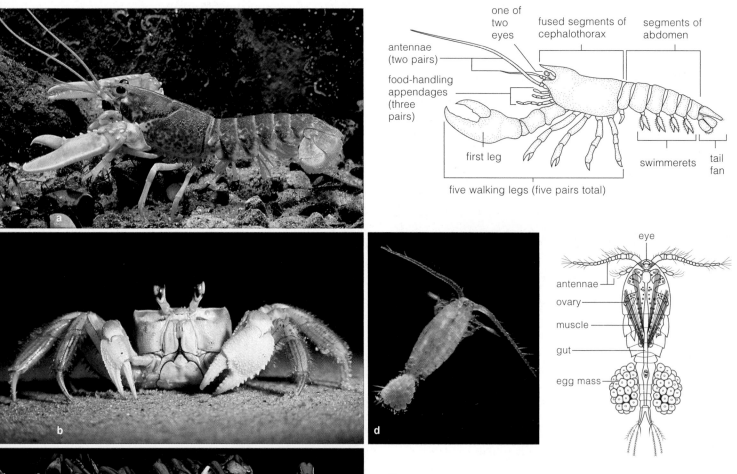

**Figure 26.33** A sampling of crustaceans and their diverse life-styles. (**a**) Photograph and body plan of a lobster. For most of their lives, lobsters are secretive (**b**) Crab. Guess why crabs skitter actively and openly across sand and rocks at night but not in the day. (**c**) Stalked barnacles. Adults cement themselves to one spot. You might mistake them for mollusks, but as soon as they open their hinged shell to filter-feed, you can observe their jointed appendages—the hallmark of arthropods. See also Figure 5.1. (**d**) Photograph and body plan of a copepod. Copepods are free-living filter feeders, predators, or parasites. This female has a pair of long antennae and is carrying her eggs with her.

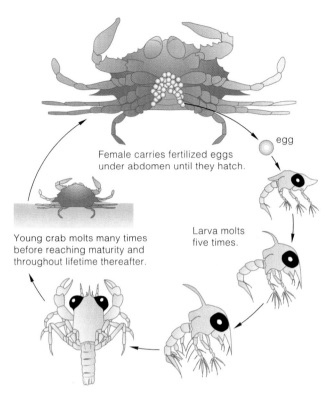

**Figure 26.34** Life cycle of a crab. The larval and juvenile stages molt repeatedly and grow in size.

Female carries fertilized eggs under abdomen until they hatch.

egg

Larva molts five times.

Young crab molts many times before reaching maturity and throughout lifetime thereafter.

shrimps, and their various relatives are equipped with five pairs of walking legs.

Of all the arthropods, only barnacles have a calcified "shell," a modified external skeleton that protects them from predators, drying winds, battering surf, and strong currents. Adult barnacles cement themselves to rocks, wharf pilings, and similar surfaces (Figure 26.33*c*). A few kinds attach themselves only to the skin of whales.

Figure 26.33*d* shows a copepod. Copepods are less than two millimeters long and are the most numerous animals in aquatic habitats, maybe even in the world. About 1,500 kinds parasitize various invertebrates and fishes. The majority—8,000 species—are consumers of phytoplankton, the "pastures" of aquatic habitats. Some also eat larval or small adult invertebrates, fish eggs, and fish larvae, which they grab with their pair of food-handling appendages. In turn, the copepods are food for different invertebrates, fishes, and baleen whales.

Like other arthropods, crustaceans molt repeatedly to shed the exoskeleton during their life cycle. Figure 26.34 shows a crab's larval stages and increases in size.

**Crustaceans differ greatly in the number and kind of their appendages. As is the case for arthropods in general, they repeatedly replace their external skeleton by molting.**

## *HOW* MANY LEGS?

The **millipedes** and **centipedes** have a long, segmented body with many legs. Of course, millipedes don't have "a thousand," as their name implies. Most have about 100 legs, although one exuberant individual grew 752. Centipedes have between 15 and 177 pairs of legs, not a nicely rounded number of "one hundred."

As the millipedes develop, pairs of segments fuse, and each segment in the cylindrical body ends up with two pairs of legs (Figure 26.35*a*). Millipedes scavenge decaying plant material in soil and forest litter.

**Figure 26.35** (**a**) Millipede. (**b**) A Southeast Asian centipede.

Centipedes have a flattened body, and all but two segments have a pair of walking legs. All species are fast-moving, aggressive predators outfitted with fangs and venom glands. They prey on insects, earthworms, and snails. The one in Figure 26.35*b* can subdue small lizards, toads, and frogs. A house centipede (*Scutigera*) often hides out in buildings, where it hunts cockroaches, flies, and other pests. Athough helpful in this respect, its vaguely terrifying body keeps it from being welcomed.

**The mild-mannered, scavenging millipedes and aggressive, predatory centipedes do not lend themselves to leg counts as they walk by.**

# A LOOK AT INSECT DIVERSITY

As a group, insects share the adaptations listed earlier in Section 26.15. Here we expand the list a bit. Insects have a head, a thorax, and an abdomen. The head has paired sensory antennae and paired mouthparts, which specialize in biting, chewing, sucking, or puncturing (Figure 26.36). Three pairs of legs and usually two pairs of wings project from the thorax. Most appendages of the abdomen are reproductive structures, such as egg-laying devices. An insect has a foregut, midgut (where most digestion proceeds), and hindgut (where water is reabsorbed). Insects get rid of waste material through **Malpighian tubules**, small tubes that connect with the midgut. When they break down proteins, the nitrogen-containing wastes diffuse from blood into the tubules and are converted into harmless crystals of uric acid. The crystals are eliminated with feces. This system lets land-dwelling insects get rid of potentially toxic wastes without losing precious water.

We've already catalogued more than 800,000 species of insects. If we use sheer numbers and distribution as the yardstick, the most successful insects are small in size and have a staggering reproductive capacity. For example, you may find some of these species growing and reproducing in great numbers on a single plant that might be only an appetizer for another animal. By one estimate, if all the progeny of a single female fly were to survive and reproduce through six more generations, that fly would have more than 5 trillion descendants!

Besides this, the most successful insect species are winged. In fact, they are the *only* winged invertebrates. They can move among food sources that are too widely scattered to be exploited by other kinds of animals. The capacity for flight contributed to their success on land.

Finally, insect life cycles commonly proceed through stages that allow exploitation of different resources at different times. As an insect embryo develops, organs required for feeding and other vital activities form and become functional. Before an insect becomes an adult (the sexually mature form of the species), it proceeds through immature, post-embryonic stages. **Nymphs** and **pupae**, as well as larvae, are examples of the stages.

Like human infants, nymphs of some insects are shaped like miniature adults. But unlike children, they undergo growth and molting (Figure 26.36a). Other insects go through post-embryonic stages of reactivated growth, tissue reorganization, and remodeling of body parts. Metamorphosis, remember, is the name for this resumption of growth and transformation into an adult form. Figure 26.37 indicates how the transformation is more drastic in some species than in others.

All the factors that contribute to insect success also make them our most aggressive competitors. Insects destroy crops, stored food, wool, paper, and timber. As some stealthily draw blood from us and from our pets,

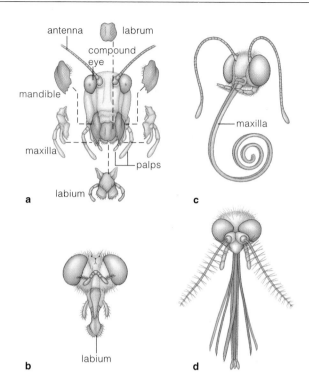

**Figure 26.36** Examples of insect appendages. Headparts of (**a**) grasshoppers, a chewing insect; (**b**) flies, which sponge up nutrients with a specialized labium; (**c**) butterflies, which siphon up nectar with a specialized maxilla; and (**d**) mosquitoes, with piercing and sucking appendages.

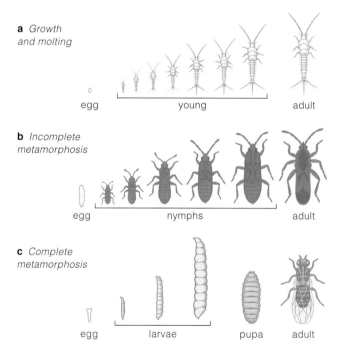

**Figure 26.37** Examples of post-embryonic development. (**a**) Young silverfish are adults in miniature, changing little except in size and proportion as they mature into adults. (**b**) True bugs show *incomplete* metamorphosis, which involves gradual, partial change from the first immature form until the last molt. (**c**) Fruit flies show *complete* metamorphosis. Tissues of immature forms are destroyed and replaced before emergence of the adult.

Further reading: Student Guide to InfoTrac on web site →

**Figure 26.38** Representative insects. (**a**) Duck louse (order Mallophaga). It eats bits of feathers and skin. (**b**) European earwig (order Dermaptera), a common household pest. (**c**) Flea (order Siphonaptera), with strong legs for jumping onto and off animal hosts. (**d**) Mediterranean fruit fly (order Diptera). Its larvae destroy citrus fruit and other crops.

(**e**) Stinkbugs (order Hemiptera), newly hatched. (**f**) At center, a honeybee (order Hymenoptera) attracting its hive mates with a dance, as described in Section 47.4. (**g**) Ladybird beetles (order Coleoptera) swarming. These beetles are commercially raised and released as biological controls of aphids and other pests. Also in this order, the scarab beetle (**h**). With more than 300,000 species, Coleoptera is the largest order of the animal kingdom. (**i**) Luna moth (order Lepidoptera) of North America. Microscopic scales cover the wings and body of most butterflies and moths, including this one. (**j**) One of the dragonflies (order Odonata). It swiftly captures and eats insects in midflight.

they often transmit pathogenic microorganisms. Yet many pollinate flowering plants, which include highly valued crop plants. And many "good" insects attack or parasitize the ones we would rather do without. Figure 26.38 highlights representatives of some major orders of insects. In the next section, we conclude our arthropod survey with a selection of the notorious species.

---

**As a group, insects show immense variation on the basic arthropod body plan. Many have wings, and their life cycles have stages that allow exploitation of different and often widely scattered food sources. Many species produce great numbers of small individuals that pass through immature, post-embryonic stages before the adult form emerges.**

---

# UNWELCOME ARTHROPODS

Given that there are now more than 6 billion of us on the planet, very, very few of us encounter the truly nasty arachnids—certain spiders, ticks, and scorpions that can, directly or indirectly, inflict memorable pain. All things considered, humans collectively do more harm to more species than, say, the spiders, most of which do great good by popping off uncountable numbers of insect pests. But the "good" ones have some "bad" relatives, and it doesn't hurt to know them when we see them.

**HARMFUL SPIDERS**  All spiders of the genus *Loxosceles* are poisonous to humans and other mammals, but we seldom meet up with most of them. One exception, the brown recluse (*L. reclusa*), favors hiding in dry, warm places. These include houses, especially near refrigerator motors and clothes dryers, and in the folds of clothing or linens in closets. You can recognize it by the violin-shaped marking on its cephalothorax (Figure 26.39*a*). A bite ulcerates and heals poorly. Some bitten people have required skin grafts; a few died. *L. laeta*, a related species in Chile, Argentina, and Peru, is similarly dangerous.

About five people die each year in the United States as a result of black widow bites (Figure 26.39*b*)—orders of magnitude less than deaths attributed to, say, drug overdoses or accidents involving SUVs. Even so, besides being painful, this spider's neurotoxin causes muscles to stiffen, so medical attention should be prompt.

**MIGHTY MITES**  Of all arachnids, mites are among the smallest, most diverse, and most widely distributed. Of an estimated 500,000 species, a few parasitic mites—ticks especially—are harmful. Figure 26.40 shows a deer tick (*Ixodes*), a vector for *Lyme disease* in the northeastern and north-central United States. (Different ticks spread the

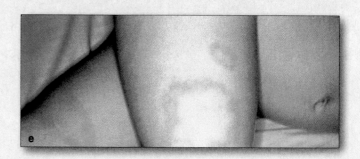

**Figure 26.39**  (**a**) Brown recluse, with a violin-shaped mark on its forebody. Its bite can be severe to fatal. (**b**) Female black widow, with a red, hourglass-shaped marking on the underside of her shiny black abdomen. Males are smaller and do not bite.

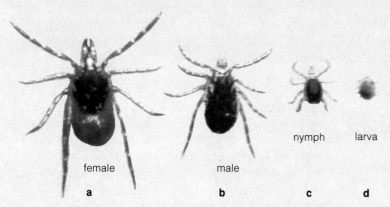

**Figure 26.40**  (**a–d**) Deer ticks (*Ixodes dammini*), the most common vector for Lyme disease. The female is about three millimeters long. All stages of the life cycle are capable of transmitting the disease agent, *Borrelia burgdorferi*, to humans. (**e**) An extreme reaction to an infection by *Borrelia burgdorferi*, a spirochete. The rash is a symptom of what is now the most common tick-borne disease in the United States: Lyme disease. Tick bites deliver the spirochete to new hosts.

(**f**) Graph showing the increasing number of cases of Lyme disease in the United States between 1982 and 1997.

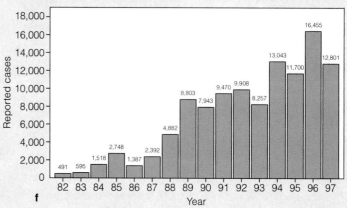

Further reading: Student Guide to InfoTrac on web site →

**Figure 26.41** *Centruroides sculpuratus*, a scorpion from Arizona that is one of the most dangerous species known.

**Figure 26.42** (**a**) Adult western corn rootworm (*Diabrotica virgifera*). (**b**) Larvae on corn roots. (**c**) Damaged corn plants.

disease in the western and southeastern states.) Normally the tick feeds on white-tail deer, mice, other mammals, and birds. *Borrelia burgdorferi*, the agent of Lyme disease, is transmitted to humans mainly by nymphal stages, probably because these are no bigger than a pinhead and escape detection. Detection is critical, because ticks usually transmit the bacterium after they have been feeding two days or more. Adults are easier to spot, so they are more likely to be removed quickly.

Ticks don't fly or jump. They crawl onto grasses and shrubs, then onto animals that brush past, then typically into hairy, hidden body parts such as the scalp and groin. Look for them after walks in the wild. Different types can transmit pathogens responsible for many serious diseases, including Rocky Mountain spotted fever (Section 22.6), scrub typhus, tularemia, babesiasis, and encephalitis. A select few can even kill directly. Their toxin interferes with motor neurons in a way that causes progressive paralysis from the lower part of the body upward. The tick has to be removed before the bitten person stops breathing.

Ending on a less scary but still irritating note, think of the house dust mites. They don't kill you, but the allergies they provoke in people can be irritating to the extreme.

**SCORPION COUNTRY**  Scorpions have that look about them. Large, prey-seizing pincers, a venom-dispensing stinger at the tip of a narrowed, jointed abdomen—what's not to love? Actually, scorpions seldom dispense venom unless their prey—other small arthropods, the occasional small lizard or mouse—puts up a mighty fight. In the wild they hunt at night and rest during the day in burrows or under logs, stones, and bark. They can stay there without starving for as long as a year.

We find scorpions in cold places, such as Canada, the southern Andes, and the southern Alps. Some live in moist forests. But most live in hot, dry places, and there they are active throughout the year. The ones that can hurt or kill people aren't the fiercest looking, and there aren't many

of them. *Centruroides sculpturatus*, an Arizona scorpion, is the most dangerous species in the United States (Figure 26.41). Its close relatives in Mexico have a reputation for causing fatalities, especially amoung small children.

When in scorpion country, it is not a good idea to walk about barefoot at night or to put on shoes without first turning them upside-down and shaking them vigorously.

**BEETLES, BEETLES EVERYWHERE**  First in America's Midwest and now in the Balkans, *Diabrotica virgifera* is making the rounds in cornfields as a representative of one of our major competitors for food. Figure 26.42 shows what it looks like and the kind of damage it causes. Its larvae feed on corn roots until the severely damaged plants simply keel over. Each year we lose about a billion dollar's worth of crops to these little critters alone.

Faced with such astronomical losses, farmers spread about 30 million pounds of pesticides a year on cornfields, which is about half the total dispensed for all row crops grown in the United States. So far rootworms are not impressed. Many have developed pesticide tolerance.

Inventive researchers are targeting the adults, which show an inordinate fondness for bitter plant juices—specifically, for curcurbitacin. This is one of the steroid terpenes that evolved as a natural pesticide in melons, squashes, cucumbers, and other members of the family Cucurbitaceae. Rootworms apparently appropriated the chemical as a defense against bird predators. (Birds learn to avoid rootworms after a taste trial and vomiting.) Robert Schroder and others are concocting to-die-for bait for the pests. For example, watermelon juice has been laced with red dye 28, a photoactive chemical. In taste trials, adult rootworms gorged themselves, turned deep red, and after five minutes in sunlight, dropped dead. Apparently sunlight triggered oxidation reactions that destroyed tissues throughout the body. Beneficial insects avoid the dyed juice. In this particular ongoing contest, chalk one up for the researchers.

# THE PUZZLING ECHINODERMS

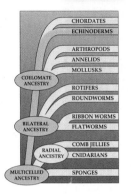

We turn, finally, to the second lineage of the coelomate animals, the deuterostomes. The major invertebrate members of this lineage are **echinoderms** (Echinodermata). The feather star, sea urchin, sea cucumber, and brittle stars shown in Figure 26.43 belong to this phylum. So do the sea lilies (crinoids), sand dollars, and sea biscuits. The sea lilies, which look a bit like stalked plants, flourished in Silurian times (Figure 21.14). About 13,000 echinoderm species are known from the fossil record, although most of these became extinct. Nearly all of the 6,000 or so existing species live in marine habitats.

An echinoderm body wall bears a number of spines, spicules, or plates made rigid with calcium carbonate. (*Echinodermata* means spiny-skinned.) These structures serve defensive functions, as you might suspect if you have ever stepped barefoot on a sea urchin. Its spines trigger painful swelling when they break off under the skin. Most echinoderms have a well-developed internal skeleton, which is composed of calcium carbonate and other substances secreted from specialized cells.

Oddly, the adult echinoderms are radial with some bilateral features. Most species even produce bilateral larvae. Did bilateral invertebrates give rise to ancestors of echinoderms, in which some radial features evolved at a later time? Maybe.

Adult echinoderms have no brain. However, their decentralized nervous system allows them to respond to information about food, predators, and so forth that is coming from different directions. For instance, any arm of a sea star that senses the shell of a tasty scallop

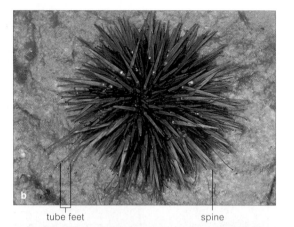

tube feet          spine

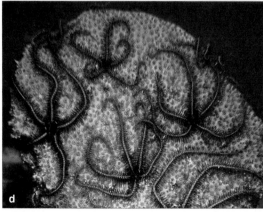

**Figure 26.43**  Representative echinoderms. (**a**) Feather star, with finely branched food-gathering appendages. (**b**) Sea urchin, which moves about on spines and tube feet. (**c**) Sea cucumber, with rows of tube feet along its body. (**d**) Brittle stars. Their arms (rays) make rapid, snakelike movements.

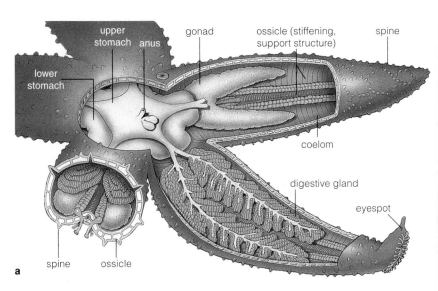

part of the water-vascular system

sieve plate

ring canal

ampulla

b

c

d

tube feet on the ventral surface of a sea star arm

**Figure 26.44** (**a**) Key aspects of the radial body plan of a sea star. Its water-vascular system, combined with great numbers of tube feet, is the basis of locomotion. (**b–d**) Five-armed sea star, with closer views of tube feet.

can become the leader, directing the rest of the body to move in a direction suitable for capturing prey.

Figures 26.43 and 26.44 show examples of tube feet. These fluid-filled, muscular structures have suckerlike adhesive disks. Sea stars use their tube feet for walking, burrowing, clinging to rocks, or gripping a clam or snail about to become a meal. Tube feet are components of a **water-vascular system** unique to echinoderms. In sea stars, that system includes a main canal in each arm. Short side canals extend from them and deliver water to the tube feet. Each tube foot has an ampulla, a fluid-filled, muscular structure shaped rather like the rubber bulb on a medicine dropper. As an ampulla contracts, it forces fluid into the foot and causes it to lengthen.

Tube feet change shape constantly as muscle action redistributes fluid through the water-vascular system. Hundreds of tube feet may move at a time. After being released, each one swings forward, reattaches to the substrate, then swings backward and is released before swinging forward again. Their motions are splendidly coordinated, so sea stars glide rather than lurch along.

On their ventral surface, sea stars have a formidable feeding apparatus, such as the one shown here:

Some eager sea stars simply swallow their prey whole. Others push part of their stomach outside the mouth and around their prey, then start digesting their meal even before swallowing it. Sea stars get rid of coarse, undigested residues through the mouth. They do have a small anus, but this is of no help in getting rid of an empty clam shell or snail shell.

With their curious traits, echinoderms are a suitable point of departure for this chapter. Even though we can identify broad trends in animal evolution, we should keep in mind that there are confounding exceptions to the perceived macroevolutionary patterns.

---

**Echinoderms are coelomate animals with spines, spicules, or plates in the body wall. From the evolutionary perspective, they are a puzzling mix of bilateral and radial features.**

---

# 26.22 SUMMARY

1. Animals are multicelled, aerobic heterotrophs that ingest or parasitize other organisms. Nearly all have diploid body cells organized into tissues, organs, and organ systems. Animals reproduce sexually and often asexually. They undergo embryonic development. Most are motile during at least part of the life cycle.

2. Animals range from structurally simple placozoans and sponges to vertebrates.

   a. By comparing major animal phyla and integrating the information with the fossil record, biologists have identified major evolutionary trends among them.

   b. Revealing aspects of body plans are the type of symmetry, gut, and cavity (if any) between the gut and body wall; whether there is a head end; and whether the body is divided into segments (Figure 26.45).

3. *Trichoplax*, the only known placozoan, is the simplest animal. It is composed of little more than two layers of cells with a fluid matrix in between.

4. The sponge body has no symmetry and it is at the cellular level of construction. Although it consists of several kinds of cells, these are not organized like the epithelia and other tissues seen in complex animals.

5. Cnidarians include the jellyfishes, sea anemones, and hydras. They have radial symmetry, and they are at the tissue level of construction. Cnidarians alone produce nematocysts, capsules with dischargeable threads that are used mainly in prey capture.

6. Nearly all animals more complex than cnidarians show bilateral symmetry, and they form tissues, organs, and organ systems. Their gut may be saclike, as it is in flatworms, but it usually is complete, with an anus and a mouth. Most animals more complex than flatworms have a coelom or false coelom (cavities between the gut and body wall). A coelom has a lining, a peritoneum.

7. Two major lineages diverged shortly after flatworms evolved. One (protostomes) gave rise to the mollusks, annelids, and arthropods. The other (deuterostomes) gave rise to echinoderms and chordates.

8. All mollusks have a fleshy, soft body and a mantle. Most have either a shell or a remnant of one. Mollusks vary greatly in size, body details, and life-styles.

9. The annelids (earthworms, polychaetes, and leeches) have a segmented body, complex organs, and a series of coelomic chambers.

10. Collectively, arthropods are the most successful of all groups in terms of diversity, numbers, distribution, defenses, and capacity to exploit food resources.

   a. All arthropods, including arachnids, crustaceans, and insects, have hardened, jointed exoskeletons. They also have modified segments, specialized appendages,

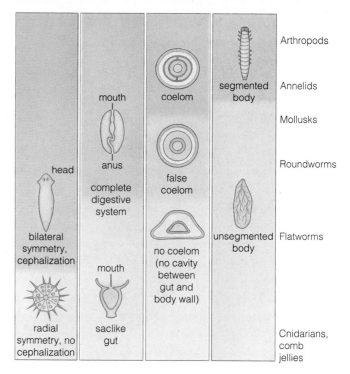

**Figure 26.45** Summary of key trends in the evolution of animals, as identified by comparing body plans of major phyla. All of these features did not appear in every group.

notably specialized respiratory, nervous, and sensory organs, and (in insects only) wings.

   b. Arthropods develop by growth in size and molting or develop through a series of immature stages, such as larvae and nymphs. Many metamorphose; the tissues of immature forms undergo major reorganization, and body parts are remodeled before the adult emerges.

11. Echinoderms have spines, spicules, or plates in their body wall. Evolutionarily, the phylum is puzzling. The larvae of most species form bilateral features, but they go on to develop into basically radial adults.

## Review Questions

1. List the six main features that characterize animals. *26.1*

2. When attempting to discern evolutionary relationships among major groups of animals, which aspects of their body plans provide the most useful clues? *26.1*

3. What is a coelom? Why was it important in the evolution of certain animal lineages? *26.1*

4. Name some animals with a saclike gut. Evolutionarily, what advantages does a complete gut afford? *26.1, 26.4–26.6*

5. Choose a species of insect that lives in your neighborhood and describe some of the observable adaptations that underlie its success. *26.15, 26.19*

## Self-Quiz (Answers in Appendix III)

1. Which is *not* a general characteristic of the animal kingdom?
   a. multicellularity; most have tissues, many form organs
   b. exclusive reliance on sexual reproduction
   c. motility at some stage of the life cycle
   d. embryonic development during the life cycle

**Figure 26.46** *Telesto*, one of the soft branching corals.

2. Most animals that are more complex than cnidarians have _____ symmetry, and _____ forms in their embryos.
   a. radial; mesoderm       c. bilateral; mesoderm
   b. bilateral; endoderm    d. radial; endoderm

3. More complex animals have a _____ between their gut and body wall.
   a. pharynx          c. coelom
   b. pseudocoelom     d. archenteron

4. Jellyfishes, sea anemones, and their relatives have _____ symmetry, and their cells form _____ .
   a. radial; mesoderm     c. radial; tissues
   b. bilateral; tissues   d. bilateral; mesoderm

5. Which phylum contains members that are notorious for causing serious diseases in humans?
   a. cnidarians       c. segmented worms
   b. flatworms        d. chordates

6. _____ have a coelom and pronounced segmentation, and a dizzying variety of modifications to the segments.
   a. arthropods    c. sponges           e. sea stars
   b. annelids      d. snails and clams   f. vertebrates

7. In sheer numbers and distribution, _____ are the most successful animals.
   a. arthropods    c. sponges           e. sea stars
   b. annelids      d. snails and clams   f. vertebrates

8. Match the terms with the appropriate groups.
   ____ sponges       a. spiny-skinned
   ____ cnidarians    b. vertebrates and kin
   ____ flatworms     c. flukes and tapeworms
   ____ roundworms    d. no tissue organization
   ____ rotifers      e. no males for some
   ____ mollusks      f. nematocysts, radial symmetry
   ____ annelids      g. hookworms, elephantiasis
   ____ arthropods    h. jointed exoskeleton
   ____ echinoderms   i. "belly-foots" and kin
   ____ chordates     j. segmented worms

## Critical Thinking

1. A carnivorous sponge was discovered in an underwater cave in the Mediterranean Sea. Unlike other sponges, it has no pores or canals. Projecting from branching outgrowths of its body surface are hooklike spicules that act like Velcro to trap shrimp and other animals. The outgrowths envelop prey, which sponge cells then digest. Which components of the sponge body had to evolve for such a feeding strategy?

2. Tapeworms are hermaphroditic. What selective advantages might this feature offer, and in what kinds of environments?

3. People who eat raw oysters or clams harvested from sewage-polluted waters develop mild to severe gastrointestinal ailments. Think about the feeding modes of these mollusks and develop a hypothesis to explain why mollusk eaters can get sick.

4. You are diving in calm, warm waters behind a large tropical reef. You observe something that looks like a red and white plant, but it has a profusion of tiny tentacles (Figure 26.46). This is a branching soft coral with a great many tiny polyps. You will never see such a coral growing on reef surfaces exposed to the open sea. Formulate two hypotheses that might explain their distribution.

5. The animal in Figure 26.47a is packed with organs and sports a crown of cilia at its head end. The marine worm in Figure 26.47b has a highly segmented body. Most of its segments are similar to one another; each has bristles on the sides that were employed to dig a burrow in sediments. To which groups do the two animals belong?

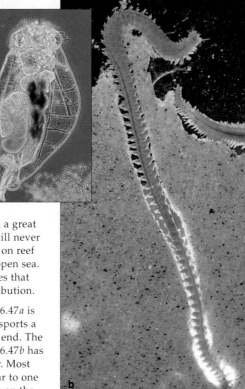

**Figure 26.47** Go ahead, name the mystery animals.

## Selected Key Terms

| | | |
|---|---|---|
| animal *26.1* | flatworm *26.7* | nerve cord *26.14* |
| annelid *26.14* | gonad *26.5* | nymph *26.19* |
| arthropod *26.15* | gut *26.1* | organ *26.7* |
| bilateral | hermaphrodite *26.7* | organ system *26.7* |
|   symmetry *26.1* | hydrostatic | placozoan *26.2* |
| brain *26.14* |   skeleton *26.4* | planula *26.5* |
| centipede *26.18* | invertebrate *26.1* | polyp *26.4* |
| cephalization *26.1* | larva *26.3* | proglottid *26.7* |
| cnidarian *26.4* | Malpighian | protostome *26.11* |
| coelom *26.1* |   tubule *26.19* | pupa *26.19* |
| collar cell *26.3* | mantle *26.12* | radial symmetry *26.1* |
| comb jelly *26.6* | medusa *26.4* | ribbon worm *26.7* |
| contractile cell *26.4* | mesoderm *26.1* | rotifer *26.10* |
| cuticle *26.8* | metamorphosis *26.15* | roundworm *26.8* |
| deuterostome *26.11* | millipede *26.18* | sponge *26.3* |
| echinoderm *26.21* | mollusk *26.12* | torsion *26.13* |
| ectoderm *26.1* | molting *26.15* | vertebrate *26.1* |
| endoderm *26.1* | nematocyst *26.4* | water-vascular |
| epithelium *26.4* | nephridium *26.14* |   system *26.21* |
| exoskeleton *26.15* | nerve cell *26.4* | |

***Readings*** *See also www.infotrac-college.com*

Kerr, E. 2 October 1998. "Tracks of Billion-Year-Old Animals?" *Science.* 282:19–21.

Pearse, V., et al. 1987. *Living Invertebrates*. Palo Alto, California: Blackwell.

Pechenik, J. 1995. *Biology of Invertebrates*. Third edition. Dubuque: W. C. Brown.

Raloff, J. July 10, 1999. "The Bitter End: Enticing Agricultural Pests to Their Last Repast." *Science News* 156:24–26.

# ANIMALS: THE VERTEBRATES

## *Making Do (Rather Well) With What You've Got*

In 1798, a few naturalists were skeptically probing a specimen delivered to the British Museum in London, looking for signs that a prankster had slyly stitched the bill of an oversized duck onto the pelt of a small furry mammal. They didn't know it, but they were examining the remains of a platypus, a web-footed mammal about half the size of a housecat. Like the other mammals, the duck-billed platypus (*Ornithorhynchus anatinus*) has mammary glands and hair. However, like birds and reptiles, it has a cloaca, an enlarged duct through which gametes, feces, and excretions from the kidneys pass. It lays shelled eggs, as birds and most reptiles do. Its young hatch pink and unfinished, as embryonic stages too helpless to fend for themselves (Figure 27.1*a*). Its fleshy bill does appear ducklike and its broad, flat, furry tail looks like the one on a beaver (Figure 27.1*b*).

With its unusual traits, the platypus invites us to challenge preconceived notions of what constitutes "an animal." This particular combination of traits started coming together when the supercontinent Pangea was breaking up. As you will see, the platypus's ancestors happened to be stuck on a huge fragment that remained isolated from the other continents for 150 million years; eventually it became Australia. The geographic separation was a springboard for genetic divergence, and unique mutations in combination with selection pressures gave rise to the platypus body. Even though that body plan is much ridiculed, we scarcely can call it a failure. It has endured far longer than the body plan we humans inherited.

The platypus inhabits streams and lagoons in remote regions of Australia and Tasmania. During the day it hides in underground burrows. At night it slips into the water, where it hunts for small invertebrates. Even when it stays in water all night, its dense fur

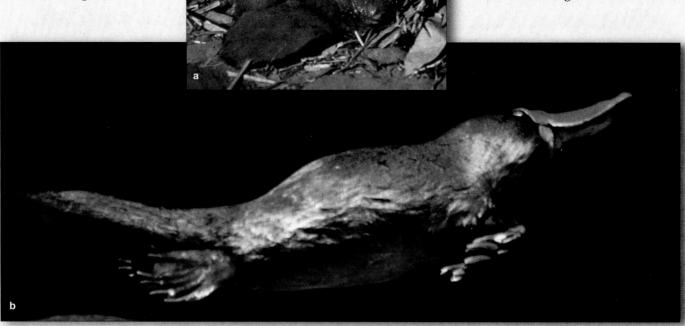

**Figure 27.1** One of evolution's success stories—the platypus, (**a**) raising its incompletely formed offspring in a burrow and (**b**) underwater, eyes and ears shut, yet homing in on prey.

helps maintain body temperature, just as it does all day long when the platypus is resting in its cool burrow. The broad, thick tail is a most excellent rudder during underwater maneuvers and a fine storehouse for

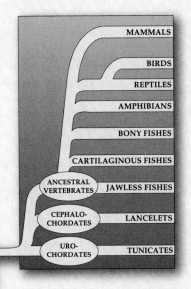

**Figure 27.2** Family tree for the vertebrates and other chordates.

energy-rich reserves of fat. The broad, flattened platypus bill is well shaped for scooping up aquatic snails, shrimps, worms, mussels, and insect larvae. The horny pads that line the jaws grind up and make short work of the shelled and hard-cuticled meals. Back at its burrow, the platypus uses the strong claws that project from its hind feet for digging out underground chambers.

Like a submarine, a platypus closes the hatches, so to speak, when it dives into the water. Its nostrils as well as a fleshy groove around the eyes and ears snap shut. No need for eyes or ears to zero in on prey; many of the 800,000 sensory receptors in its bill detect tiny oscillations in water pressure as prey swim past. Other receptors detect changes in electric fields as weak as those initiated by the flicks of a shrimp tail.

The platypus is one of 4,500 kinds of mammals that now occupy the vertebrate branch of the animal family tree (Figure 27.2). Its vertebrate relatives include fishes, amphibians, reptiles, and birds, which are described in this chapter. This chapter surveys all vertebrate groups. It concludes with a look at the evolutionary history of the human species, starting with its mammalian and primate stocks. As you consider each group, keep this key concept in mind: *Each animal is a mosaic of traits, many conserved from remote ancestors and others unique to its branch on the animal family tree.*

## KEY CONCEPTS

**1.** The chordate branch of the family tree for animals includes invertebrate and vertebrate species. All are bilateral animals. While they are embryos or larvae, each typically develops a supporting rod for the body (notochord), a dorsal nerve cord, a pharynx, gill slits in the pharynx wall, and a tail that extends past the anus. Some or all of these features persist in adults.

**2.** Existing invertebrate chordates include the tunicates and lancelets.

**3.** Of eight classes of vertebrates, seven have living representatives. These are jawless fishes, cartilaginous fishes, bony fishes, amphibians, reptiles, birds, and mammals. The other class, the placoderms, became extinct early in vertebrate history.

**4.** Four major trends occurred when certain vertebrate lineages evolved. Support and movement of the body came to depend less on the notochord and more on a backbone. After jaws evolved, the nerve cord evolved into a spinal cord and brain. When some lineages moved onto land, gills became less important than lungs for gas exchange, a trend enhanced by the evolution of more efficient circulatory systems. Among pioneers on land, fleshy fins with skeletal supports evolved into limbs, which evolved in diverse ways among the amphibians, reptiles, birds, and mammals.

**5.** In the primate branch of the mammalian lineage are prosimians, tarsioids, and anthropoids (monkeys, apes, and humans). Apes and humans are hominoids. Humans and some extinct species with a mosaic of apelike and humanlike traits are further classified as hominids.

**6.** A long-term cooling trend started during the Miocene. It led to seasonal changes in habitats and food sources. As food sources became scarcer, early hominoids spread out through Africa and, later, entered Europe and southern Asia. One lineage gave rise to the hominids.

**7.** Unlike earlier hominids, *Homo erectus* and *H. sapiens* displayed remarkable behavioral flexibility and creative experimentation with their environment, as when they started using fire. This characteristic allowed them to survive the challenges of dispersing into novel and often harsh environments around the world.

### Characteristics of Chordates

The preceding chapter left off with echinoderms, one of the most ancient lineages of the deuterostome branch of the animal family tree. Dominating this branch are their more recently evolved relatives, the **chordates** (phylum Chordata). Of these bilateral animals, about 2,100 are, like the echinoderms, invertebrates. The vast majority—about 48,000 species—are vertebrates.

**Vertebrates** are chordates with a backbone, of either cartilage or bone, and a brain located inside a protective chamber of skull bones. The "invertebrate chordates" share certain features with these dominant members of the phylum, but a backbone isn't one of them.

Four features are evident in chordate embryos, and in many species these features persist into adulthood. *First*, a **notochord**, a long rod of stiffened tissue (not cartilage or bone), helps support the body. *Second*, the nervous system of chordate embryos develops from a tubular, dorsal **nerve cord**, which runs parallel to the notochord and gut. The cord's anterior end increases in mass and undergoes modifications to form the brain. *Third*, the embryos have distinctive slits in the wall of their **pharynx**, which is a muscularized tube. A pharynx functions in feeding, respiration, or both. *Fourth*, a tail forms in embryos and extends past the anus.

### Chordate Classification

Biologists group nearly all of the chordates into three subphyla. These are named Urochordata (tunicates and their kin), Cephalochordata (lancelets), and Vertebrata (vertebrates). There are eight classes of vertebrates:

| | |
|---|---|
| Agnatha | *Jawless fishes* |
| Placodermi | *Jawed, armored fishes (extinct)* |
| Chondrichthyes | *Cartilaginous fishes* |
| Osteichthyes | *Bony fishes* |
| Amphibia | *Amphibians* |
| Reptilia | *Reptiles* |
| Aves | *Birds* |
| Mammalia | *Mammals* |

Appendix I has an expanded classification scheme for the vertebrates. Unit VI provides details of their body plans and functions. Here we become acquainted with major trends in their evolution. We find clues to those trends among the invertebrate chordates.

---

The embryos of chordates alone have this combination of features: a notochord, a tubular dorsal nerve cord, a pharynx with slits in its wall, and a tail extending past the anus.

### Tunicates

The 2,000 species of existing urochordates are baglike animals one to several centimeters long. More often they are called the **tunicates**, after the gelatinous or leathery "tunic" that adults secrete around themselves. Adults of the most common species, the "sea squirts," squirt water through a siphon when something irritates them.

Tunicates live in marine habitats from the intertidal zone to surprising depths. Most adults remain attached to rocks and other suitably hard substrates. Some live solitary lives; others are colonial. Figure 27.3a shows an adult sea squirt. It develops from a bilateral, swimming larva that resembles a tadpole (Figure 27.3b,c). A larva, recall, is an immature stage between the embryonic and adult stages of an animal life cycle. The firm, flexible notochord of a tunicate larva is a series of cells that acts like a torsion bar. As muscles on one side of the tail or the other contract, the notochord bends, then it springs back as muscles relax. The strong, side-to-side motion propels the larva forward. Most fishes use muscles and their backbone for the same kind of motion.

Like some other animals, tunicates are **filter feeders**: they filter food from a current of water that is directed through part of their body. Water flows in through one siphon and passes through **gill slits**, or openings in the thin pharynx wall. Water flows out through a different siphon. Their pharynx also acts as a respiratory organ. Dissolved oxygen is less concentrated in blood vessels adjoining the pharynx than in the surrounding water. Therefore, it diffuses from water into the blood. Carbon dioxide's concentration gradient is such that it diffuses from blood into the water leaving the pharynx.

Sea squirts undergo metamorphosis. By this process of development, recall, remodeling and reorganization of body tissues transform an immature stage into the adult (Section 26.19). As a sea squirt larva develops, its tail and notochord disappear, and a tunic forms. The pharynx enlarges, and perforations in its wall become subdivided into many slits. The nerve cord regresses, so all that remains is a very simplified nervous system.

### Lancelets

About twenty-five species of fish-shaped, translucent animals called cephalochordates live in nearshore marine sediments around the world. Most are shorter than your little finger. Much of the time they are buried, almost up to their mouth, in sand and sediments. Their common name, **lancelets**, refers to the sharp tapering of their body at both ends. The lancelet body plan sketched in Figure 27.4a clearly shows the four chordate features. Notice the segmented pattern of muscles on both sides of the notochord. Like tunicate larvae, lancelets use the

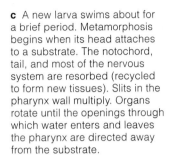

**b** Body plan of the tadpole-like larva, shown midsagittal section.

**c** A new larva swims about for a brief period. Metamorphosis begins when its head attaches to a substrate. The notochord, tail, and most of the nervous system are resorbed (recycled to form new tissues). Slits in the pharynx wall multiply. Organs rotate until the openings through which water enters and leaves the pharynx are directed away from the substrate.

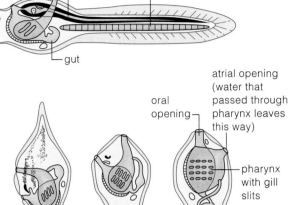

**Figure 27.3** Example of a tunicate. (**a**) Adult form of a sea squirt. (**b**) Diagram of a sea squirt larva. (**c**) Metamorphosis of the larva into the adult.

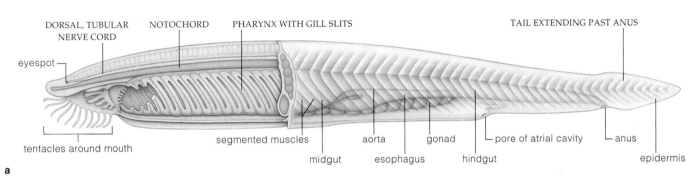

a

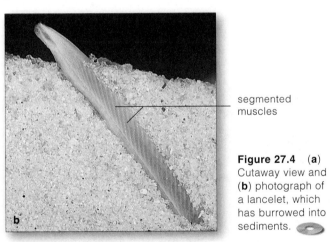

segmented muscles

**Figure 27.4** (**a**) Cutaway view and (**b**) photograph of a lancelet, which has burrowed into sediments.

Like tunicates, lancelets are filter-feeding animals. Cilia line their mouth cavity and create a current that draws water through it. The water then moves through the pharynx, where food becomes trapped in mucus. The trapped food is delivered to the rest of the gut. By itself, the collective beating of tiny cilia cannot deliver sufficient food to a filter-feeding animal. Delivery also depends on having a large food-trapping surface area. In lancelets, the pharynx provides the area. It is large relative to the overall body length (Figure 27.4). Besides this, as many as 200 ciliated, food-trapping gill slits perforate its wall. As you will read in sections to follow, the pharynx turned out to be an organ with interesting evolutionary possibilities for the vertebrates.

notochord and muscles to produce swimming motions. Unlike tunicates, they have a closed circulatory system (but no red blood cells). A complex brain is nowhere in sight, but the anterior end of the dorsal nerve cord is expanded, and pairs of nerves extend into each muscle segment. The mode of respiration is simple yet effective for such a small body. Dissolved oxygen and carbon dioxide diffuse across the body's thin skin.

Tunicate larvae and the lancelets use their notochord and muscles for fishlike swimming movements. Their pharynx, a muscularized tube, has finely divided openings across its thin wall.

The tunicates and lancelets are filter feeders. They use their pharynx to filter microscopic food from a current of water, which they draw in through the body. Tunicate larvae also use the pharynx as a respiratory organ, for gas exchange.

# EVOLUTIONARY TRENDS AMONG THE VERTEBRATES

## Puzzling Origins, Portentous Trends

Did vertebrates arise from tadpole-shaped chordates? The hemichordates, an obscure phylum, suggest this may well have happened (*hemi-* meaning half, as in "halfway to chordates"). In morphology, these marine invertebrates are midway between echinoderms and chordates. They don't have a notochord, but they have a gill-slitted pharynx and a dorsal, tubular nerve cord. Their larvae resemble echinoderm *and* tunicate larvae. What if gene mutations in an echinoderm lineage accelerated the rates at which sex organs developed? If those organs started functioning earlier, in tadpole-shaped *larvae*, then metamorphosis—and the original adult form—could be dispensed with. This idea is not far-fetched. A few modern tunicates resemble larvae but have sex organs. Some amphibians become sexually mature even though they are larvae in some respects, and they can reproduce generation after generation.

Assuming tadpole-shaped chordates emerged, how did they evolve into vertebrates? The key trends started in fishes (Figures 27.5 and 27.6). One involved a shift from the notochord to reliance on a column of separate, hardened segments: **vertebrae** (singular, vertebra). This was the start of an endoskeleton—an internal skeleton —that muscles could work against and initiate motion.

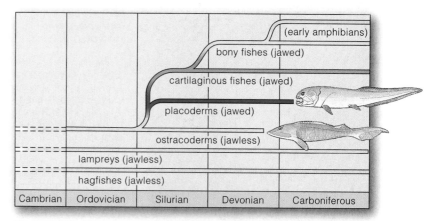

**Figure 27.5**  Evolutionary history of the fishes.

*A vertebral column evolved in fast-moving predators, some of which were ancestral to all other vertebrates.*

In a related trend, part of the nerve cord expanded into a brain. It did so after **jaws** evolved from the first of a series of structures that lent support to the gill slits. Those structural elements were a key innovation; they meant new feeding possibilities and greater competition among predators. Fishes able to smell or see predators or food farther away now had the advantage—provided that the brain also had evolved enough to process and respond to the new sensory information. *A trend toward complex sensory organs and nervous systems did indeed begin among fishes, and it continued in land vertebrates.*

Another trend began when paired fins evolved. **Fins** are appendages that help propel, stabilize, and guide the body through water (compare Figure 27.9). In some lineages, ventral fins became fleshy and equipped with skeletal supports that turned out to be forerunners of limbs. *Paired, fleshy fins were the starting point for all of the legs, arms, and wings that evolved among amphibians, reptiles, birds, and mammals.*

A major change in respiration was another trend. Think of the lancelet. Except when it is buried in sediments, oxygen and carbon dioxide merely diffuse across its body surface. But **gills** evolved in most vertebrate lineages. Gills are respiratory organs having a moist, thin, greatly folded surface. They are richly endowed with blood vessels, and they offer a large surface area for gas exchange. For example, five to seven pairs of a shark's gill slits are continuous with gills extending from the pharynx to the body's surface. When a shark opens its mouth and closes its external gill openings, the pharynx expands. Oxygen diffuses

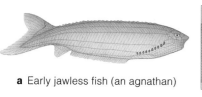

**a** Early jawless fish (an agnathan)

**b** Early jawed fish (a placoderm)

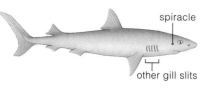

spiracle

other gill slits

**c** Modern jawed fish (a shark)

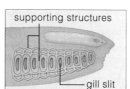

supporting structures

gill slit

jaw

spiracle (small gill slit)   jaw support

jaw

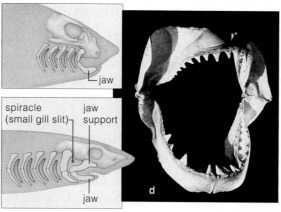

d

**Figure 27.6**  Comparison of gill-supporting structures in jawless fishes and jawed fishes. In the placoderms and other early jawed vertebrates, cartilage supported the rim of the mouth. In modern jawed fishes, a gill slit between the jaws and a nearby supporting element is a spiracle, an opening through which water is drawn. In (**a**), gill supports are just under the skin. In (**b**) and (**c**), they are internal to the gill surface. (**d**) Jaws of a modern shark.

*from* water in the mouth into the gills, while carbon dioxide diffuses *into* that water. Muscles constrict the pharynx and force the oxygen-depleted, carbon dioxide-enriched water out through the gill slits.

As some fishes became larger and more active, gills became more efficient. But gills can't work out of water; they stick together unless water flows through them and keeps them moist. In fishes ancestral to the land vertebrates, pouches developed from the gut wall. The pouches evolved into **lungs**—internally moistened sacs for gas exchange. In a related trend, modifications to the heart enhanced the pumping of oxygen and carbon dioxide through the body. *Ancestors of land vertebrates relied less on gills and more on lungs. And more efficient circulatory systems accompanied the evolution of lungs.*

### The First Vertebrates

Figure 27.5 has a time frame for vertebrate evolution. Free-swimming species originated in Cambrian times, and they gave rise to two categories of fishes—those without and those with jaws. Among them were the ancestors of all vertebrate lineages that followed. The earliest jawless fishes (Agnatha) included **ostracoderms**. Figure 27.6a shows one of those bottom-dwelling filter feeders. Probably the skeleton was a notochord plus a protective case for the enlarged brain. Armorlike plates on the body consisted of bony tissue and dentin, a hardened tissue still present in vertebrate teeth. The armor might have been useful against the pincers of giant sea scorpions but apparently wasn't much good against jaws. Ostracoderms disappeared when jawed fishes began their first adaptive radiation.

**Placoderms** were among the earliest fishes with jaws and paired fins (Figure 27.6b). Those bottom-dwelling scavengers or predators had a notochord reinforced with bony elements. The first gill-supporting structures were enlarged and had bony projections something like teeth, and they functioned as jaws. Before placoderms, feeding strategies had been limited to filtering, sucking, or rasping away at food material. When placoderms started biting and tearing off large chunks of prey, they set off an evolutionary race of offensive and defensive adaptations that has continued to the present.

Placoderms diversified, but they all became extinct in the Carboniferous period. New kinds of predators, the cartilaginous and bony fishes, replaced them in the seas. We will consider the living descendants of those new predators after a look at existing jawless fishes.

---

**The vertebral column, jaws, paired fins, and lungs were pivotal developments in the evolution of lineages that gave rise to fishes, amphibians, reptiles, birds, and mammals.**

---

MAMMALS
BIRDS
REPTILES
AMPHIBIANS
BONY FISHES
CARTILAGINOUS FISHES
ANCESTRAL VERTEBRATES     JAWLESS FISHES

The **hagfishes** and **lampreys** are living descendants of early jawless fishes. All seventy-five species have a cylindrical body and a skeleton of cartilage (Figure 27.7). None has the paired fins seen in jawed fishes (Figure 27.6). Most species are less than 1 meter (3.3 feet) long.

Hagfishes live together in groups on sediments of continental shelves. They prey on polychaete worms or scavenge for weakened or dead organisms. Even though they lack jaws, hagfishes eat quite well. They have sensory tentacles around the mouth and a tongue that rasps soft tissues from their prey. They defend themselves by secreting copious amounts of sticky, smelly, slimy mucus from a series of glands along their body.

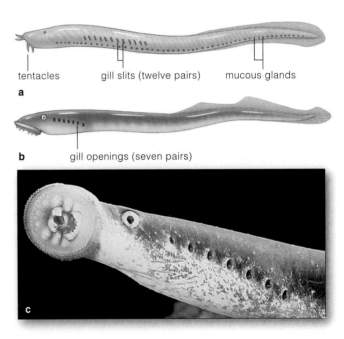

**a** tentacles     gill slits (twelve pairs)     mucous glands

**b** gill openings (seven pairs)

**c**

**Figure 27.7** Body plan of (**a**) a hagfish and (**b**) a lamprey. (**c**) A lamprey's oral disk, pressed against aquarium glass.

Lampreys are specialized predators that are almost parasitic. Their suckerlike oral disk has horny, toothlike parts that rasp flesh from prey. Some types latch onto salmon, trout, and other commercially valuable fishes, then suck out juices and tissues. Just before the turn of the century, lampreys began invading the Great Lakes of North America. Their introduction resulted in the collapse of populations of lake trout and other large, valued fishes, as it has done in other aquatic habitats.

---

**Hagfishes and lampreys get along without jaws; they latch onto prey with their efficient, specialized mouthparts.**

---

# EXISTING JAWED FISHES

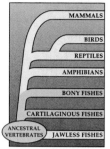

Few of us make a career of studying life beneath the surface of the seas and other bodies of water, so fishes are not widely recognized as being the world's dominant vertebrates. Their numbers exceed those of all other vertebrate groups combined. Fishes also show far more diversity than other vertebrates; we know of more than 21,000 species of bony fishes alone.

The form and behavior of a fish tell us something about the challenges it faces in water. For example, being about 800 times more dense than air, water resists fast motion through it. As an adaptation to this constraint, many marine fishes are streamlined for pursuit or for escape (Figure 27.8*a*). Sharks are an example. Their long, trim body reduces friction, and their tail muscles are organized for great propulsive force and forward motion. By contrast, some bottom-dwelling fishes, such as the rays, have a flattened body (Figure 27.8*b*). Theirs is not a high-speed body plan; it is good for hiding out from predators or prey.

A motionless trout, suspended in shallow water, is another example of adaptation to water's density. Like many fishes, it maintains neutral buoyancy with a **swim bladder**, an adjustable flotation device that exchanges gases with blood. When a fish gulps air at the water's surface, it is adjusting the volume of its swim bladder.

## Cartilaginous Fishes

**Cartilaginous fishes** (Chondrichthyes) include about 850 species of skates, sharks, and chimaeras, as shown in Figure 27.8. These marine predators are equipped with prominent fins, a skeleton of cartilage, and five to seven gill slits on both sides of their pharynx. Most species have a few scales or many rows of them. **Scales** are small, bony plates at the body surface. Fish scales typically protect the body without weighing it down.

Skates and rays are mainly bottom dwellers with flattened teeth suitable for crushing hard-shelled prey. Both have enlarged fins that extend onto the side of the head. The largest species, the manta ray, measures up to six meters from fin tip to fin tip. A venom gland in the tail of stingrays probably helps to deter predators. Other rays have electric organs in the tail or fins that can stun prey with as much as 200 volts of electricity.

At fifteen meters from head to tail, some sharks are among the largest living vertebrates. As Figure 27.6*d* shows, sharks have formidable jaws. They continually shed and replace sharp-edged teeth (modified scales), which they use to grab prey and rip off chunks of flesh. The relatively few shark attacks on humans have given the group a bad reputation. Surfboards with human legs dangling over the sides are a new temptation for some

**Figure 27.8** Cartilaginous fishes: (**a**) shark, (**b**) blue-spotted reef ray, and (**c**) chimaera, sometimes called a ratfish.

sharks, which have been hunting invertebrates, fishes, and marine mammals for many millions of years.

The thirty or so species of chimaeras feed mostly on mollusks. With their bulky body and slender tail, they look a bit like a rat; hence the common name, ratfishes. They have a venom gland in front of the dorsal fin.

## Bony Fishes

**Bony fishes** (Osteichthyes) are the most numerous and diverse vertebrates. They make up all but 4 percent of modern fish species. Their ancestors arose in Silurian times and gave rise to three lineages: ray-finned fishes, lobe-finned fishes, and lungfishes. Descendants of the early forms radiated into nearly all aquatic habitats.

Body plans vary greatly. Marine predators typically have a torpedo shape, a flexible body, and strong tail fins used in swift pursuit. Many reef dwellers are small finned and box shaped; they move easily in narrow spaces. Those with an elongated, flexible body, such as the moray eel, wriggle through mud and into concealing crevices. The cryptic body shape of sea horses and many bottom-dwelling species helps conceal them from prey, predators, or both. Figure 27.9 shows examples.

In ray-finned fishes, outwardly projecting rays that are derived from dermis (skin's inner layer) support paired fins. Most species have highly maneuverable fins

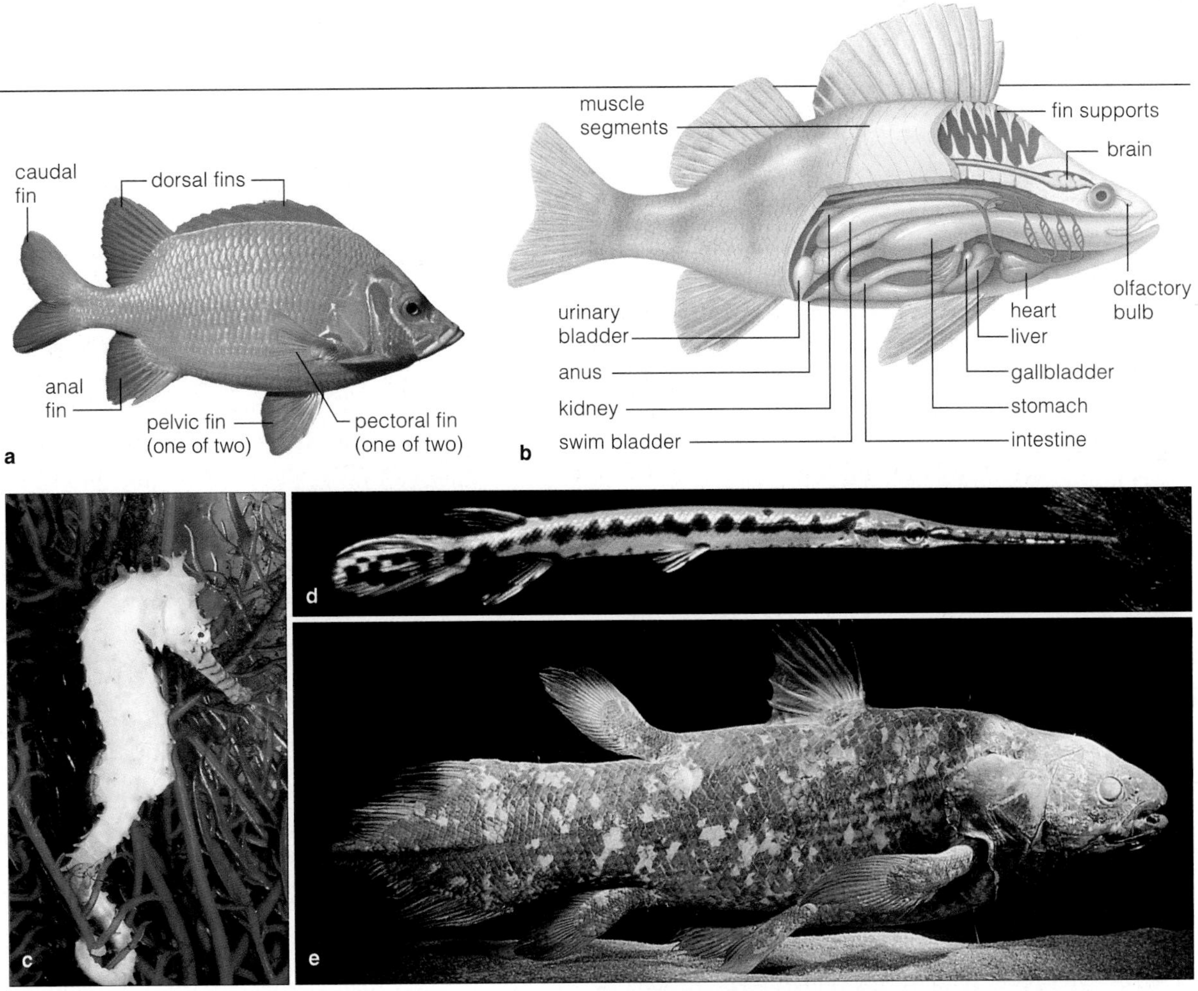

**Figure 27.9** Some common features of the teleosts, the most diverse bony fishes: (**a**) Fins of a soldierfish. (**b**) Internal organs of a perch.

Just a few of the variations on the basic body plan: (**c**) Sea horse, which uses its tail to attach to substrates. (**d**) Long-nose gar. (**e**) Coelacanth (*Latimeria*), a "living fossil." It shares traits with early lobe-finned fishes. (**f**) Sketch of a moray eel. See Section 28.3.

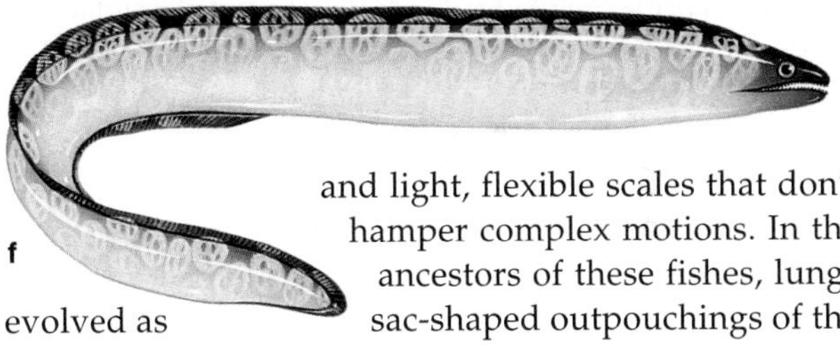

**a** — caudal fin, dorsal fins, anal fin, pelvic fin (one of two), pectoral fin (one of two)

**b** — muscle segments, fin supports, brain, olfactory bulb, heart, liver, gallbladder, stomach, intestine, swim bladder, kidney, anus, urinary bladder

**c**   **d**   **e**   **f**

**Figure 27.10** Evolution of swim bladders and lungs. Respiratory surfaces are coded *pink* and the esophagus (a tube leading to the stomach), *gold*. Lungs evolved as outpouchings of the esophagus. Being in close contact with blood vessels, they increased the surface area for gas exchange in oxygen-poor habitats. The lung sacs became swim bladders (buoyancy devices) in some species.

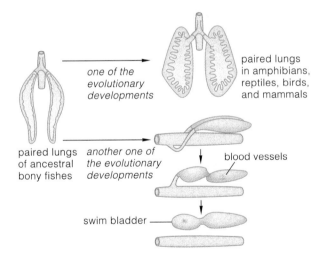

one of the evolutionary developments — paired lungs in amphibians, reptiles, birds, and mammals

paired lungs of ancestral bony fishes — another one of the evolutionary developments — blood vessels — swim bladder

Trout and other less specialized fishes surface and gulp air, which enters the swim bladder through a duct from the esophagus. Most bony fishes have no such duct. Gases dissolved in blood diffuse into a swim bladder that has a dense mesh of arteries and veins. Blood flow through the vessels increases the concentrations of gases in the swim bladder. Another area of the swim bladder enhances reabsorption of gases by cells in body tissues.

and light, flexible scales that don't hamper complex motions. In the ancestors of these fishes, lungs evolved as sac-shaped outpouchings of the wall of the esophagus, a tube to the stomach. The sacs supplemented gills for gas exchange (Figure 27.10). In certain species, those sacs evolved into swim bladders, which help a fish hold its position at different depths.

One group of ray-finned fishes resembles their early ancestors. It includes the sturgeons and paddlefishes of the Mississippi River basin. Teleosts, the most abundant fishes, include salmon, tuna, rockfish, catfish, perch, minnows, moray eels, flying fish, sculpins, blennies, scorpionfish, and pikes. All are thin scaled or scaleless.

By contrast, only one species of lobe-finned fishes and three genera of lungfishes made it to the present. As their name suggests, a **lobe-finned fish** is unique in having paired fins that incorporate fleshy extensions from the body. As you will see next, neither it nor the lungfishes have changed much from their ancient stock.

Of all existing vertebrates, the bony fishes are the most spectacularly diverse and the most abundant.

MAMMALS
BIRDS
REPTILES
AMPHIBIANS
BONY FISHES
CARTILAGINOUS FISHES
ANCESTRAL VERTEBRATES  JAWLESS FISHES

### Origin of Amphibians

Take another look at the coelacanth, the lobe-finned fish shown in Figure 27.9*e*. It is a "living fossil," a relic of an early time when some vertebrates first moved onto land. Remember the earlier description of living conditions in the Devonian? Sea levels rose and fell repeatedly, and the swamps fringing the coasts flooded and drained many times. Ancestors of the lobe-finned fishes evolved during those trying times. And they probably used their lobed fins to pull themselves from dried-up ponds to ones that still had water and were habitable. What else made their pond-to-pond lurchings possible? They gulped air, and they had lungs.

Existing lungfishes provide more clues to how the ancestral forms might have made it through stressful times. They live in stagnant water but surface to gulp air. In dry seasons, when streams dwindle to mud, they encase themselves in a slathering of mud and slime that keeps them from drying out until the next rainy season.

Devonian lobe-finned fishes lurched over land only as a way of reaching more hospitable ponds. Yet the very act of traveling out of water favored the evolution of stronger fins and more efficient lungs. Among the evolving forms were the ancestors of amphibians. An **amphibian** (Amphibia) is a vertebrate with a body plan and reproductive mode somewhere between fishes and reptiles. Most kinds have a largely bony endoskeleton, and they have four legs (or four-legged ancestors).

Early amphibians were spending time on land by the close of the Devonian (Figure 27.11*a*). For them, life in those drier habitats was dangerous—and promising. Temperatures shifted more on land than in the water, air didn't support the body as well as water did, and water was not always plentiful. However, air is richer in oxygen. Amphibian lungs continued to be modified in ways that enhanced oxygen uptake. Also, circulatory systems became better at rapidly distributing oxygen to living cells throughout the body. Both modifications increased the energy base for more active life-styles.

New sensory information also challenged the early amphibians. Swamp forests supported vast numbers of edible insects and other invertebrate prey. Animals with good vision, hearing, and balance—the senses that are most advantageous on land—were favored. And brain regions concerned with those senses expanded.

All of the salamanders, frogs, toads, and caecilians alive today are descended from those first amphibians. None has escaped the water entirely. Even when gills or lungs are present, amphibians can use their thin skin as a respiratory surface (that is, for gas exchange). But respiratory surfaces must be kept moist, and skin dries easily in air. Some species still spend their entire lives in water; others lay their eggs in water or produce aquatic larvae. Even species adapted to land lay eggs in moist places or their embryos develop inside the moist body.

a

**Figure 27.11** (**a**) *Ichthyostega*, one of the early Devonian amphibians. Fossils of this species have been recovered in Greenland. The skull, deep tail, and fins were decidedly fishlike. Unlike fishes, this species had four limbs adapted for moving on land, and a short neck intervened between its head and the rest of the body. Its vertebral column and rib cage were adapted to support the body's weight out of water.

(**b**,**c**) Proposed evolution of skeletal elements inside the lobed fins of certain fishes into the limb bones of early amphibians.

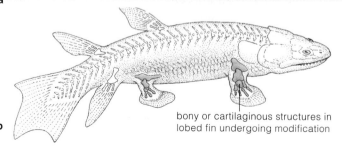

b

bony or cartilaginous structures in lobed fin undergoing modification

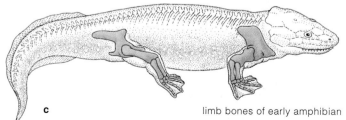

c

limb bones of early amphibian

Further reading: Student Guide to InfoTrac on web site →

## Salamanders

About 380 species of salamanders and their kin, the newts, live in north temperate zones and tropical areas of Central and South America. Most are not even fifteen centimeters long. Most have legs at right angles to the body, with forelimbs and hindlimbs of about the same size (Figure 27.12a). Like fishes and early amphibians, salamanders bend from side to side when they walk:

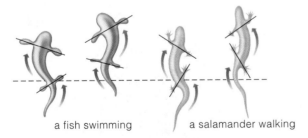

a fish swimming      a salamander walking

Probably the first four-legged vertebrates also walked this way. Larval and adult salamanders are carnivores. Adults of some species retain several larval features. For example, the Mexican axolotl retains the external gills of the larval form, and the development of its teeth and bones is arrested at an early stage. As in some other species, its larvae are sexually precocious; they breed.

## Frogs and Toads

With close to 4,000 species, the frogs and toads are the most successful amphibians (Figure 27.12b,c). Their long

**Figure 27.12** Amphibians. (**a**) Terrestrial stage in the life cycle of a red-spotted salamander. (**b**) A frog, splendidly jumping. (**c**) An American toad. (**d**) A caecilian.

hindlimbs and powerful muscles allow them to catapult into the air or barrel through the water. Most frogs flip a sticky-tipped, prey-capturing tongue from their mouth. An adult eats just about any animal it can catch; only its head size limits prey sizes. Like all amphibians, frogs have mucous and poison glands in their skin. Notably poisonous types often have bright warning coloration. Frog skin contains antibiotics that provides protection against many pathogens in aquatic habitats. In Unit VI, you will have opportunities to take a look at how frogs are put together and how they function.

## Caecilians

The ancestors of caecilians lost their limbs and most of their scales. They gave rise to decidedly worm-shaped amphibians (Figure 27.12d). Nearly all of the 160 or so existing species burrow through soft, moist soil as they pursue insects and earthworms. A few live in shallow freshwater habitats. Adults of most species are blind.

---

**In body form and behavior, amphibians show resemblances to fishes and certain reptiles. Regardless of how far they venture onto land, most have not fully escaped dependency on aquatic or moist habitats to complete the life cycle.**

---

# THE RISE OF REPTILES

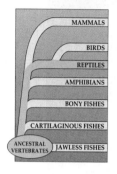

MAMMALS
BIRDS
REPTILES
AMPHIBIANS
BONY FISHES
CARTILAGINOUS FISHES
ANCESTRAL VERTEBRATES    JAWLESS FISHES

Late in the Carboniferous, a divergence from the amphibian lineage gave rise to the **Reptiles** (Reptilia). Of all vertebrates, reptiles were first to escape dependency on standing water. They did so mainly by four adaptations, as listed below.

First, reptiles have tough, dry, scaly skin that restricts loss of water from the body (Figure 27.13*a*). Second, fertilization is internal (as it is in some amphibians). A copulatory organ deposits sperm inside a female's body; sperm do not require free water to reach eggs. Third, reptilian kidneys are good at conserving water. Fourth, **amniote eggs** form during the life cycle of most species. Embryos in such eggs develop to an advanced stage before being hatched or born into dry habitats. Amniote eggs have membranes that conserve water and protect or provide metabolic support to the embryo. Most have a leathery or calcified shell (Figure 27.13*b,c* and Section 27.9). We will consider the structure and development of amniote eggs in Chapter 45.

Compared to most amphibians, early reptiles chased prey with far greater cunning and speed. With their well-muscled jawbones and formidable teeth, reptiles could seize and apply sustained, crushing force on prey. Their prey included other vertebrates as well as insects. Also, reptilian limbs generally were more efficient at supporting the trunk of the body on land. For many species, the nervous system increased in complexity. A reptile's brain is small compared to the rest of the body mass, but it governs complex forms of behavior that are unknown among amphibians. For example, the cerebral cortex is the most complex part of the forebrain. Here, information from diverse sensory organs is integrated and stored, and commands for responses are issued. The cerebral cortex evolved first among the reptiles.

b

c

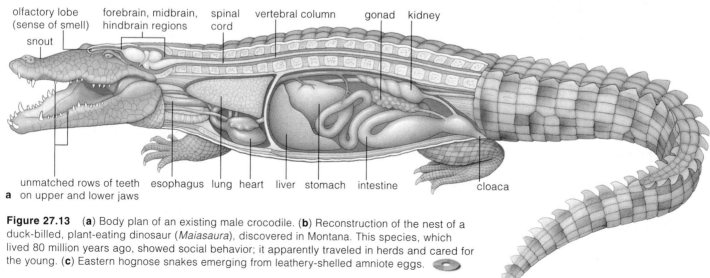

olfactory lobe (sense of smell)    forebrain, midbrain, hindbrain regions    spinal cord    vertebral column    gonad    kidney

snout

unmatched rows of teeth on upper and lower jaws    esophagus    lung    heart    liver    stomach    intestine    cloaca

a

**Figure 27.13** (**a**) Body plan of an existing male crocodile. (**b**) Reconstruction of the nest of a duck-billed, plant-eating dinosaur (*Maiasaura*), discovered in Montana. This species, which lived 80 million years ago, showed social behavior; it apparently traveled in herds and cared for the young. (**c**) Eastern hognose snakes emerging from leathery-shelled amniote eggs.

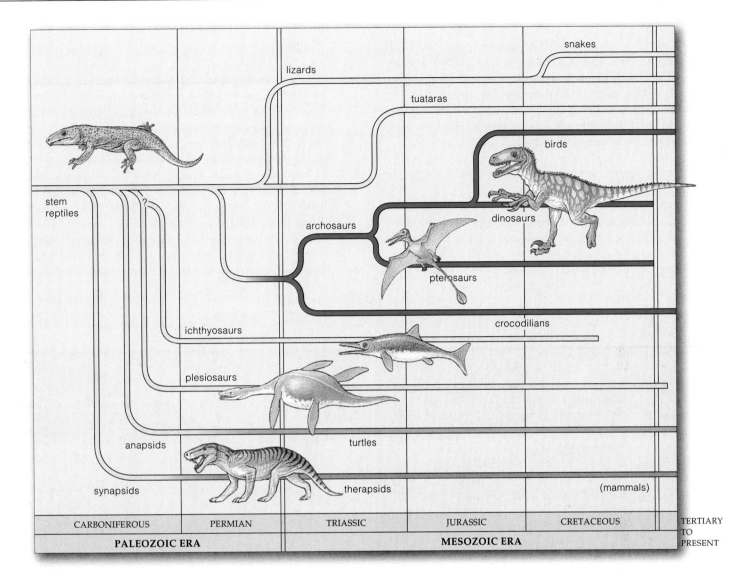

**Figure 27.14** Evolutionary history of the reptiles.

Crocodilians, which diverged from ancestral reptiles, were the first animals with a muscular, four-chambered heart fully separated into two halves. (Blood enters the first chamber of each half; the second chamber pumps it out.) As Chapter 39 describes, such separation permits oxygen-rich blood to travel from the lungs to the rest of the body, and oxygen-poor blood from the body to the lungs, in two independent circuits. Also, gas exchange across the skin, so vital for amphibians, was abandoned by reptiles. Nearly all reptiles depend on lungs.

Early reptiles underwent adaptive radiations that led to fabulously diverse forms. In the seas, some of the marine reptiles called masosaurs evolved into the largest lizards ever known; some were ten meters long. Monitor lizards, such as the Komodo dragon, are their living relatives (compare Figure 28.3). The group called dinosaurs evolved during the Triassic. For the next 125 million years they were the dominant land vertebrates. The fossilized nests of one kind (*Maiasaura*) contain not only eggs but also juveniles a few months old, at most

(Figure 27.13*b*). Such evidence indicates that at least some dinosaurs showed parental behavior; that is, they took care of their young during a period of dependency. (*Maiasaura* means "good mother lizard.")

When the Cretaceous ended abruptly, so did the last of the dinosaurs (Sections 21.6 and 21.7). As you can see from Figure 27.14 and the next section, the crocodilians, turtles, tuataras, snakes, and lizards are reptilian groups that made it to the present.

---

**Reptiles, with their tough, scaly skin, reliance on internal fertilization, water-conserving kidneys, and amniote eggs, were the first vertebrates to escape dependency on free-standing water in their habitats.**

**The move onto land also required major modifications in the nervous, circulatory, and respiratory systems.**

---

The name of class Reptilia is derived from the Latin *repto*, meaning "to creep." Maybe all of the first reptiles did creep about slowly in muddy swamps and on dry land. But the capacity to move swiftly and with agility evolved among some of their early descendants, such as the bipedal (two-legged) *Velociraptors* of the Mesozoic. Existing descendants race, lumber, and slither about.

### Crocodilians

Modern crocodiles and alligators, the closest relatives of birds and dinosaurs, live in or near water. Among them are the largest living reptiles. Crocodiles and alligators have powerful jaws, a long snout, and sharp teeth, as in Figure 27.15*a*. The feared "man-eaters" of southern Asia

### Turtles

The 250 existing species of turtles live inside a shell that is attached to the skeleton. When most are threatened, they pull the head and limbs inside (Figure 27.15*b,c*). It is a body plan that works well; it has been around since Triassic times. Only among the sea turtles and other highly mobile types is the shell reduced in size.

Instead of teeth, turtles have tough, horny plates that are suitable for gripping and chewing food. They have strong jaws and often a fierce disposition that helps keep predators at bay. All turtles lay eggs on land, then leave them. Predators eat most of the eggs, so few new turtles hatch. Sea turtles have been hunted to the brink of extinction. Section 41.8 describes one of these.

**Figure 27.15** Representative reptiles. (**a**) Spectacled caiman. As is true of other crocodilians, its peglike upper teeth do not match up with peglike lower ones. The body plan and life-styles of crocodilians have not changed much for nearly 200 million years. Now their future is not rosy. Housing developments are encroaching on many of their habitats, and their belly skin is in demand for wallets, shoes, and handbags. (**b**) A heavy-shelled Galápagos tortoise. (**c**) Section through the skeleton and shell of a turtle, with head withdrawn and extended. *Facing page:* (**d**) Frilled lizard, flaring a ruff of neck skin as defensive behavior. (**e**) Rattlesnake in mid-strike. (**f**) Tuatara (*Sphenodon*).

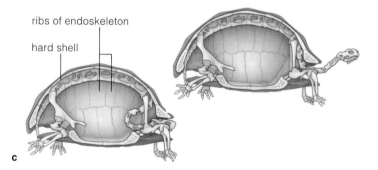

ribs of endoskeleton

hard shell

c

and Nile crocodiles weigh as much as 1,000 kilograms. They drag a mammal or bird into water, tear it apart by violently spinning about, then gulp down the chunks.

Like the other reptiles and birds, crocodilians adjust body temperature with behavioral and physiological mechanisms. They are like birds in displaying complex social behaviors, as when parents guard nests and assist hatchlings into water. This trait and others suggest that crocodilians and birds share a common ancestor.

### Lizards and Snakes

Lizards and snakes, which make up 95 percent or so of living reptiles, are distant relatives of dinosaurs. Most are small, but the Komodo monitor lizard is big enough to capture a young water buffalo. The longest modern snake would stretch across ten yards of a football field.

Most of the 3,750 kinds of lizards are insect eaters of deserts and tropical forests. They include aggressive, adhesive-toed geckos and iguanas; the photograph of the Unit VI introduction shows one of the more colorful types. Most lizards grab prey with small, peglike teeth. Chameleons capture prey with accurate flicks of their tongue, which is longer than their body (Section 33.4).

Being small themselves, most lizards are prey for many other animals. Some attempt to startle predators and intimidate rivals by flaring their throat fan (Figure 27.15*d*). Many give up their tail when a predator grabs them. The detached tail wriggles for a bit and might be distracting enough to permit a getaway.

During the early Cretaceous, certain short-legged, long-bodied lizards gave rise to the elongated, limbless snakes. Most of the 2,300 existing species slither in S-shaped waves, much like salamanders. "Sidewinders"

make J-shaped movements and leave distinctive trails across loose sand and sediments. Some species have bony remnants of ancestral hindlimbs (Section 17.1).

All snakes are carnivores. They have flexible skull bones and highly movable jaws; some swallow animals wider than they are. Pythons and boas coil around prey and suffocate it into submission before gulping it down. Fanged types, including coral snakes and rattlesnakes (a pit viper) bite and subdue prey with venom (Figure 27.15*e*). Snakes usually do not act aggressively toward people. Even so, during an average year in the United States, the rattlesnakes and other poisonous types bite 8,000 or so people and kill about 12 of them. In India, king cobras and other snakes bite 200,000 or so people and kill about 9,000 of them.

Even the most feared snakes are vulnerable during their life cycle; birds and other predators relish snake eggs. The female snakes store sperm and lay several clutches of fertilized eggs at intervals after they mate, which improves the odds that at least some will hatch.

### Tuataras

Besides having reptilian traits, tuataras are like modern amphibians in some aspects of their brain and in their way of walking. The two existing species live on small, windswept islands near New Zealand (Figure 27.15*f*). Their body plan has not changed much for the past 140 million years. Like lizards, tuataras have a third "eye," with retina, lens, and nerves to the brain. Being covered with skin, it can only register changes in daylength and light intensity. Does it have roles in hormonal controls over reproduction? Maybe (compare Section 37.8). The tuataras, like turtles, may live longer than sixty years. They engage in sex only after they are twenty years old.

---

**Existing crocodilians are the closest relatives of dinosaurs and birds. Turtles, like tuataras, have changed little in body plan for millions of years. Lizards and snakes represent about 95 percent of all living reptiles.**

---

# BIRDS

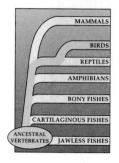

Of all organisms, only **birds** (Aves) grow feathers. **Feathers**, lightweight structures derived from skin, are used for flight, body insulation, or both. Judging from the fossil record, birds descended from tiny reptiles that ran around on two legs 160 million years ago. *Archaeopteryx* was in or near the lineage leading to modern birds. It had reptilian *and* avian traits—including feathers (Section 17.4). In fact, feathers evolved as highly modified reptilian scales.

In many respects, birds still resemble reptiles. For example, they have scales on their legs and a number of the same internal structures. They, too, lay eggs and commonly engage in parenting behavior, as when they guard the nest (Figure 27.16). Thomas Huxley, one of Darwin's supporters, was the first to argue that birds are glorified reptiles. Today, most biologists do indeed classify birds as a branching of the reptilian lineage.

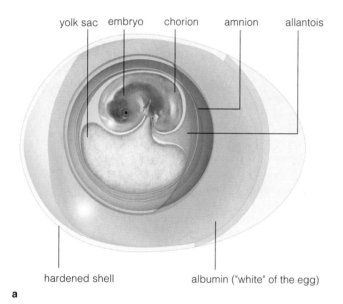

yolk sac   embryo   chorion   amnion   allantois

hardened shell        albumin ("white" of the egg)

a

**Figure 27.16**   Characteristics of birds. (**a**) Generalized sketch of an amniote egg. All birds lay these eggs. (**b**) Flight. Of all living vertebrates, only birds and bats fly by flapping their wings. (**c**) Feathers, the key defining characteristic of birds. The flamboyant plumage of this male pheasant is an outcome of sexual selection. This bird is a native of the Himalaya Mountains of India, and it is an endangered species. As is the case for many other kinds of birds, its bright, jewel-colored feathers end up adorning humans—in this case, on the caps of native tribespeople.

(**d**) Many birds, including these Canada geese, display migratory behavior. *Animal migration* is a recurring pattern of movement between two or more places in response to environmental rhythms. For instance, seasonal change in daylength is a cue that acts on internal timing mechanisms (biological clocks), which trigger physiological and behavioral changes in individuals. Such changes induce migratory birds to make round trips between distant regions that differ in climate. Canada geese spend the summer nesting in marshes and lakes in the northern United States and Canada. Their wintering grounds are in New Mexico and other parts of the southern United States.

(**e**) Speckled, hard-shelled eggs of a magpie, unhatched in the nest.

e

There are almost 9,000 named species of birds. They show stunning variation in their body size, proportions, coloration, and capacity for flight. One of the smallest hummingbirds barely tips the scales at 2.25 grams (0.08 ounce). The largest existing bird, the ostrich, weighs about 150 kilograms (330 pounds). Ostriches cannot fly, but they are impressively long-legged sprinters (Figure 17.2). Many birds, such as warblers and other perching types, differ markedly in feather coloration and in their territorial songs. Bird songs and other complex social behaviors are topics of later chapters.

Flight demands high metabolic rates, which require a good flow of oxygen through the body. Like mammals, birds have a large, durable, four-chambered heart. The heart pumps oxygen-enriched blood to the lungs and to the rest of the body along separate routes, as it probably did in reptilian ancestors of birds. Also, a bird's respiratory system incorporates a unique ventilating apparatus that enormously enhances oxygen uptake. Bird respiration is described in Section 41.3.

Flight also demands an airstream, low weight, and a powerful downstroke that can provide lift (a force at right angles to the airstream). The bird wing, a modified forelimb, is composed of feathers and lightweight bones attached to powerful muscles. The bones do not weigh much, owing to profuse air cavities in the bone tissue.

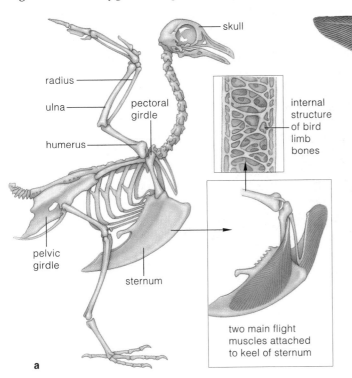

For example, the skeleton of a frigate bird, with its seven-foot wingspan, weighs only four ounces. That is less than the feathers weigh! A bird's flight muscles attach to an enlarged breastbone (sternum) and to the upper limb bones adjacent to it (Figure 27.17). Muscle contraction creates the powerful downstroke required for flight. Wings, with their long flight feathers, serve as airfoils. Usually, a bird spreads out long feathers on a downstroke and thus increases the size of the surface pushing against air (Figure 27.16b). On the upstroke, it folds the feathers somewhat, so each wing presents the least possible resistance to air.

**Figure 27.17** (**a**) Body plan of a typical bird. Flight muscles attach to the large, keeled breastbone (sternum). A bird wing is a complicated system of lightweight bones and feathers. (**b**) Feathers gain strength from a hollow central shaft and from tiny barbules that are interlocked in a latticelike array.

**Of all animals, birds alone have feathers, which they use in flight, in heat conservation, and in socially significant communication displays.**

# THE RISE OF MAMMALS

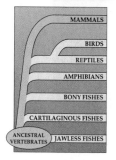

## *Distinctly Mammalian Traits*

**Mammals** are vertebrates with hair and mammary glands (hence the name of the class Mammalia, from the Latin *mamma*, which means breast). Of all organisms, mammals are the only ones having these traits. Females feed the young with milk, a nutritious fluid that mammary glands secrete into ducts that lead to openings at the body's ventral or anterior surface (Figure 27.18*a*). Mammals also are highly diverse. For example, in size alone, existing species range from the 1.5-gram Kitti's hognosed bat to 100-ton great whales. They also differ greatly in the amount, distribution, and type of hair.

A few aquatic mammals, including the whales, lost most of their hair after their land-dwelling ancestors returned to the seas. But look closely and you see that even the whale snout has some "whiskers"—modified hairs that serve sensory functions, just as they do in dogs and cats. More typically, mammals have a furry coat of underhair (a dense, soft, insulative layer that traps heat) and coarser, longer guard hairs that protect the insulative layer from wear and tear (Figure 27.18b). When wet, the guard hairs of platypuses, otters, and other aquatic mammals become matted down like a protective blanket, and the underhair stays dry.

Mammals also are distinctive in that most care for the young for an extended period, and adults serve as models for their behavior. Young mammals are born with a capacity to learn and to repeat behaviors that have survival value. It is among mammals that we find the most stunning shows of **behavioral flexibility**—a capacity to expand on basic activities with novel forms of behavior—although the trait is far more pronounced in some species than in others. Behavioral flexibility coevolved with expansion of the brain, especially the cerebral cortex. Remember, this outermost layer of the forebrain receives, processes, and stores information from sensory structures, and it issues commands for complex responses. We find the most highly developed cerebral cortex among the mammals called primates. We will soon turn to their story.

Unlike reptiles, which generally swallow their prey whole, most mammals secure, cut, and sometimes chew food before swallowing. They also differ from reptiles in **dentition** (that is, in the type, number, and size of teeth). Mammals have four distinctive types of upper and lower teeth that match up and work together to crush, grind, or cut food (Figure 27.18*c*). Their incisors (flat chisels or cones) nip or cut food. Horses and other grazing mammals have large incisors. Canines, with piercing points, are longest in meat-eating mammals. Premolars and molars (cheek teeth) are a platform with surface bumps (cusps); they crush, grind, and shear food. If a mammal has large, flat-surfaced cheek teeth, you can safely bet its ancestors evolved in places where tough, fibrous plants were abundant foods. As you will see later on in the chapter, fossilized jaws and teeth from the early primates as well as from species that apparently were on the road to modern humans offer clues to their life-styles.

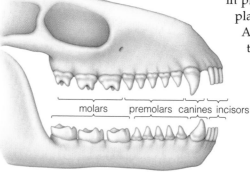

**Figure 27.18** Three distinctly mammalian traits. (**a**) A human baby busily demonstrating a key defining feature: It derives nourishment from mammary glands. (**b**) A pair of juvenile raccoons displaying their fur coat. (**c**) Unlike the teeth of their reptilian ancestors, the upper and lower rows of mammalian teeth match up. (Compare Figure 27.15*a*.)

molars premolars canines incisors

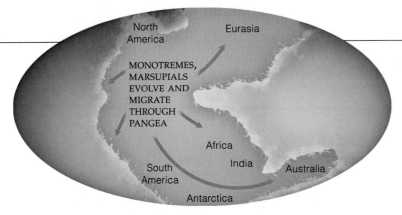

**a** About 150 million years ago, during the Jurassic

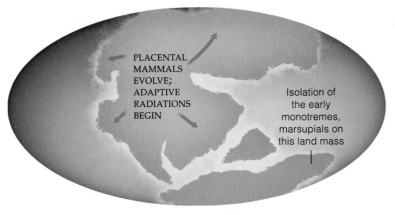

**b** Between 100 and 85 million years ago, during the Cretaceous

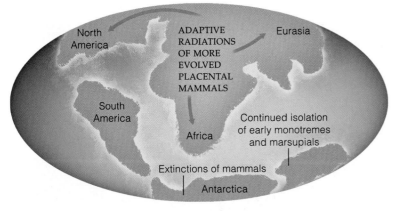

**c** About 20 million years ago, during the Miocene

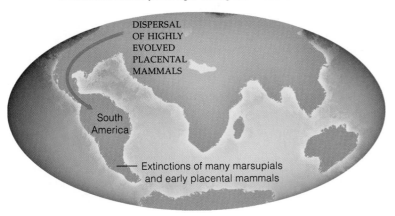

**d** About 5 million years ago, during the Pliocene

**Figure 27.19** Adaptive radiations of mammals.

## Mammalian Origins and Radiations

More than 200 million years ago, during the Triassic, a genetic divergence from small, hairless reptiles called synapsids gave rise to the **therapsids**, the ancestors of mammals (Figure 27.14). And by Jurassic times, diverse plant-eating and meat-eating mammals called **therians** had evolved. Most were the size of a mouse, and they were endowed with hair and major changes in the jaws, teeth, and body form. For instance, their four limbs were positioned upright under the body's trunk. This skeletal arrangement made it easier to walk erect, but a trunk higher from the ground was not as stable. At that time the cerebellum, a brain region dealing with the body's balance and spatial positioning, started expanding.

The therians coexisted with dinosaurs through the Cretaceous. When the last of the dinosaurs vanished, diverse adaptive zones opened up for three lineages of those previously inconspicuous mammals (Sections 19.5 and 21.8). New opportunities, differences in traits, and key geologic events put those lineages—the **monotremes** (egg-laying mammals), **marsupials** (pouched mammals), and **eutherians** (placental mammals)—on very different paths to the present. Compared with monotremes and marsupials, both of which retained many archaic traits, placental mammals had the competitive edge. They had higher metabolic rates, more precise ways of regulating body temperature, and a new way of nourishing their developing embryos. You will read about those traits in later chapters. For now, it is enough to know that the placental mammals radiated into many new adaptive zones at the expense of their less competitive relatives.

As Figure 27.19a shows, ancestors of monotremes and marsupials had entered the southern part of the supercontinent Pangea by the late Jurassic. Following the breakup of Pangea, those on the land mass that would become Australia were already isolated from the ancestors of placental mammals, which were evolving elsewhere (Figure 27.19b). In the future South America, monotremes were replaced by marsupials and early placental mammals that had evolved independently of their relatives on other continents. When a land bridge rejoined South and North America during the Pliocene, highly evolved placental mammals radiated southward and rapidly drove many of those previously isolated mammals to extinction. Only opossums and a few other species successfully invaded land in the other direction.

Mammals alone have hair and mammary glands. They have distinctive dentition, a highly developed nervous system, and a notable capacity for behavioral flexibility.

Much of mammalian history was a matter of luck, of species with particular traits being in the right or wrong places on a changing geologic stage at particular times.

# PORTFOLIO OF EXISTING MAMMALS

Evolutionarily distant, geographically isolated lineages have sometimes evolved in similar ways in similar habitats, so that they come to resemble each other in form and function. We call this **convergent evolution**, as defined in Section 20.4. With their interesting history, the three lineages of mammals offer classic examples, as the lists in Table 27.1 suggest.

The only living monotremes are two species of spiny anteaters and the duck-billed platypus described at the start of the chapter. One spiny anteater (*Tachyglossus*, in Figure 27.20*a*) is common in Australia. The other lives in mountains of New Guinea. These small, burrowing mammals feed mostly on ants. Like porcupines, they bristle with protective spines that are modified hairs. As is the case for platypuses, females lay eggs. Unlike platypuses, they do not dig out nests. They incubate a single egg, and the hatchling suckles and completes its development in a skin pouch that forms *temporarily* (by muscle contractions) on the mother's ventral surface.

Of the 260 existing species of marsupials, most are native to Australia and nearby islands; only a few live in the Americas (Figure 27.20*b,c*). The tiny, blind, and hairless newborns suckle and finish their development in a *permanent* pouch on the mother's ventral surface.

**Table 27.1  Convergences Among Mammalian Groups**

| Life-Style | Home | Mammalian Family |
|---|---|---|
| Aquatic invertebrate eater | North America<br>Central America<br>Australia | Water shrew (Soricidae)<br>Water mouse (Cricedidae)<br>Platypus (Ornithorhynchidae) |
| Land-dwelling carnivore | North America<br>Australia | Wolf (Canidae)<br>Tasmanian wolf (Thylacinedae) |
| Land-dwelling anteater | South America<br>Africa<br>Australia | Giant anteater (Myrmecophagidae)<br>Aardvark (Orycteropodidae)<br>Spiny anteater (Tachyglossidae) |
| Ground-dwelling leaf, tuber eater | North America<br>South America<br>Eurasia | Pocket gopher (Geomyidae)<br>Tuco-tuco (Ctenomyidae)<br>Mole rat (Spalacidae) |
| Tree-dwelling leaf eater | South America<br>Africa<br>Madagascar<br>Australia | Howler monkey (Cebidae)<br>Colobus monkey (Cercopithecidae)<br>Woolly lemur (Indriidae)<br>Koala (Phascolarctidae) |
| Tree-dwelling nut, seed eater | Southeast Asia<br>Africa<br>Australia | Flying squirrel (Sciuridae)<br>Flying squirrel (Anomaluridae)<br>Flying squirrel (Phalangeridae) |

**Figure 27.20**  A representative monotreme: (**a**) Spiny anteater (*Tachyglossus*). Two Australian marsupials: (**b**) A female koala and her albino offspring living in relative safety in the San Diego Zoo. Their ancestors evolved millions of years ago, when the climate turned drier. Drought-resistant plants evolved, and koalas came to depend on eucalyptus (gum trees) for food and, often, for shelter. Koalas were abundant until Europeans arrived. Eucalyptus forests were cleared for farmland; millions of the slow-moving mammals were shot for their pelts. Koalas became a protected species in the 1930s. They still declined in numbers because nothing protected the eucalyptus trees. This still is the case through much of their home range. (**c**) Tasmanian devil, the largest carnivorous marsupial. It is about the size of a small dog. Its fierce show of fangs (a threat display used as a communication signal) as well as its hair-raising screeches, coughs, and snarls have given it an undeserved bad reputation. The Tasmanian devil is a nocturnal scavenger, a bit famous for its rowdy communal feeds at carcasses.

Every other existing descendant of therians is a placental mammal. A **placenta** is a spongy tissue of maternal and fetal membranes (Figure 27.21). It forms inside a pregnant female's uterus, the chamber in which an embryo develops in relative freedom from harsh conditions in the outer world. The female sends nutrients and oxygen to her embryo across the placenta and removes its

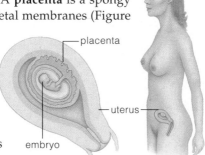

**Figure 27.21**  Location of the placenta in a human female.

**Figure 27.22** A few placental mammals. (**a**) Herd of camels traversing an extremely hot desert with ease. (**b**) One of the bats, the only truly flying mammals, which dominate the night sky vacated by birds. (**c**) Manatee. It lives in the sea and eats submerged seaweed. (**d**) Arctic fox, with thick, insulative fur that camouflages it from prey. In summer, its light-brown fur blends with golden-brown grasses. In winter, the fur turns white and blends with the snow-covered ground. (**e**) Walruses swim in frigid waters after their prey and sunbathe on ice.

metabolic wastes. Section 45.9 provides a look at the structure and function of the placenta. For now, it is enough simply to remember that placental mammals generally have a developmental advantage. They grow faster in the uterus than marsupials do in their pouch, and many species are fully formed (and less vulnerable to predators) at birth. Figure 27.22 shows a few species; Appendix I lists the major groups.

The body form and function, behavior, and ecology of the mammals will occupy our attention later in the book. Chapter 28, for example, will provide insight into why many existing mammalian species are threatened by a mass extinction that humans are bringing about.

**Convergent evolution occurred among the families of all three lineages of mammals that now occupy similar habitats in different regions of the world.**

Having traversed the mammalian branch of the animal family tree, we are now ready to follow it along the branchings leading to primates, then to humans. As we explore the branchings, keep this key point in mind: *"Uniquely" human traits emerged through modification of traits that evolved earlier, in ancestral forms.*

## Primate Classification

In the order **Primates** are prosimians, tarsioids, and anthropoids (Appendix I). Figure 27.23 shows a few. The first prosimians were arboreal (tree-dwelling). They dominated northern forests for millions of years before monkeys and apes evolved and almost displaced them. Small tarsiers of Southeast Asia, the only living tarsioids,

on their sense of smell and more on daytime vision. *Second*, skeletal changes led to upright walking, which freed hands for novel tasks. *Third*, changes in bones and muscles led to refined hand movements. *Fourth*, teeth became less specialized. *Fifth*, evolution of the brain, behavior, and culture became interlocked.

## Overview of Key Trends

**ENHANCED DAYTIME VISION**   Early primates observed the world through one eye on each side of their head. Later ones had forward-directed eyes, an arrangement that is far better for sampling shapes and movements in three dimensions. Other modifications allowed eyes to respond to variations in color and light intensity (dim

| Table 27.2 | Primate Classification |
|---|---|
| Taxon | Representatives |
| **PROSIMIANS** | |
| Lemuroids | Lemur, loris |
| **TARSIOIDS** | |
| Tarsioids | Tarsier |
| **ANTHROPOIDS** | |
| Ceboids (New World monkeys) | Spider monkey, howler monkey |
| Cercopithecoids (Old World monkeys) | Baboon, langur, macaque |
| Hominoids: | |
| Hylobatids | Gibbon, siamang |
| Pongids | Bonobo, orangutan, chimpanzee, gorilla |
| Hominids | Humans, extinct humanlike forms |

are somewhere between prosimians and anthropoids in traits. Monkeys, apes, and humans are all anthropoids. In biochemistry and body form, the apes are closer to humans than to monkeys. Apes, humans, and extinct species of their lineages are classified as hominoids. **Hominids** include all humanlike and human species of a line of descent that started with its divergence from the last shared ancestor of apes and humans.

Most primates live in tropical or subtropical forests, woodlands, or savannas (open grasslands with sparse stands of trees). Like their ancestors, most are *arboreal*, or tree dwellers. No one feature sets them apart from other mammals, and each lineage has its own defining traits. Five trends that help define the lineage leading to humans were set in motion when certain primates started adapting to life in the trees. *First*, there was less reliance

**Figure 27.23**   Representative primates. (**a**) Bonobos, our closest primate relatives. Like us, they walk upright. (**b**) Gibbon, with body and limbs well adapted for swinging arm over arm through trees. (**c**) Spider monkey, a four-legged climber, leaper, and runner. (**d**) Tarsiers, vertical climbers and leapers.

**Figure 27.24** Comparison of the skeletal organization and stance of a monkey, ape, and human. Monkeys climb and leap. Gorillas and chimps use their forelimbs to climb and help support their weight, but most of the time they walk on all fours on the ground. Humans are two-legged walkers. Differences in modes of locomotion arose through modifications of the basic skeletal plan of mammals.

to bright light). Being able to interpret and respond swiftly to the novel stimuli proved advantageous for life in the trees.

**UPRIGHT WALKING** How a primate walks depends on its arm length and the shape and positioning of its shoulder blades, pelvic girdle, and backbone. With arm and leg bones of about the same length, a monkey can climb, leap, and run on four legs, but not two (Figures 27.23c and 27.24). A gorilla walks on two legs and on the knuckles of its two long arms. Only humans and bonobos show **bipedalism**: they can routinely stride and run about on two legs. Compared with an ape or monkey backbone, the human backbone is shorter, S-shaped, and flexible. Bipedalism was a key innovation that evolved in the ancestors of hominids.

**POWER GRIP AND PRECISION GRIP** How did we get our versatile hands? The first mammals spread their toes apart to support the body as they walked or ran on four legs. Primates still spread fingers or toes. Many also cup their fingers, as when monkeys lift food to the mouth. Among ancient tree-dwelling primates, modifications to handbones allowed them to wrap fingers around objects (*prehensile* movements) and to touch a thumb to the tip of a finger (*opposable* movements). In time, hands became freed from load-bearing functions. Later, when hominids evolved, so did power and precision grips:

power grip          precision grip

These hand positions gave early humans a capacity to make and use tools. They were a foundation for unique technologies and cultural development.

**TEETH FOR ALL OCCASIONS** Before hominids evolved, modifications in primate jaws and teeth accompanied a shift from eating insects to fruits and leaves, and on to a mixed diet. Later, rectangular jaws and long canines

**Figure 27.25** Trend toward longer life spans and greater dependency of offspring on adults among primates.

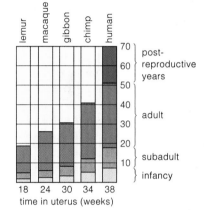

came to be more defining features of monkeys and apes. Along the road that led to modern humans, a notably bow-shaped jaw and smaller teeth of about the same length evolved.

**THE BRAIN, BEHAVIOR, AND CULTURE** Among some of the early primates, an arboreal life-style required shifts in reproductive and social behavior. In many lineages, parents put more effort in fewer offspring. They formed stronger bonds with offspring, maternal care grew more intense, and the time for dependency and learning grew longer (Figure 27.25).

Expansion of brain regions became interlocked with selection for more complex behavior. Novel behaviors promoted new neural connections in brain regions that dealt with sensory inputs, a brain with more intricate wiring favored more novel behavior, and so on. We find evidence of such interlocking in the parallel evolution of the human brain and culture. We define **culture** as the sum of behavior patterns of a social group, handed down among successive generations through learning and symbolic behavior, especially language. A capacity for language arose in ancestral human species mainly as an outcome of changes in skull bones and the brain.

---

These key adaptations evolved along the pathway from arboreal primates to modern humans: complex, forward-directed vision; bipedalism; refined hand movements; generalized dentition; and an interlocked elaboration of brain regions, behavior, and culture.

---

# FROM EARLY PRIMATES TO HOMINIDS

## Origins and Early Divergences

Primates evolved from mammals more than 60 million years ago, in tropical forests of the Paleocene. The first ones resembled small rodents or tree shrews (Figures 27.26 and 27.27). Like tree shrews, they probably had huge appetites and foraged at night for insects, seeds, buds, and eggs beneath the trees. They had a long snout and a refined sense of smell, suitable for snuffling food and predators. They clawed their way up stems, although not with much speed or grace.

During the Eocene (between 54 and 38 million years ago), certain primates were staying up in the trees. They had a shorter snout, better daytime vision, a larger brain, and a capacity for refined grasping motions. How did those traits evolve?

Consider the trees. Arboreal life-styles had advantages: abundant food and safety from ground-dwelling predators. However, they also were habitats of uncompromising selection. Visualize an Eocene morning—dappled sunlight, boughs swaying in the breeze, colorful, tasty fruit hidden among the leaves, perhaps predatory birds. A long, odor-sensitive snout would not have been much good in a dense forest canopy, where

**Figure 27.26** A tree shrew of Indonesia.

air currents disperse odors. But a brain that could assess movement, depth, shape, and color would have been a definite plus. So would a brain that could work fast when its owner was running and leaping (especially!) from branch to branch. Body weight, wind speed, and the distance and suitability of a destination had to be estimated swiftly, all at the same time. Adjustments for miscalculations had to be quick.

By 36 million years ago, before the dawn of the Oligocene epoch, the tree-dwelling anthropoids had evolved in the forests. One form, the squirrel-size *Catopithecus*, was on or near the evolutionary road leading to monkeys and apes. It had forward-directed eyes. Given its snoutless, flattened face and upper jaw with its shovel-shaped front teeth, it must have grabbed fruit and insects with its hands.

Some of the anthropoids lived in swamplands that were infested with fearsome, predatory reptiles. They rarely ventured down to the ground. Maybe that's why it was imperative to think fast and grip strongly. Many primates still fall out of the trees, so we can assume early anthropoids also slipped up.

Between 25 and 5 million years ago, in the Miocene, apelike forms—*the first hominoids*—evolved. They underwent an adaptive radiation into Africa, Europe, and southern Asia. While these early hominoids were evolving, continents were on the move, and ocean circulation patterns were shifting in a major way. The shift triggered a long-term cooling trend that would continue into the Pliocene (Figure 27.28).

The climate in eastern and southern Africa became cooler, drier, and more seasonal. Tropical forests—with their bounty of edible soft fruits, leaves, and abundant insects—started to give way to dense woodlands, then to grasslands. Food was drier and harder (and harder to find). Hominids that had evolved in the forests had two options: move into new adaptive zones or die out.

Not all of the early hominoids made the transition; most were extinct by the time the Miocene ended. But

**Figure 27.27** Comparison of the skull shape and teeth of three early primates. (**a**) From the Paleocene, *Plesiadapis* was as tiny as a tree shrew and had rodentlike teeth. (**b**) The monkey-sized *Aegyptopithecus*, an Oligocene anthropoid, probably predates the divergence that gave rise to Old World monkeys and apes. (**c**) One of the apelike dryopiths that lived during the Miocene. Some dryopiths were as large as a chimpanzee.

a b c

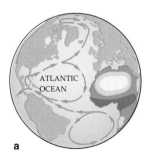

ATLANTIC OCEAN

a

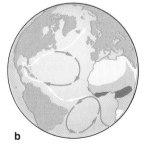

b

**Figure 27.28** Model of a long-term shift in global climate. (**a**) Before the Isthmus of Panama formed, salinity levels of currents near the ocean surface were much the same around the world. Circulation patterns kept Arctic waters warm. (**b**) After the isthmus formed, North Atlantic surface currents became saltier and heavier, so they sank before they reached the Arctic region. With colder water in polar regions, the Arctic ice cap formed and ushered in a long-term trend toward a cooler, drier climate in Africa.

**Figure 27.29** (a) Remains of "Lucy" (*Australopithecus afarensis*), who lived 3.2 million years ago. (b) At Laetoli in Tanzania, Mary Leakey found these footprints, made in soft damp volcanic ash 3.7 million years ago. (c,d) The arch, big toe, and heel marks of such footprints are signs of bipedal hominids. Unlike apes, early hominids didn't have a splayed big toe, as the chimpanzee in (c) obligingly demonstrates.

*Ardipithecus ramidus*
4 million years old

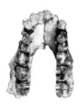

*Australopithecus anamensis*
4.2–3.9 million years
(walked upright)

*A. afarensis*
3.6–2.9 million years

*A. africanus*
3.2–2.3 million years

*A. garhi*
2.5 million years
(first tool user?)

*A. boisei*
2.3–1.4 million years
(huge molars)

*A. robustus*
1.9–1.5 million years

**Figure 27.30** Representative fossils of African hominids that lived between 4.4 and 1.4 million years ago.

genetic analyses suggests that two lineages branched from a common ancestor some time between 6 million and 4 million years ago. One lineage gave rise to the great apes, and the other gave rise to the first *hominids*.

## The First Hominids

Fossil fragments of early hominids, some older than 5 million years, have been found in eastern and southern Africa (Figures 27.29 and 27.30). One species, *Ardipithecus ramidus*, lived 4.4 million years ago. It was more apelike than humanlike. Its small molars, large canines, and thin tooth enamel suggest that its diet consisted of easily chewed fruits and vegetables. Paleontologist Tim White puts it in the minor leagues of brain development.

Diverse hominids evolved over the next 2 million years. We still don't know how they are related, so we informally call them **australopiths**, or "southern apes." At this writing, the oldest known australopith species is *Australopithecus anamensis*. Like *A. afarensis* and *A. africanus*, it was gracile (slightly built). The ones called *A. boisei* and *A. robustus* were robust (muscular and heavily built). Like apes, australopiths had a large face, protruding jaws, and a small skull (and brain) size. Yet they differed in key respects from the hominoids. For

example, their molars had thicker enamel and could grind harder foods. They could walk upright. We know this from fossilized hip bones and limb bones and from footprints. Around 3.7 million years ago, *A. afarensis* walked on newly fallen volcanic ash, which a light rain turned into quick-drying cement (Figure 27.29b).

Bipedal hominids had started to evolve in the late Miocene. In the trees, their hands had become adapted to grip objects strongly and with precision. When their descendants left the trees for life on the ground, rather than becoming specialized in running fast on all fours, they used manipulative skills to advantage. They kept their hands free to hold offspring and probably to carry precious food during their foraging expeditions.

---

Primates evolved from small, rodentlike mammals that had moved into arboreal habitats more than 60 million years ago. The earliest hominoids evolved between 25 million and 5 million years ago in Africa, Europe, and southern Asia.

Between 6 million and 4 million years ago, lineages that would give rise to the great apes and to the first hominids branched from a common hominoid ancestor.

Early hominids called australopiths were apelike in many skeletal details but humanlike in walking upright.

---

# EMERGENCE OF EARLY HUMANS

What can the fossilized fragments of early hominids tell us about our own origins? The record is still too sketchy for us to know how the diverse australopiths were related to one another, let alone which ones may have been ancestral to humans. Besides, which traits do characterize **humans**—members of the genus *Homo*?

Well, what about the brain? The brain of modern humans is the basis of great analytical and verbal skills, complex social behavior, and technological innovation. It clearly sets us apart from apes, which have a much smaller skull volume and brain size (Figure 27.31). Yet this feature alone cannot tell us when certain hominids made the leap to being human. Why? Their brain size almost certainly fell within the ape range. Even though they made simple tools, so do chimps and some parrots. We have no clues to their social behavior.

We are left to speculate on a continuum of physical traits among a number of fossils—a skeleton adapted for bipedalism; manual dexterity; larger brain volume; a smaller face; and smaller, more thickly enameled teeth. Those traits, which originated in the late Miocene, were evident in what may have been the earliest humans—*Homo habilis*, a name that means "handy man."

**Figure 27.32** Olduvai Gorge stone tools: (**a**) a crude chopper, (**b**) a more refined chopper, (**c**) a hand ax, and (**d**) a cleaver,

Between 2.4 and 1.6 million years ago, early forms of *Homo* lived in dry woodlands adjoining the savannas of eastern and southern Africa. Their dentition implies they could handle hard-shelled nuts and seeds as well as soft fruits, leaves, and insects. Those food supplies were seasonal. Most likely, *H. habilis* had to think ahead, plan when to venture about to gather and maybe store foods that would help it survive the cold dry seasons.

*H. habilis* shared its habitat with saber-tooth cats and other predators. The cats' teeth could impale prey and shear off flesh but couldn't crush open marrow bones. So carcasses of their kills, with shreds of meat clinging to marrow bones, were concentrated stores of nutrients in nutrient-stingy places. *H. habilis* was a forager, not a full-time carnivore. But it opportunistically enriched its diet by scavenging carcasses (Figure 27.31c).

Fossil hunters have found a great many stone tools that date to the time of *H. habilis*. They cannot say with certainty that *H. habilis* was the only species that made them. Possibly australopiths as well as *H. habilis* used sticks and other perishable materials for tools before then, as modern apes do, but we have no way of knowing.

Maybe ancestors of modern humans started down a toolmaking road by picking up rocks to crack marrow bones. Maybe they scraped flesh from bones with sharp flakes split naturally from rocks. However it happened, at some point they started *shaping* stone implements. Paleontologist Mary Leakey was first to find evidence of stone toolmaking at Africa's Olduvai Gorge, which cuts through many sedimentary layers. The oldest tools there are crudely chipped pebbles in the deepest layers (Figure 27.32). Early humans might have used them to smash marrow bones, dig for roots, and poke insects from bark. More recent layers have more complex tools. Where shorelines of ancient lakes prevailed, we observe numerous bones and tools. Why? In arid habitats, lakes would have beckoned plenty of thirsty prey animals.

At such sites we find fossils of an early form of *Homo* that was twice as brainy as australopiths and obviously

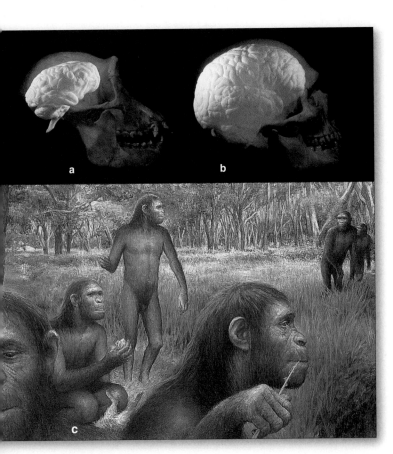

**Figure 27.31** Comparison of brain size for (**a**) a chimpanzee and (**b**) a modern human. (**c**) Artist's view of *Homo habilis* in an East African woodland. Australopiths are in the distance.

Further reading: Student Guide to InfoTrac on web site →

| *Homo rudolfensis* | *H. habilis* | *H. erectus* | *H. neanderthalensis* | *H. sapiens* |
| 2.4–1.8 million years | 1.9–1.6 million years | 2 million–53,000? years | 200,000–30,000 years | 100,000–? |

**Figure 27.33** Representative fossils of early humans (*Homo*).

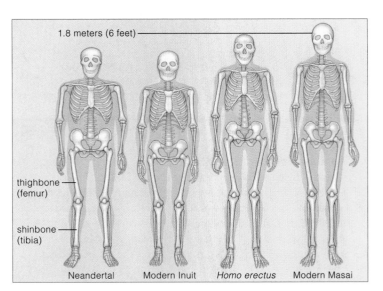

1.8 meters (6 feet)

thighbone (femur)

shinbone (tibia)

Neandertal    Modern Inuit    *Homo erectus*    Modern Masai

**Figure 27.34** Body build correlated with climate. Humans adapted to cold climates have a heat-conserving body (stockier, shorter legs, compared with humans adapted to hot climates).

ate well. There apparently was no selection pressure for more creativity in securing food resources; stone tools did not change much for the next 500,000 years.

Apparently, the ancestors of modern humans stayed put in Africa until about 2 million years ago. Then, a genetic divergence gave rise to *Homo erectus*, a species clearly related to fully modern humans (Figure 27.33). The name means upright man. Its forerunners also were upright, two-legged walkers, but *H. erectus* populations did the name justice. They trekked out of Africa, turned left into Europe, and right into Asia. Some traversed the 14,000 kilometers to China; fossils from Southeast Asia and the former Soviet republic of Georgia are 1.8 million and 1.6 million years old. More than once, *H. erectus* survived when ice sheets advanced down into northern Europe, southern Asia, and North America.

Whatever pressures triggered the far-flung travels, this was a time of physical changes, as in skull size and leg length. It also was a time of cultural lift-off for the human lineage. *H. erectus* had a larger brain and was

a more creative toolmaker. Its social organization and communication skills must have been well developed. How else can we explain its successful dispersal? From southern Africa to England, different populations used the same kinds of hand axes and other tools to pound, scrape, shred, chop, cut, and whittle material. *H. erectus* withstood environmental challenges by building fires and using furs for clothing. Clear evidence of fire use dates from an ice age in the early Pleistocene.

As Middle Eastern fossils show, *Homo sapiens* evolved by 100,000 years ago. Early *H. sapiens* had smaller teeth and jaws than *H. erectus* (Figure 27.33). Many had a novel feature: a chin. Facial bones were smaller, the skull higher and rounder, and the brain larger. They might have developed complex language. But their origin and geographic dispersal are hotly debated (Section 27.15).

One group of early humans, the Neandertals, lived in Europe and in the Near East from 200,000 to 30,000 years ago. Massively built and large brained, some were the first to adapt to the coldest regions (Figure 27.34). Their disappearance coincided with the appearance of anatomically modern humans in the same regions about 40,000 to 30,000 years ago. We have no evidence that Neandertals warred or interbred with the later arrivals. We still do not know what happened to them. We do know that Neandertal DNA has unique sequences, so they might not have contributed to the gene pools of modern European populations.

From 40,000 years ago to today, human evolution has been almost entirely cultural, not biological. And so we leave the story with these conclusions: Humans spread rapidly through the world by devising *cultural* means to deal with a broader range of environments. Compared to their predecessors, they developed rich, varied cultures. Although hunters and gatherers persist in parts of the world, others moved from "stone-age" technology to the age of "high tech," attesting to the great plasticity and depth of human adaptations.

---

**Cultural evolution has outpaced the biological evolution of the only remaining human species, *H. sapiens*. Today, humans everywhere rely on cultural innovation to adapt rapidly to a broad range of environmental challenges.**

---

**27.15**

# OUT OF AFRICA—ONCE, TWICE, OR . . .

If researchers are interpreting the fossil record of human evolution correctly, then it would seem Africa was the cradle for us all. At this writing, at least, no one has any fossils of humans older than 1.8 million years *except* in Africa. *H. erectus* coexisted for a time with earlier humans (*H. habilis*) before dispersing from Africa to the cooler grasslands, forests, and mountains of Europe and Asia. This form apparently left Africa in waves between about 2 million and 500,000 years ago. Judging from *H. erectus* fossils from Java, isolated populations may have endured until some time between 53,000 and 37,000 years ago.

Where, on the larger geologic stage, do we place the origin of *H. sapiens*? *Here we find a good example of how the same body of evidence can be interpreted in different ways.* The interpretations are called the **multiregional model** and **African emergence model** for modern human origins. Both attempt to explain the world distribution of fossils of particular ages and measured genetic distances among existing human populations. For example, biochemical and immunological studies imply that the greatest genetic distance separates *H. sapiens* populations native to Africa from populations everywhere else; the next greatest distance separates Southeast Asia and Australia from everywhere else (Figure 27.35). As still another example, Figure 27.36 correlates *H. sapiens* fossils to specific times.

**MULTIREGIONAL MODEL** By this model, *H. erectus* populations spread through much of the world by about 1 million years ago. The geographically separated groups faced different selection pressures, so their traits evolved in regionally distinctive ways. Then the subpopulations ("races") of *H. sapiens* evolved from them in different places. They differed phenotypically but did not evolve into separate species because gene flow continued among them, even to the present. Thus, for example, while the armies of Alexander the Great were sweeping eastward, they also contributed "blue-eye genes" from Greeks to

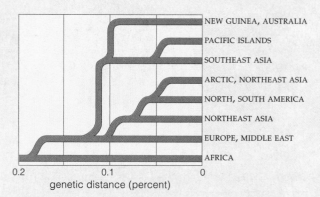

**Figure 27.35** One proposed family tree for populations of modern humans (*Homo sapiens*) that are native to different regions. Branch points show presumed genetic divergences. The tree is based on nucleic acid hybridization studies of many genes (including those for mitochondrial DNA and the ABO blood group) and immunological comparisons.

the allele pool of generally brown-eyed subpopulations in Africa, the Near East, and Asia.

**AFRICAN EMERGENCE MODEL** This model does not dispute fossil evidence that populations of *H. erectus* evolved in different ways in different regions. However, it holds that *H. sapiens*—modern humans—originated in sub-Saharan Africa somewhere between 200,000 and 100,000 years ago. Only later did *H. sapiens* populations move out of Africa, into regions along the routes shown in the Figure 27.36 map. In each region that populations of *H. sapiens* settled, they replaced the archaic *H. erectus* populations that had preceded them. Only then did regional phenotypic differences become superimposed on the original *H. sapiens* body plan.

In support of this model, the oldest known *H. sapiens* fossils are from Africa. Also, in Zaire, a finely wrought barbed-bone harpoon and other exquisite tools suggest that African populations were as skilled at making tools as *Homo* populations known earlier from Europe. Also, in 1998, researchers at the University of Texas and in China announced the findings from the Chinese Human Genome Diversity Project. Detailed analysis of gene patterns from forty-three ethnic groups in Asia suggest that modern humans moved from central Asia, along the coast of India, then on into Southeast Asia and southern China. Populations later moved north and northwest into China, then into Siberia, and then on down into the Americas.

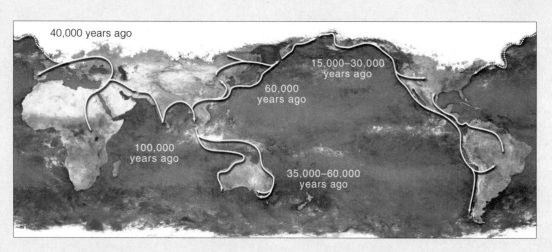

**Figure 27.36** Estimated times when populations of early *H. sapiens* were colonizing different regions of the world, based on radiometric dating of fossils. The presumed dispersal routes (white lines) seem to support the African emergence model.

1. Almost all chordate embryos have a notochord, a dorsal hollow nerve cord, a pharynx with gill slits (or hints of these), and a tail that extends past the anus. Some or all of these traits persist in adults. Chordates with a backbone are vertebrates. Tunicates (including sea squirts) and lancelets are invertebrate chordates.

2. The eight classes of vertebrates are jawless fishes, jawed armored fishes (now extinct), cartilaginous fishes, bony fishes, amphibians, reptiles, birds, and mammals.

3. The earliest vertebrates, the jawless fishes that arose in the Cambrian, included ostracoderms. Their modern descendants are lampreys and hagfishes. Jawed fishes also arose in Cambrian times. A once-dominant lineage, the placoderms, became extinct. Other lineages gave rise to the cartilaginous fishes and bony fishes.

4. Four evolutionary trends are associated with certain vertebrate lineages:

   a. A vertebral column supplanted the notochord as a structural element against which muscles can act. This development led to fast-moving predatory animals.

   b. Jaws evolved from gill-supporting elements. This led to increased predator–prey competition; it favored more efficient nervous systems and sensory organs.

   c. In one lineage of bony fishes, paired fins evolved into fleshy lobes having internal structural elements. The lobes were the forerunners of paired limbs.

   d. In some fish lineages, respiration by gills came to be supplemented by paired lungs, which proved to be adaptive during the invasion of the land. In a related development, the circulatory system became far more efficient at distributing oxygen through the body.

5. Amphibians were the first vertebrates to invade the land, but they never fully escaped the water. Their skin dries out, and aquatic stages persist in most life cycles.

6. Unlike amphibians, reptiles escaped dependence on standing free water. The key adaptations that allowed them to do so include toughened, scaly skin that can restrict loss of water from the body; a reliance upon internal fertilization; more efficient, water-conserving kidneys; and amniote eggs, often leathery or shelled, which protect and metabolically support the embryos.

7. Reptiles, and birds and mammals (which descended from certain reptilian lineages), have highly efficient circulatory and respiratory systems. They also have well-developed nervous system and sensory organs.

8. Of all vertebrates, birds alone have feathers, which they use in flight, heat conservation, and social displays.

9. Of all animals, mammals alone have milk-producing mammary glands, and they have hair or thick skin that functions in insulation. They have distinctive dentition and a highly developed cerebral cortex, the brain's most complex region. Most adults nurture offspring through an extended period of dependency and learning.

   a. Therians, the first mammals, evolved from small, hairless reptiles (therapsids) during the Triassic. At the end of the Cretaceous, new adaptive zones opened up and, combined with geologic events, set the stage for three previously inconspicuous mammalian lineages.

   b. Those lineages still persist. They are monotremes (egg-laying mammals), marsupials (pouched mammals), and eutherians (placental mammals). In similar habitats in different parts of the world, some of their members converged evolutionarily.

   c. In many places, placental mammals outcompeted the monotremes and marsupials, which retained many archaic traits. Placental mammals had higher metabolic rates, more precise control of body temperature, and an efficient way to nourish and protect embryos.

10. Primates include the prosimians (lemurs and related forms), the tarsioids, and the anthropoids (which include monkeys, apes, and humans). Apes and humans alone are hominoids. Only the anatomically modern humans (*H. sapiens*) and others of the lineage (from *Ardipithecus* and australopiths to *H. erectus*) are hominids.

11. The first primates were small, rodentlike mammals that evolved by 60 million years ago in tropical forests.

12. Apelike forms (the first hominoids) were evolving in Miocene times, 25 to 5 million years ago. In eastern and southern Africa, some gave rise to hominids by at least 5 million years ago.

   a. The origin and early evolution of hominids might coincide with selection pressures that emerged during a long-term change in climate, brought about by shifts in land masses and ocean circulation patterns. Tropical forest gave way to dense woodlands, then grasslands.

   b. Some hominids became adapted to the changes. They had physical modifications that allowed upright walking (bipedalism), a more diverse diet (changes in dentition), and a capacity to find scarcer and seasonally available food (possible with a more complex brain).

13. *H. habilis*, the earliest known species of the genus *Homo*, evolved by 2.5 million years ago. It might have been the first stone toolmaker. *Homo erectus*, possibly ancestral to modern humans, evolved by 2 million years ago. *H. erectus* populations radiated out of Africa, into Asia and Europe. The oldest known fossils of the early modern humans (*H. sapiens*) date from about 100,000 years ago. About 40,000 years ago, cultural evolution outstripped biological evolution of the human form.

14. Modern humans are adapted to a broad range of environments. This capacity resulted from evolutionary modifications in certain primate lineages. Starting with arboreal primate ancestors, there was less reliance on the sense of smell and more on enhanced daytime vision.

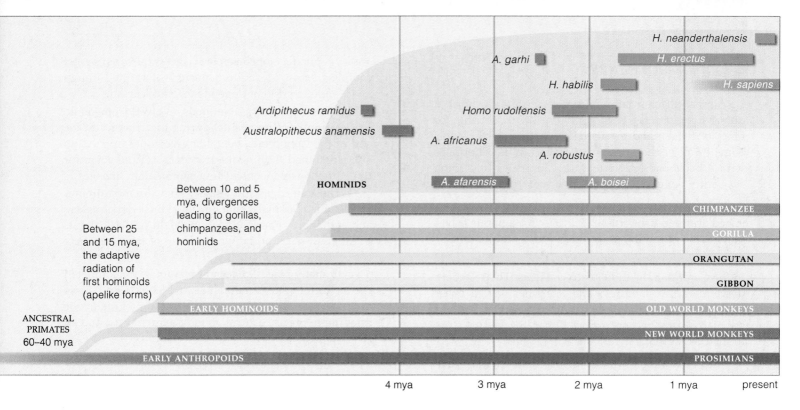

Figure 27.37 shows the timeline with the following labels:

H. neanderthalensis
A. garhi
H. erectus
H. habilis
H. sapiens
Ardipithecus ramidus
Homo rudolfensis
Australopithecus anamensis
A. africanus
A. robustus
A. afarensis
A. boisei

HOMINIDS

Between 10 and 5 mya, divergences leading to gorillas, chimpanzees, and hominids

Between 25 and 15 mya, the adaptive radiation of first hominoids (apelike forms)

CHIMPANZEE
GORILLA
ORANGUTAN
GIBBON

EARLY HOMINOIDS
OLD WORLD MONKEYS

ANCESTRAL PRIMATES 60–40 mya

NEW WORLD MONKEYS

EARLY ANTHROPOIDS
PROSIMIANS

4 mya     3 mya     2 mya     1 mya     present

Also among the arboreal primates, manipulative skills increased as hands began to be freed from load-bearing functions. Starting with Miocene apes, there was a shift from four-legged climbing to bipedalism, a shift from specialized to omnivorous eating habits, and increases in brain complexity and behavior.

15. Figure 27.37 summarizes what we currently know or suspect about primate origins and evolution.

Figure 27.37 Summary of presumed evolutionary branchings leading to modern humans in the primate family tree.

## Review Questions

1. Which features distinguish chordates from other groups of animals? *27.1*

2. List four major trends that occurred during the evolution of at least some vertebrate lineages. *27.3*

3. Describe some features of jawless fishes, cartilaginous fishes, and bony fishes. Which group is the most abundant? *27.3*

4. Which traits distinguish reptiles from amphibians? *27.6, 27.7*

5. Birds, crocodilians, and dinosaurs share what traits? *27.7–27.9*

6. List some characteristics that distinguish each of the three mammalian lineages from the reptiles. *27.10, 27.11*

7. What is the difference between a "hominoid" and a "hominid"? *27.12*

8. Briefly describe some of the conserved physical traits that connect anatomically modern humans with their mammalian ancestors, then their primate ancestors. *27.12, 27.13*

## Self-Quiz *(Answers in Appendix III)*

1. Only _____ have a notochord, a tubular dorsal nerve cord, a pharynx with slits in the wall, and a tail extending past the anus.
   a. echinoderms
   b. tunicates and lancelets
   c. vertebrates
   d. both b and c
   e. all of the above

2. Gills function in _____ .
   a. respiration
   b. circulation
   c. food trapping
   d. water regulation
   e. both a and c

3. A shift from a reliance on _____ to reliance on _____ was pivotal in the evolution of all vertebrates.
   a. the notochord; a backbone
   b. filter feeding; jaws
   c. gills; lungs
   d. all of the above

4. The first vertebrates probably were _____ .
   a. bony fishes
   b. jawless fishes
   c. lobe-finned fishes
   d. both a and b

5. Of all existing vertebrates, _____ are the most diverse.
   a. cartilaginous fishes
   b. bony fishes
   c. amphibians
   d. reptiles
   e. birds
   f. mammals

6. Generally, the only amphibian groups that entirely escaped dependency on free-standing water are _____ .
   a. salamanders
   b. toads
   c. caecilians
   d. none of the above

7. Reptiles moved fully onto land owing to _____ .
   a. tough skin
   b. internal fertilization
   c. good kidneys
   d. amniote eggs
   e. none of the above
   f. all of the above

8. A four-chambered heart evolved first in _____ .
   a. bony fishes
   b. amphibians
   c. birds
   d. mammals
   e. crocodilians
   f. both c and d

9. _____ have highly efficient circulatory and respiratory systems, and a complex nervous system and sensory organs.
   a. Reptiles
   b. Birds
   c. Mammals
   d. all of the above

10. Birds use feathers in _____ .
   a. flight
   b. heat conservation
   c. social functions
   d. all of the above

11. Various mammals _____ .
    a. hatch
    b. complete their embryonic development in pouches
    c. complete their embryonic development in the uterus
    d. both b and c
    e. all of the above

12. Early humans _____ .
    a. adapted to a wide range of environments
    b. adapted to a narrow range of environments
    c. had flexible bones that cracked easily
    d. were limber enough to swing through the trees

13. Match the organisms with the appropriate features.
    ____ jawless fishes          a. complex cerebral cortex,
    ____ cartilaginous              thick skin or hair
         fishes                  b. respiration by skin and lungs
    ____ bony fishes             c. include coelacanths
    ____ amphibians              d. include hagfishes
    ____ reptiles                e. include sharks and rays
    ____ birds                   f. complex social behavior,
    ____ mammals                    feathers
                                 g. first with amniote eggs

## Critical Thinking

1. Describe the factors that might contribute to the collapse of native fish populations in a lake following the introduction of a novel predator, such as the lamprey.

2. Think about the flight muscles of birds and their demands for oxygen and ATP energy. What type of organelle would you expect to be profuse in these muscles? Explain your reasoning.

3. Kathie and Gary, two amateur fossil hunters, have unearthed the complete fossilized remains of a mammal. How can they determine whether that mammal was a herbivore, carnivore, or omnivore?

4. In Australia, many species of marsupials are competing with recently introduced placental mammals (such as rabbits) for resources—but they are not winning. Explain how it is that placental mammals that did not even evolve in the Australian habitats show greater fitness than the native mammals.

5. About seventy-five species of coral snakes live in the rain forests, grasslands, mountains, and desertlike regions of North, Central, and South America. Coral snakes of family Colubridae are harmless; those of family Elapidae are poisonous. In both families, the most poisonous types bite mostly to kill prey, yet moderately poisonous types can be extremely vicious. All coral snakes, including the deadliest (*Micrurus*, Figure 27.38), have distinctive color banding. Even snake experts (herpetologists) may have trouble knowing which is which. However, relatives of the two families in other parts of the world have no such banding; neither do bigger snakes.

Thus, color banding in the two families can't be explained in terms of environmental differences, phylogeny, or lethality. What brought it about? According to a hypothesis proposed by herpetologist R. Mertens, the ancestors of *Micrurus* migrated across a land bridge that connected Asia with North America more than 60 million years ago. Because predators learned to avoid them by associating the color banding with "taste trials," the ancestral snakes must *not* have been as deadly as some of their descendants. (If the snakes were *too* deadly, all trials might have ended in death.) Therefore, the protective function of the color banding must depend on unpleasant trials with only *moderately* poisonous snakes.

Among the harmless snakes (Colubridae), identical warning coloration also evolved. This was a case of *mimicry*, in which one group (mimics) came to bear deceptive resemblance to another

**Figure 27.38** One of the extremely poisonous coral snakes (*Micrurus*).

group (models) that enjoy a survival advantage. For this to work, of course, mimics can't outnumber models. That said, explain the following numbers of trapped snakes that were brought to the Butantan Institut in São Paulo for the years shown (trappers did not distinguish among poisonous and nonpoisonous types):

|            | 1950 | 1951 | 1952 | 1953 |
|------------|------|------|------|------|
| *Micrurus*: | 59   | 41   | 58   | 56   |
| Others:    | 218  | 246  | 265  | 284  |

## Selected Key Terms

*Vertebrates*

amniote egg *27.7*
amphibian *27.6*
behavioral
    flexibility *27.10*
bird *27.9*
bony fish *27.5*
cartilaginous fish *27.5*
chordate *27.1*
convergent
    evolution *27.11*
dentition *27.10*
eutherian *27.10*
feather *27.9*
filter feeder *27.2*
fin *27.3*
gill *27.3*
gill slit *27.2*

hagfish *27.4*
jaw *27.3*
lamprey *27.4*
lancelet *27.2*
lobe-finned
    fish *27.5*
lung *27.3*
mammal *27.10*
marsupial *27.10*
monotreme *27.10*
nerve cord *27.1*
notochord *27.1*
ostracoderm *27.3*
pharynx *27.1*
placenta *27.11*
placoderm *27.3*
Primates
    (order) *27.12*

reptile *27.7*
scale (fish) *27.5*
swim bladder *27.5*
therapsid *27.10*
therian *27.10*
tunicate *27.2*
vertebra *27.3*
vertebrate *27.1*

*Human Evolution*

African emergence
    model *27.15*
australopith *27.13*
bipedalism *27.12*
culture *27.12*
hominid *27.12*
human *27.14*
multiregional
    model *27.15*

*Readings* *See also www.infotrac-college.com*

Gould, S. J. (general editor). 1993. *The Book of Life*. New York: Norton. Splendid, easy-to-read essays, gorgeous illustrations.

Romer, A. S., and T. S. Parsons. 1986. *The Vertebrate Body*. Sixth edition. Philadelphia: Saunders.

Strickberger, M. 1996. *Evolution*. Second edition. Boston: Jones and Bartlett.

Tattersall, I. 1998. *Becoming Human: Human Evolution and Human Uniqueness*. Pennsylvania: Harvest Books.

Wickler, W. 1968. *Mimicry in Plants and Animals*. New York: McGraw-Hill. Paperback.

# BIODIVERSITY IN PERSPECTIVE

## *The Human Touch*

In 1722, on Easter morning, a European explorer landed on a small volcanic island and found a few hundred skittish, hungry Polynesians living in caves. He also found more than 200 massive stone statues near the coast and 700 unfinished, abandoned ones in inland quarries (Figure 28.1). There were no trees whatsoever on the island, only dry, withered grasses and scorched, shrubby plants. Without trees for timber and fibrous plants to make ropes, how did the islanders erect the statues? Some statues weighed fifty tons. And there were no wheeled carts or strong animals. How did the statues get from the quarries to the coast?

Two years later James Cook visited the island. When islanders paddled out to his ships, he noted that they had neither the knowledge nor the materials to keep their canoes from leaking; they spent half the time bailing water. He saw only four canoes on the whole island. Nearly all the statues had been tipped over, often onto spikes that shattered the faces on impact.

Later, researchers solved the mystery of the statues. Easter Island, as it came to be called, has a mere 165 square kilometers (64 square miles) of surface land. Voyagers from the Marquesas discovered this eastern outpost of Polynesia around A.D. 350, possibly after being blown off course by a series of storms.

The place was a paradise. Its fertile volcanic soil supported dense palm forests, hauhau trees, toromino shrubs, and lush grasses. The new arrivals built large, buoyant, ocean-going canoes by using long, straight palms strengthened with rope made of fibers from the hauhau trees. They used toromino wood as fuel to cook fishes and dolphins; they cleared forests to plant crops of taro, bananas, sugarcane, and sweet potatoes.

By 1400, between 10,000 and 15,000 descendants were living on the island. Society had become highly structured, given the need to cultivate, harvest, and distribute food for so many. Crop yields had declined by then, for every crop withdrew precious nutrients from the soil, as did ongoing, neglected erosion. Edible marine species vanished from the safe waters around the island, so fishermen had to build larger canoes to sail out ever farther, into the open sea.

**Figure 28.1** On Easter Island, a few of the massive stone tributes to the gods. Apparently islanders erected the statues as a plea for divine intervention after their once-large population wiped out the biodiversity of their tropical paradise. The plea didn't have any effect whatsoever on reversing the losses on land and in the surrounding sea. The human population didn't recover, either.

Survival was at stake. Those in power appealed to the gods. They redirected community resources into carving divine images of unprecedented size and power, and to moving the newly carved statues over miles of rough terrain to the coast. As Jo Anne Van Tilburg recently demonstrated, they probably lashed the statues to canoe-shaped rigs, then simply rolled them along a horizontal "ladder" made of greased logs. (She got the idea after observing how existing islanders haul canoes from water onto land.)

By now, islanders had all of the arable land under cultivation. They had eaten all the native birds, and no new birds came to nest. They were raising and eating rats, the descendants of hitchhikers on the first canoes. They coveted palm seeds, now a scarce delicacy.

Wars broke out over dwindling food and space. By about 1550 no one was venturing offshore to harpoon fishes or dolphins. They couldn't build canoes because the once-rich forests were gone. They had cut down all of the palms. Hauhau trees were extinct; islanders had used every last one as firewood for cooking.

The islanders turned to the only remaining source of animal protein. They started to hunt and eat one another.

Central authority crumbled, and gangs replaced it. As gang wars raged, those on the rampage burned the remaining grasses to destroy hideouts. The rapidly dwindling population retreated to caves and launched raids against perceived enemies. Winners ate the losers and tipped over the statues. Even if the survivors had wanted to leave the island, they no longer had a way to do so. What possibly could they have been thinking when they chopped down the last palm?

Easter Island initially sustained a society that rose to greatness amid abundant resources, then abruptly fell apart. In life's greater evolutionary story, its loss of biodiversity might not even warrant a footnote. After all, compare the loss to the uncountable numbers of species that appeared and then became extinct in the distant past!

Yet there is a lesson here. The near-total destruction that 15,000 Polynesians wrought on biodiversity was confined to a small bit of land in the vastness of the Pacific Ocean. What are the requirements and whims of *6 billion* people doing to global biodiversity today? Is the loss of species on Easter Island an isolated case? Or should we take what happened there as a warning of a worldwide extinction crisis of our own making, one that is even now under way? Let's take a look.

KEY CONCEPTS

1. The current range of global biodiversity is largely an outcome of an overall pattern of abrupt extinction crises and slow recoveries. It takes tens of millions of years to reach the level of biodiversity that prevailed before such mass extinctions.

2. The loss of individual species as well as extinctions of major groups may have complicated causes, the understanding of which requires scientific analysis.

3. Species found only in a restricted geographic region and extremely vulnerable to extinction are classified as endangered species. Over the past four decades, rates of extinction have been rising through a combination of habitat loss and fragmentation, species introductions, overharvesting, and illegal trading in wildlife.

4. The rapidly growing human population is expected to reach 9 billion by 2050. Its demands for food, materials, and living space are threatening biodiversity around the world and are the basis of a new extinction crisis.

5. Conservation biology takes a three-pronged approach to preserving biodiversity. It includes the systematic survey of the full range of biodiversity, analysis of its evolutionary and ecological origins, and identification of methods to maintain and use biodiversity for the benefit of the human population.

6. Systematic surveys focus initially on identifying hot spots, which are limited areas where many species are in danger of extinction owing to human activities. At the next research level, broader or multiple hot spots are researched. Data are combined into a global picture of ecoregions where biodiversity is most threatened.

# ON MASS EXTINCTIONS AND SLOW RECOVERIES

Based on many lines of evidence accumulated over the past few centuries, an estimated 99 percent of all species that have ever lived are extinct. Even so, the full range of biodiversity is greater now than it has ever been at any time in the past.

Reflect on the evolutionary stories of the preceding chapters in this unit. For at least the first 2 billion years of life's history, single-celled prokaryotes dominated the evolutionary stage. They did so until the Cambrian period, when atmospheric oxygen started to approach current levels. At that time, some of the single-celled eukaryotes began their genetic divergences and intricate species interactions, which led to the origin of diverse protistans, plants, fungi, and animals.

You saw how mass extinctions reduced biodiversity on land and in the seas. Each global episode spurred evolutionary change and radiations into newly vacated adaptive zones. However, recovery to the same level of biodiversity was exceedingly slow, requiring 20 to 100 million years. Figure 28.2a is a review of the pattern of five major extinctions and slow recoveries.

That pattern is only a composite of what happened to the major taxa. Lineages, remember, differ in their time of origin, the extent to which the member species evolved, and how long they persisted. If you consider ongoing survival and reproduction to be the measures of success, then each became a loser or a winner when environmental conditions changed in drastic or novel ways. To appreciate this point, reflect on Figure 28.2b, which shows the evolutionary history of representative lineages. Many such histories were combined to give us the overall pattern of Figure 28.2a.

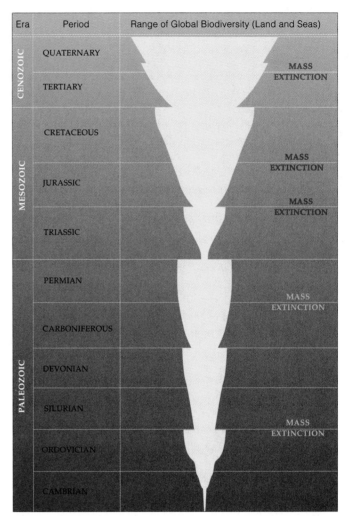

a

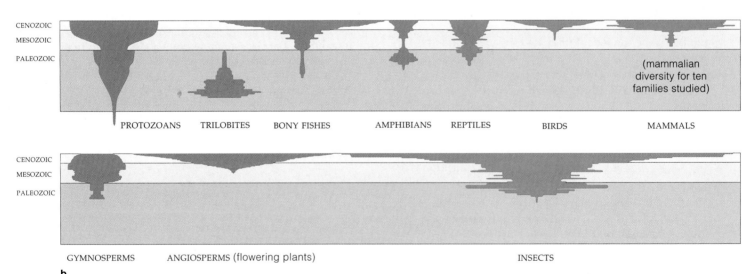

b

**Figure 28.2** (**a**) Review of the range of global diversity through geologic time, on land and in the seas, taking into account five of the greatest mass extinctions and subsequent slow recoveries. (**b**) Within the framework of this generalized graph, patterns of extinction and recoveries differed significantly for different groups of organisms, as these selected examples indicate.

Further reading: Student Guide to InfoTrac on web site ➞

a

**Figure 28.3** Two extinct and one threatened species. (**a**) Charles Knight's magnificent painting of *Tylosaurus*, one of the mosasaurs. This powerful marine lizard flourished in shallow, nearshore waters of the Cretaceous seaways. Indonesia's Komodo dragon, which grows 10 feet (3 meters) long, is a living relative. We think the Komodo dragon is a giant, yet imagine meeting up with a mosasaur. Some were 40 feet (12 meters) long.

(**b**) The dodo (*Raphus cucullatus*), a large, flightless bird that evolved on Mauritius, then vanished more than 300 years ago. Certain trees (*Calvaria major*) also evolved on that island, as did tortoises. Did the tree coevolve with dodos, tortoises, or both? Either may have fed on the tree's large fruits. For example, the large dodo gizzard could have partially digested the thick-walled coat of seeds inside fruits, just enough to help the seeds germinate after leaving the gut. Either way, after the dodo became extinct, the seeds stopped germinating. Only thirteen trees were still standing by the mid-1970s, and by some estimates they are more than 300 years old. Each year they still produce seeds, but these apparently cannot break out of the seed coats without help. Today, botanists employ turkeys or gem polishers as substitute grinders.

b

What causes mass extinctions? The answer may not be obvious. For example, considerable evidence suggests that many lineages never made it past the K–T boundary (Section 21.1). Many believe "the asteroid did it." It did deliver the coup de grace for some lineages, including the dinosaurs and mosasaurs (Figure 28.3*a*). But other factors were at work. Biodiversity had been declining for 10 million years. Pterosaurs were gone before the K–T event. The event had little effect on insects. Was tectonic change a factor? When a land mass destined to become Australia was being rafted away from Antarctica, deep, cold currents from the south were able to move into the warmer, equatorial seas—where many groups were hit. Seawater composition, sea levels, and the climate itself shifted. The consequences affected life on land as well as in the seas. The K–T asteroid impact did indeed end some lineages abruptly. But for others, it may simply have been the final blow in a long streak of bad luck.

Now extend this thought about hidden causes of extinctions to individual species. Long ago, Dutch sailors clubbed to death every last dodo, a flightless bird that lived only on Mauritius. After that, a tree species also native to the island simply stopped reproducing. It was not until the 1970s that a hypothesis emerged: If the tree depended intimately on a coevolved species that became extinct, then the tree would be vulnerable to extinction, also. Was its partner the dodo? Maybe (Figure 28.3*b*).

Biodiversity is greater now than it has ever been in the past.

The current range of global biodiversity is an outcome of an overall pattern of mass extinctions and slow recoveries in the history of life. Within that pattern, lineages differ in which member species persisted and which became extinct.

The loss of individual species, as well as mass extinctions, may have obvious or complicated causes.

# THE NEWLY ENDANGERED SPECIES

No biodiversity-shattering asteroids have hit the Earth for 65 million years. *Yet the sixth major extinction event is under way.* Throughout the world, human activities are swiftly driving many species to extinction. Think of the mammalian lineage, which outlasted the dinosaurs. About 2 million years ago, early humans started to hunt mammals in earnest. About 11,000 years ago, they started to encroach on mammalian habitats in a big way; with agriculture, domesticated species became favored at the expense of wild stock. Think of the once-stupendous herds of bison, which nearly vanished in the 1800s. As adventurers moved westward, they shot too many bison for sport, sometimes from the platforms of trains.

Only 4,500 or so species of mammals made it to the present. Of those, 300 (including most whales, wild cats, otters, and primates other than humans) have the bad luck to be enrolled in the endangered species club. An **endangered species** is an endemic species extremely vulnerable to extinction. *Endemic* means it originated in only one geographic region and lives nowhere else.

We are just one species among millions. Yet because of our rapid population growth, we are threatening others by way of habitat losses, species introductions, overharvesting, and illegal wildlife trading. For the past forty years, combinations of these factors have raised extinction rates. Figure 28.4 shows the most threatened regions. In time, nature probably will again heal itself. But recoveries after mass extinctions are known to take millions of years. We certainly will not be around to congratulate ourselves on not having caused irrevocable harm, if and when healing is complete.

## Habitat Loss and Fragmentation

Edward O. Wilson defines **habitat loss** as the physical reduction in suitable places to live, as well as the loss of suitable habitat as an outcome of chemical pollution. Habitat loss is one of the major threats to more than 90 percent of endemic species facing extinction.

Biodiversity is greatest in the tropics. There, habitats on land are assaulted daily by deforestation, a topic you will read about in Sections 51.4 and 51.5. Grasslands, wetlands, and freshwater and marine habitats are also under attack. Section 28.3 gives one example. As other examples, 98 percent of the tallgrass prairie, 50 percent of the wetlands, and 85–95 percent of old-growth forests of the United States are gone.

Habitats also may become chopped up into isolated patches. Such **habitat fragmentation** has three effects on biodiversity. First, it increases a habitat's boundaries (edges), so species are more vulnerable to predators, temperature changes, winds, fires, and disease. Second, the patches might not be large enough to support the population sizes required for breeding. (Generally, each tenfold decrease in habitat area leads to a 50 percent reduction in the number of species.) Third, patches may not have enough food and other resources to sustain a population of the species.

**ISLAND BIOGEOGRAPHY AND HABITAT ISLANDS** Section 48.10 gives insight into factors that shape biodiversity on islands, which represent but a fraction of the Earth's surface. Nevertheless, about half of the plant and animal

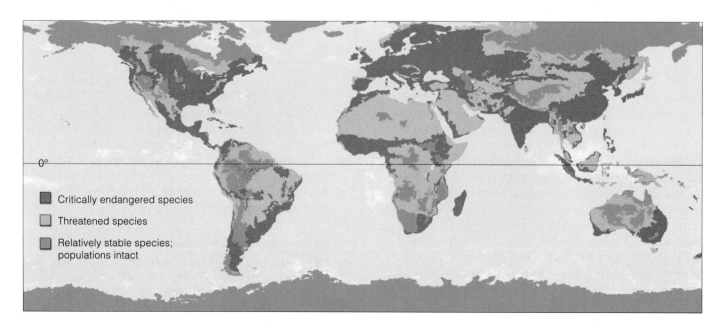

Critically endangered species

Threatened species

Relatively stable species; populations intact

**Figure 28.4** Threats to biodiversity: major regions of the world in which species are critically endangered, threatened, or relatively stable, as projected from 1998 through 2018.

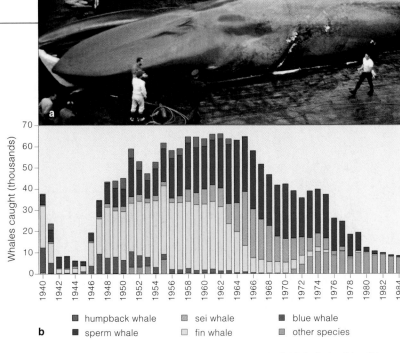

**Figure 28.5** (**a**) Blue whale being butchered in British Columbia. (**b**) Graph of whale harvesting that reflects the declining numbers of great whales, now at the brink of extinction. (**c**) Confiscated products made of endangered species. All but 10 percent of the illegal wildlife trade, which threatens more than 600 species of animals and plants, goes undetected. North American examples and their black market prices: live large saguaro cactus ($5,000–15,000), bighorn sheep head ($10,000–60,000), polar bear ($6,000), grizzly ($5,000), bald eagle ($2,500), and peregrine falcon ($10,000). Other examples: live chimpanzee ($50,000), live mountain gorilla ($150,000), live imperial Amazon macaw ($30,000), Bengal tiger hide ($100,000), snow leopard hide ($14,000), and rhinoceros horn ($28,600 per kilogram).

species that have become extinct since 1600 were native to islands, mainly because they had nowhere else to go.

R. MacArthur and Wilson have developed a model of island biogeography as a way to estimate the number of current and future extinctions in regions surrounded by logging, mining, urbanization, and other destructive activities. National parks, tropical forests, reserves, and lakes are examples. Think of them as **habitat islands** in a sea of unsuitable habitat. By the MacArthur–Wilson model, destruction of 50 percent of an island habitat or a habitat island will drive about 10 percent of endemic species to extinction. A 90 percent loss will drive about 50 percent of endemic species to extinction.

**INDICATOR SPECIES** Conservation biologists often study **indicator species**, which provide warning of changes in habitat and impending widespread loss of biodiversity. Birds are an example. Different types live in all major land regions and climate zones, they respond quickly to changes in habitats, and they are relatively easy to track.

Think of the 700 migratory species that make annual trips to and from their summer breeding grounds in North America. Many live most of the year in tropical forests. Biologists surveyed sixty-four species between

1978 and 1987. The population sizes of forty-four insect-eating, migratory songbird species declined. For twenty species, it plummeted 25 to 45 percent. Factors in the decline were deforestation of winter habitats and fragmentation of summer habitats. Intrusion of farms, freeways, and suburbs also cut forests into patches. In North America, this made it easier for skunks, opossums, squirrels, raccoons, and blue jays to feast on eggs and juveniles of migratory songbirds. About 69 percent of the 9,600 known species of birds are now facing habitat loss.

## Other Contributing Factors

Whether by accident or deliberate importation, a species sometimes moves from its home range and successfully insinuates itself into a different habitat. Each habitat has only so many resources, and its occupants compete for their share. Introduced species have adaptations that make them highly competitive, and they displace one or more endemic species. You will read about these *exotic* species in Sections 48.7 and 48.8. For now, it is enough to know that species introductions are a major factor in almost 70 percent of the cases where endemic species are being driven to extinction.

Overharvesting also is reducing biodiversity. Whales are an example (Figure 28.5*a*). Humans have killed them for food, lubricating oil, fuel, cosmetics, fertilizer, even corset stays. Substitutes exist for all whale products. Yet fishermen of several nations still ignore a moratorium on slaughtering large whales, and they routinely cross boundaries of sanctuaries set aside for their recovery.

Finally, in a sad commentary on human nature, the more rare a wild animal becomes, the more its value soars in the black market. Figure 28.5*b* gives examples.

---

An endangered species is any endemic species extremely vulnerable to extinction.

For the past forty years, human activities have raised rates of extinction through habitat losses, species introductions, overharvesting, and illegal wildlife trading.

Habitat loss is the physical reduction in suitable living spaces and habitat closure by chemical pollution.

---

Focus on the Environment

# 28.3 CASE STUDY: THE ONCE AND FUTURE REEFS

**Coral reefs** are wave-resistant formations that develop mainly in clear, warm waters between latitudes 25° north and south. All show spectacular biodiversity. All suffer losses from hurricanes, warming of ocean currents, and other natural calamities, but usually they recover within a few decades. Reefs also suffer from human assaults and are less likely to recover from them.

Each coral reef began with the accumulated remains of countless organisms. Hard parts of corals became the structural foundation. Coralline algae contributed to it; deposits of calcium and magnesium carbonate hardened the cell walls of these red algae, including *Corallina* (Figure 28.6). Secretions from other organisms helped cement things together.

The massive, pocketed spine of a present-day reef is home to hundreds of species of corals, and a staggering variety of red algae and other organisms. Figure 28.6 merely hints at the wealth of warning colors, spines, tentacles, and stealthy behavior—clues to dangers and fierce competition for resources among species that are packed together in limited space.

Dinoflagellates often live as symbionts in tissues of reef-building corals (Section 26.5). In return for protection, they provide a coral polyp with oxygen and recycle its mineral wastes. When stressed, the polyps expel their protistan symbionts. When stressed for more than a few months, they die, and only bleached hard parts remain.

Abnormal, widespread bleaching in the Caribbean and tropical Pacific began in the 1980s. So did increases in sea surface temperature, which may be a major stress factor. Is the damage one outcome of human activities that are contributing to global warming (Section 49.9)? If so, as marine biologists Lucy Bunkley-Williams and Ernest Williams suggest, the future looks grim for reefs.

Human activities are destroying reefs in more direct ways than this. For example, pollutants such as raw

**Figure 28.6** A small sampling of the stunning biodiversity of tropical reefs, such as the one shown at lower left.

sewage are being discharged into the nearshore waters around populated islands. Massive oil spills, as occurred during the Persian Gulf war, have calamitous effects. So do dredging operations and mining for coral rock.

Where commercial fishermen from Japan, Indonesia, and Kenya move in, reef life is being decimated. No simple nets for these fellows. They drop dynamite in the water, so fish hiding among the coral are blasted out and float dead to the surface. They also squirt sodium cyanide into the water to stun fish, which float to the surface. Some survivors end up as tropical fish in pet stores. The endangered species are transported to Asian restaurants, where they are dispatched and served up as exorbitantly expensive status symbols. On small, native-owned islands, fishing rights are traded for paltry sums. Then fishermen destroy the reefs, which can no longer sustain the small human populations that once depended on them for survival.

Reef biodiversity is in great danger around the world, in places extending from Australia and Southeast Asia to such distant places as the Hawaiian Islands, Galápagos Islands, the Gulf of Panama, Florida, and Kenya. Reef formations off Florida's Key Largo have already been diminished by one-third, mostly since 1970.

coral reef    island    lagoon    open ocean

**490**    Further reading: Student Guide to InfoTrac on web site →

LIONFISH

CRAB

MORAY EEL

CHAMBERED NAUTILUS

SEA ANEMONE

DAISY CORAL

PILLAR CORAL

CROWN-OF-THORNS SEA STAR

# 28.4  RACHEL'S WARNING

In 1951 Rachel Carson (Figure 28.7), an oceanographer and marine biologist, wrote a book about human impact on the ocean. *The Sea Around Us* was translated into thirty-two languages and sold more than 2 million copies. Seven years later, officials happened to spray mosquito-controlling DDT near the home and private bird sanctuary of one of her friends, who became agitated over the subsequent agonizing deaths of several birds. That friend begged Carson to find someone who might study the effects of pesticides on birds and other wildlife.

Carson discovered that independent, critical research on the subject was almost nonexistent. She conducted an extensive survey of the literature and then methodically developed information about the harmful effects of the widespread use of pesticides. In 1962, she published her findings. *Silent Spring*, the book's title, was an allusion to the silencing of "robins, catbirds, doves, jays, wrens, and scores of other bird voices" following pesticide exposure.

Many scientists, politicians, and policymakers read *Silent Spring*, and the public embraced it. Manufacturers of chemicals viewed Carson as a threat to their booming pesticide sales, and they swiftly mounted a campaign to discredit her. Critics and industry scientists claimed the book was full of inaccuracies, made selective and biased use of research findings, and failed to balance the picture with an account of the benefits of pesticides. Some even claimed that, as a woman, Carson simply was incapable of understanding such a highly technical subject. Others charged she was a hysterical woman and a radical nature lover trying to scare the public in order to sell books.

During those intense attacks, Carson knew she had terminal cancer. Yet she strongly defended her research and successfully countered her critics. She died in 1964—eighteen months after publication of *Silent Spring*—without knowing that her efforts would be the start of what is now known as the environmental movement in the United States. The new field of conservation biology is one outgrowth of that movement.

**Figure 28.7** Rachel Carson, who helped awaken the public's interest in human impact on nature. She died without knowing she helped start the environmental movement.

# 28.5  CONSERVATION BIOLOGY

All countries can count their wealth in three forms: material, cultural, and biological. Until recently, many undervalued their biological wealth—*their biodiversity*, which is a source of food, medicine, and other goods. No one knows how much countries have already lost.

For instance, in the 1970s, a Mexican college student poking around in Jalisco discovered *Zea diploperennis*, a wild species of maize. Unlike domesticated corn, it resists disease and lives for more than one growing season. Gene transfers from the wild species to *Z. mays* have the potential to enormously boost production of corn for hungry people in Mexico and elsewhere. That species was growing in a mountain habitat no larger than ten hectares (twenty-five acres). The student saved it in the nick of time. A week later, humans started to clear the habitat with machetes and controlled burns. They would have pushed *Z. diploperennis* into oblivion.

There is a monumental problem. In the next fifty years, human population size may reach 9 billion, with most growth proceeding in the developing countries. Like all organisms, each of those 9 billion people must have raw materials, energy, and living space. How many species will they crowd out? In Wilson's words, "An awful symmetry . . . binds the rise of humanity to the fall of biodiversity: the richest nations preside over the smallest and least interesting biotas, while the poorest nations, burdened by explosive population growth and little scientific knowledge, are stewards of the largest."

Awareness of the impending extinction crisis gave rise to **conservation biology**. We define this field of pure and applied research as (1) a systematic survey of the full range of biological diversity, (2) an attempt to decipher the evolutionary and ecological origins of diversity, and (3) an attempt to identify methods that might maintain and use biodiversity for the good of the human population. Its goal is to conserve and use, in sustainable fashion, as much biodiversity as possible.

## The Role of Systematics

Realistically, we simply don't have enough time, money, or people to complete a global survey. Possibly 99 percent of existing species have not even been discovered, and we don't know where most of them live. Thus initial work has focused on identifying **hot spots**. These are habitats with many species found nowhere else and in greatest danger of extinction because of human activities.

At the first survey level, researchers target a limited area, such as an isolated valley. A complete survey is out of the question, so they inventory birds, mammals, fishes, butterflies, and other indicator species for the habitat. At the next level, broader areas that are major or multiple hot spots are systematically explored. The widely separated forests of Mexico are an example of a

Further reading: Student Guide to InfoTrac on web site →

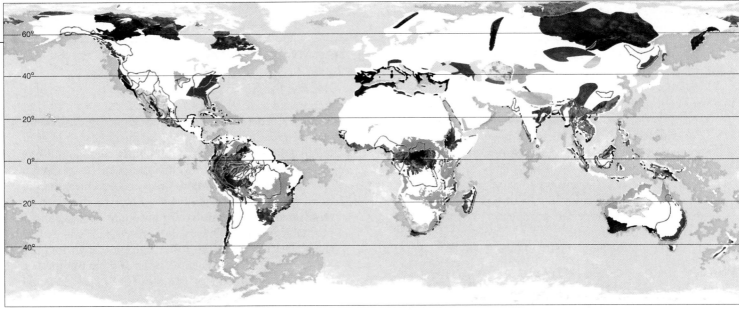

**Figure 28.8** One map of the most vulnerable areas of land and seas, compiled by the World Wildlife Fund.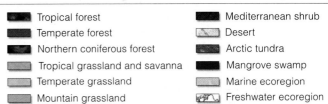

| | | | |
|---|---|---|---|
| Tropical forest | | Mediterranean shrub |
| Temperate forest | | Desert |
| Northern coniferous forest | | Arctic tundra |
| Tropical grassland and savanna | | Mangrove swamp |
| Temperate grassland | | Marine ecoregion |
| Mountain grassland | | Freshwater ecoregion |

major hot spot. Research stations are set up at different latitudes and elevations across the region.

At the highest survey level, hot spot inventories are combined with existing information on biodiversity of ecoregions. An **ecoregion** is a broad land or ocean region defined by climate, geography, and producer species. For example, Figure 28.8 is a work in progress by the World Wildlife Fund. Currently this organization has identified 233 regions crucial to global biodiversity (136 land, 36 fresh water, and 61 marine). Of these, it selected 25 for immediate conservation efforts. Such efforts on behalf of biodiversity will be refined over the coming decades.

### Bioeconomic Analysis

The systematic expansion of species inventories serves as a baseline for economic analysis—that is, for assigning future value to ecoregions. This concept is something new in human history. For example, when the ancestors of Polynesians journeyed eastward from Southeast Asia, they literally ate their way across the Pacific, through New Zealand, Tonga, and Easter Island, all the way to the Hawaiian Islands. Flightless birds, eagles, and other endemic species that had no evolutionary experience with humans didn't stand a chance. Another wave of extinction accompanied the expansion of Paleo-Indians throughout the Americas. What counted in the past, as counts today, is food on the table for self and family. It counted for the Mexican truckdriver who said, after shooting one of the two remaining imperial woodpeckers on Earth, "It was a great piece of meat."

Probably, short-sighted thinking can be countered only by showing economically pressed individuals that sustaining biodiversity has more economic value than destroying it. Should governments—and individuals— protect a habitat, withdraw resources in a sustainable

way, or obliterate it? Answers will require analysis of what a threatened species may offer. For example, some may be sources of new medicines and other chemical products. Developing countries do not have advanced laboratories for chemical analysis. Large pharmaceutical companies do. One of the largest has been paying Costa Rica's National Institute of Biodiversity to collect and identify promising species and send in chemical samples extracted from them. If natural chemicals end up being marketed, Costa Rica will get a share of the royalties, which are earmarked for conservation programs.

### Sustainable Development

Ultimately, biodiversity will be best protected when its species can be used over the long term for the good of local economies. Identifying and implementing such uses is easier said than done. The technical problems and social barriers are huge. However, there are a few success stories, as the next section describes.

---

Conservation biology entails a systematic survey of the full range of biodiversity, analysis of its evolutionary and ecological origins, and identification of methods to maintain and use biodiversity for the benefit of humans.

Researchers identify local hot spots (limited areas with many endemic species in greatest danger of extinction by human activities), then broader or multiple hot spots, which they combine into a global picture of biodiversity.

To counter economic demands of the human population, the future economic value of biodiversity must be determined.

---

# RECONCILING BIODIVERSITY WITH HUMAN DEMANDS

The fact remains that, when people locked in poverty are confronted with the choice of saving endangered species or themselves, the species will lose. Deprived of the education and resources that people in developed countries take for granted, they destroy tropical forests. They hunt animals and dig up plants in "protected" parks and reserves. They try to raise crops and herds on marginal land. They will take the last tree, whale, or primate, the last imperial woodpecker. We can expect people to stop only when they are shown ways to earn a living from the biodiversity in threatened habitats.

## A Case for Strip Logging

Section 51.4 describes the staggering pace of destruction of once-great tropical rain forests, the richest sources of biodiversity on land. For now, simply think about one concept of how to minimize the destruction.

Besides providing wood for local economies, trees of tropical rain forests yield diverse, exotic woods prized by developed countries. What if they could be logged in a profitable, sustainable way that preserved biodiversity? Gary Hartshorn was the first to propose **strip logging** in portions of forests that are sloped and have a number of streams (Figure 28.9). The idea is to clear a narrow corridor that parallels contours of the land, and use the upper part of it as a road to haul away logs. After a few years, saplings start to grow in the corridor. Another corridor is cleared above the road. Precious nutrients that runoff leaches from the exposed soil trickle into the first corridor. There they are taken up by saplings, which benefit from all the nutrient input by growing more rapidly. Later, a third corridor is cut above the second one —and so on in a profitable cycle of logging, which the habitat sustains over time.

uncut forest

cut 1 year ago

dirt road

cut 3–5 years ago

cut 6–10 years ago

uncut forest

stream in watershed

**Figure 28.9** Strip logging, a practice that might sustain biodiversity even as it enhances the regeneration of economically valued forest trees.

## Ranching and the Riparian Zones

Look again at the Figure 28.8 map, and you sense at once that the biodiversity crisis is not confined to Third World countries. Also take a look at Figure 28.10, which shows a riparian zone in the western United States. Each **riparian**

**Figure 28.10** Riparian zone along the San Pedro River in Arizona, shown before and after restoration.

zone is a relatively narrow corridor of vegetation along a stream or river. Its plants afford a major line of defense against flood damage by sponging up water during spring runoffs and summer storms. Shade from the canopy of taller shrubs and trees helps conserve water during droughts. Riparian zones also offer food, shelter, and shade for wildlife, especially in arid and semiarid regions. For example, in the western United States, 67 to 75 percent of the endemic species must spend all or part of their life cycle in riparian zones. They include more than 136 species of songbirds, some of which nest only in riparian vegetation.

Cattle raised in the American West provide beef for much of the human population. Compared with wild ungulates, cattle drink a lot more water, so they tend to congregate at rivers and streams. There they trample and feed until grasses and herbaceous shrubs are gone. It takes only a few head of cattle to destroy riparian vegetation. All but 10 percent of the riparian vegetation of Arizona and New Mexico has already disappeared, mainly into the stomachs of grazing cattle.

Restricted access and development of watering sites for cattle away from streams can help conserve riparian zones. So can the rotation of livestock and provision of supplemental feed at different grazing areas. Cattle ranchers resist implementing such measures. Putting in more fencing, for example, is costly.

Jack Turnell is one rancher who knows almost as much about riparian biodiversity as he knows about cattle. Turnell runs cattle on his 32,000-hectare (80,000-acre) ranch south of Cody, Wyoming, and on 16,000 hectares (40,000 acres) of Forest Service land for which he has grazing rights. Ten years after he took over the ranch, he decided to raise cattle in an unconventional way. He began systematically rotating cattle away from the riparian areas. He also crossed Hereford and Angus cattle with a breed from France that does not consume as much water. He made most of his ranching decisions in consultation with specialists in range and wildlife management.

Willows and other plants are sustaining diversity in the restored zones. And Turnell's ranching approach is profitable; grasses maintained in the riparian zones help put more meat on his cattle. His may be only a small step toward sustainable ranching, but it appears to be a step in the right direction.

---

**Throughout much of the world, the unprecedented rate of human population growth is the elephant in the room that no one wants to talk about.**

**Protecting biodiversity depends on finding ways for people to make a living from it without destroying it. Can such a balance be struck among so many billions of people?**

---

# SUMMARY

1. Global biodiversity is greater now than ever, but the current rate of species losses is high enough to suggest that an extinction crisis is under way.

2. The history of life indicates that, after global mass extinctions, 20 million to 100 million years pass before biodiversity recovers to the preceding level.

   a. Different taxa have different evolutionary histories. Some were eliminated by major extinction events. Others passed through them relatively unscathed.

   b. Mass extinctions, and extinction of single species, may have obvious or hidden, complicated causes.

3. Growth of the human population, which is projected to reach 9 billion within the next fifty years, is causing the sixth extinction crisis. That growth is most rapid in regions with the richest, most vulnerable biodiversity.

4. Habitat losses, habitat fragmentation, introduction of species into novel habitats, overharvesting, and illegal wildlife trading threaten endemic species. "Endemic" means it originated in and is presently restricted to one geographic region. An endangered species is defined as an endemic species extremely vulnerable to extinction.

5. Habitat loss refers to physical reduction or chemical pollution of suitable places for species to live. Habitat fragmentation is carving a habitat into isolated patches. It puts species at risk, as by chopping up populations to sizes that cannot promote successful breeding.

6. Island biogeography models help predict the number of current and future extinctions. A habitat island is a habitat in a sea of possibly destructive human activities, such as logging. Generally, destruction of 50 percent of an island habitat (or habitat island) will drive one-tenth of its species to extinction. Destruction of 90 percent will drive one-half to extinction.

7. Indicator species are birds and other easily tracked species that can provide warning of changes in habitats and impending widespread loss of biodiversity.

8. Humans are directly destroying coral reefs and other regions, as when commercial fishermen dynamite reefs to harvest fish. They may be destroying them indirectly, as by contributing to global warming (with concurrent rises in sea surface temperatures and sea levels).

9. Conservation biology is a field of pure and applied research. Its goal is to conserve and use biodiversity in sustainable ways. The field encompasses:

   a. A systematic survey at three ever more inclusive levels of the full range of biodiversity.

   b. Analysis of biodiversity's origins in evolutionary and ecological terms.

   c. Identification of methods that might maintain and use biodiversity for the good of the human population, which may otherwise destroy it.

10. Today the systematic survey of conservation biology is proceeding at three levels:

a. Local hot spots (for example, an isolated valley) are identified and indicator species inventoried. The hot spots are habitats with many species in greatest danger of extinction because of human activities.

b. Major hot spots (or multiple ones) are inventoried. Research stations are set up across these broader areas to gather data by latitude and elevation.

c. Data gathered at the first two levels are combined with data on ecoregions; these are the most vulnerable, broad regions of land and seas throughout the world.

d. Data gathered for such maps will be refined into an increasingly detailed picture of global biodiversity.

11. Protecting biodiversity depends on finding ways for people to make a living from it without destroying it. To counter growing economic demands, biodiversity's future economic value must be determined. Methods by which local economies can exploit that biodiversity in sustainable ways must be developed.

12. When people are given a choice between saving a species or themselves, that species will lose. Currently, millions of people are making such choices because of their demands for food, shelter, and material goods.

## Review Questions

1. How many major mass extinctions have occurred, including the one that is now under way? 28.1

2. List four human activities that are major contributors to the current extinction crisis. Which threaten coral reefs? 28.2, 28.3

3. Define endangered, endemic, and indicator species. 28.2

4. Distinguish between island habitat and habitat island. 28.2

5. State the goal of conservation biology and briefly describe its three-pronged approach to achieving that goal. 28.5

6. Define hot spot. Why are conservation biologists focusing on hot spots rather than quickly completing a global survey? 28.5

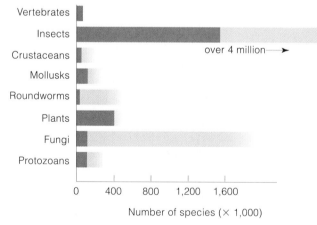

**Figure 28.11** Current species diversity for a few major taxa. *Red* indicates the number of named species; *gold* indicates the estimated number of species not yet discovered.

## Self-Quiz (Answers in Appendix III)

1. Following mass extinctions in the past, recovery to the same level of biodiversity has taken many _____ of years.
   a. hundreds     b. millions     c. billions

2. Species may be driven to extinction by _____ .
   a. obvious factors              c. human activities
   b. complicated, hidden factors  d. all of the above

3. The goal(s) of conservation biology is (are) to _____ .
   a. conduct a three-level, systematic survey of all biodiversity
   b. analyze biodiversity's evolutionary and ecological origins
   c. identify ways to maintain and use biodiversity for people
   d. all of the above

4. Strip logging _____ .
   a. destroys wild habitats     c. sustains forests
   b. is profitable              d. both b and c

## Critical Thinking

1. Figure 28.11 compares the current estimated number of species for some major taxa. Given what you learned about taxa from chapters in this unit, which are most vulnerable in the current extinction crisis? Which are likely to pass through it with much of their biodiversity intact? List some reasons why (for example, compare the distribution of member species).

2. Visit or study a riparian zone near where you live. Imagine visiting it five years from now, then ten years after that. Given its location, what kinds of changes do you predict for it?

3. Mentally transport yourself to a tropical rain forest of South America. Imagine you don't have a job. There are no jobs in sight, not even in overcrowded cities some distance away. You have no money or contacts to get you anywhere else. Yet you are the sole supporter of your large family. Today a stranger approaches you. He tells you he will pay good money if you can capture alive a certain brilliantly feathered parrot in the forest. You know the parrot is rare; it is seldom seen. However, you have an idea of where it lives. What will you do?

4. In his review of this chapter, ecologist Robin Tyser offered these comments: Students might find the current biodiversity crisis too overwhelming to contemplate. But people are making a difference. For example, they helped reestablish viable bald eagle populations in the continental United States and wolves in northern Wisconsin. Daniel Janzen is working to recreate a dry forest ecosystem in Costa Rica. Do a literature search or computer research and report on one or two success stories that you personally might consider inspiring.

## Selected Key Terms

conservation          endangered species 28.2    hot spot 28.5
  biology 28.5        habitat fragmentation 28.2  indicator species 28.2
coral reef 28.3       habitat island 28.2         riparian zone 28.6
ecoregion 28.5        habitat loss 28.2           strip logging 28.6

## Readings  See also www.infotrac-college.com

Cox, G. 1999. *Alien Species in North America and Hawaii: Impacts on Natural Ecosystems.* Washington, D.C.: Island Press.

Wilson, Edward O. 1992. *The Diversity of Life.* Cambridge, Massachusetts: Belknap Press. This slender book proves that Wilson is one of the most eloquent defenders of life's diversity.

FACING PAGE: *A flowering plant (Prunus) busily doing what it does best: producing flowers for the fine art of reproduction.*

# APPENDIX I.   CLASSIFICATION SCHEME

The classification scheme that follows is a composite of several that microbiologists, botanists, and zoologists use. Major groupings are agreed upon, more or less. There is not always agreement on what to name a particular grouping or where it may fit within the overall hierarchy. There are several reasons for the lack of total consensus.

First, the fossil record varies in its quality and in its completeness. Therefore, the phylogenetic relationship of one group to others is sometimes open to interpretation. Comparative studies at the molecular level are firming up the picture, but this work is still under way.

Second, ever since the time of Linnaeus, classification schemes have been based on the perceived morphological similarities and differences among organisms. Although some original interpretations are now open to question, we are so used to thinking about organisms in certain ways that reclassification often proceeds slowly.

For example, birds and reptiles traditionally have been placed in separate classes (Reptilia and Aves), yet there are many compelling arguments for grouping the lizards and snakes in one class and the crocodilians, dinosaurs, and birds in a separate class. Some favor six kingdoms but others favor three domains: archaebacteria, eubacteria, and eukaryota (alternatively, archaea, bacteria, and eukarya).

Third, researchers in microbiology, mycology, botany, zoology, and the other fields of biological inquiry have inherited a wealth of literature, based on classification schemes that were developed over time in each of those fields. Many see no good reason to give up the established terminology and thereby disrupt access to the past.

For instance, many microbiologists and botanists use *division*, and zoologists *phylum*, for taxa that are equivalent in the hierarchy of classification. Also, opinions are still polarized with respect to the kingdom Protista, certain members of which could just as easily be grouped with plants, fungi, or animals. Indeed, the term protozoan is a holdover from an earlier scheme in which some single-celled organisms were ranked as simple animals.

Given the problems, why do we even bother imposing artificial frameworks on the history of life? We do this for the same reason that a writer might decide to break up the history of civilization into several volumes, a number of chapters, and many paragraphs. Both efforts are attempts to impart obvious structure to what might otherwise be an overwhelming body of knowledge and to enhance the retrieval of information from it.

Finally, bear in mind that we include this classification scheme primarily for your reference purposes. Besides being open to revision, it also is by no means complete. It does not include the most recently discovered species, as from the mid-ocean. Many existing and extinct organisms of the so-called lesser phyla are not represented here. Our strategy is to focus mainly on organisms mentioned in the text. A few examples of organisms also are listed.

**SUPERKINGDOM PROKARYOTA**.   Prokaryotes. Almost all microscopic species. DNA organized at nucleoid (a region of cytoplasm), not inside a membrane-bound nucleus. All are bacteria, either single cells or simple associations of cells. Autotrophs and heterotrophs. Table A on the following page lists representative types. Reproduce by prokaryotic fission, sometimes by budding and by bacterial conjugation.

The authoritative reference in bacteriology, *Bergey's Manual of Systematic Bacteriology*, calls this "a time of taxonomic transition." It groups bacteria mostly by numerical taxonomy (Section 22.6), not on phylogeny. The scheme presented here reflects strong evidence of evolutionary relationships for at least some bacterial groupings.

**KINGDOM EUBACTERIA**.  Gram-negative, gram-positive forms. Peptidoglycan present in cell wall. Photosynthetic autotrophs, chemosynthetic autotrophs, and heterotrophs.

PHYLUM GRACILICUTES.   Typical Gram-negative, thin wall. Autotrophs (photosynthetic and chemosynthetic) and heterotrophs. *Anabaena* and other cyanobacteria. *Escherichia, Pseudomonas, Neisseria, Myxococcus.*

PHYLUM FIRMICUTES.   Typical Gram-positive, thick wall. Heterotrophs. *Bacillus, Staphylococcus, Streptococcus, Clostridium, Actinomycetes.*

PHYLUM TENERICUTES.   Gram-negative, wall absent. Heterotrophs (saprobes, pathogens). *Mycoplasma.*

**KINGDOM ARCHAEBACTERIA**.  Methanogens, extreme halophiles, extreme thermophiles. Evolutionarily closer to eukaryotic cells than to eubacteria. Strict anaerobes. Distinctive cell wall, membrane lipids, ribosomes, RNA sequences. *Methanobacterium, Halobacterium, Sulfolobus.*

**SUPERKINGDOM EUKARYOTA**.   Eukaryotes. Both single-celled and multicelled species. Cells start out life with a nucleus (encloses the DNA) and usually other membrane-bound organelles. Chromosomes have many histones and other proteins attached.

**KINGDOM PROTISTA**.   Diverse single-celled, colonial, and multicelled eukaryotic species. Existing species are unlike bacteria in characteristics and are most like the earliest, structurally simple eukaryotes. Autotrophs, heterotrophs, or both (Table 23.3). Reproduce sexually and asexually (by meiosis, mitosis, or both). Many related evolutionarily to plants, fungi, and possibly animals.

PHYLUM CHYTRIDIOMYCOTA.   Chytrids. Heterotrophs; saprobic decomposers or parasites. *Chytridium.*

PHYLUM OOMYCOTA.   Water molds. Heterotrophs. Decomposers, some parasites. *Saprolegnia, Phytophthora, Plasmopara.*

PHYLUM ACRASIOMYCOTA.   Cellular slime molds. Heterotrophs with free-living, phagocytic amoeboid cells and spore-bearing stages. *Dictyostelium.*

PHYLUM MYXOMYCOTA.   Plasmodial slime molds. Heterotrophs with free-living, phagocytic amoeboid cells and spore-bearing stages. Aggregate into streaming mass of cells that discard plasma membranes. *Physarum.*

## Table A Representative Eubacteria and Archaebacteria Grouped on the Basis of Numerical Taxonomy

| Some Major Groups | Main Habitats | Characteristics | Representatives |
|---|---|---|---|
| **EUBACTERIA** | | | |
| *Photoautotrophs:* | | | |
| Cyanobacteria, green sulfur bacteria, and purple sulfur bacteria | Mostly lakes, ponds; some marine, terrestrial habitats | Photosynthetic; use sunlight energy, carbon dioxide; cyanobacteria use oxygen-producing noncyclic pathway; some also use cyclic route | *Anabaena, Nostoc, Rhodopseudomonas, Chloroflexus* |
| *Photoheterotrophs:* | | | |
| Purple nonsulfur and green nonsulfur bacteria | Anaerobic, organically rich muddy soils, and sediments of aquatic habitats | Use sunlight energy; organic compounds as electron donors; some purple nonsulfur may also grow chemotrophically | *Rhodospirillum, Chlorobium* |
| *Chemoautotrophs:* | | | |
| Nitrifying, sulfur-oxidizing, and iron-oxidizing bacteria | Soil; freshwater, marine habitats | Use carbon dioxide, inorganic compounds as electron donors; influence crop yields, cycling of nutrients in ecosystems | *Nitrosomonas, Nitrobacter, Thiobacillus* |
| *Chemoheterotrophs:* | | | |
| Spirochetes | Aquatic habitats; parasites of animals | Helically coiled, motile; free-living and parasitic species; some major pathogens | *Spirochaeta, Treponema* |
| Gram-negative aerobic rods and cocci | Soil, aquatic habitats; parasites of animals, plants | Some major pathogens; some fix nitrogen (e.g., *Rhizobium*) | *Pseudomonas, Neisseria, Rhizobium, Agrobacterium* |
| Gram-negative facultative anaerobic rods | Soil, plants, animal gut | Many major pathogens; one bioluminescent (*Photobacterium*) | *Salmonella, Escherichia, Proteus, Photobacterium* |
| Rickettsias and chlamydias | Host cells of animals | Intracellular parasites; many pathogens | *Rickettsia, Chlamydia* |
| Myxobacteria | Decaying organic material; bark of living trees | Gliding, rod-shaped; aggregation and collective migration of cells | *Myxococcus* |
| Gram-positive cocci | Soil; skin and mucous membranes of animals | Some major pathogens | *Staphylococcus, Streptococcus* |
| Endospore-forming rods and cocci | Soil; animal gut | Some major pathogens | *Bacillus, Clostridium* |
| Gram-positive nonsporulating rods | Fermenting plant, animal material; gut, vaginal tract | Some important in dairy industry, others major contaminators of milk, cheese | *Lactobacillus, Listeria* |
| Actinomycetes | Soil; some aquatic habitats | Include anaerobes and strict aerobes; major producers of antibiotics | *Actinomyces, Streptomyces* |
| **ARCHAEBACTERIA (ARCHAEA)** | | | |
| Methanogens | Anaerobic sediments of lakes, swamps; animal gut | Chemosynthetic; methane producers; used in sewage treatment facilities | *Methanobacterium* |
| Extreme halophiles | Brines (extremely salty water) | Heterotrophic; also, unique photosynthetic pigments (bacteriorhodopsin) form in some | *Halobacterium* |
| Extreme thermophiles | Acidic soil, hot springs, hydrothermal vents | Heterotrophic or chemosynthetic; use inorganic substances as electron donors | *Sulfolobus, Thermoplasma* |

PHYLUM SARCODINA.   Amoeboid protozoans. Heterotrophs, free-living or endosymbiotic, some pathogens. Soft-or shelled bodies, locomotion by pseudopods. The rhizopods (naked amoebas, foraminiferans), *Amoeba proteus, Entomoeba*. Also actinopods (radiolarians, heliozoans).

PHYLUM CILIOPHORA.   Ciliated protozoans. Heterotrophs, predators or symbionts, some parasitic. All have cilia. Free-living, sessile, or motile. *Paramecium, Didinium*, hypotrichs.

PHYLUM MASTIGOPHORA.   Animal-like flagellated protozoans. Heterotrophs, free-living, many internal parasites. All with one to several flagella. *Trypanosoma, Trichomonas, Giardia*.

APICOMPLEXA.   Heterotrophs, sporozoite-forming parasites. Complex structures at head end. Most familiar members called sporozoans. *Cryptosporidium, Plasmodium, Toxoplasma*.

PHYLUM EUGLENOPHYTA.   Euglenoids. Mostly heterotrophs, some autotrophs (photosynthetic), some switch depending on environmental conditions. Most with one short, one long flagellum; red, green, or colorless. *Euglena, Peranema*.

PHYLUM PYRRHOPHYTA.   Dinoflagellates. Photosynthetic, mostly, but some heterotrophs. *Fiesteria, Gymnodinium breve*.

PHYLUM CHRYSOPHYTA.   Golden algae, yellow-green algae, diatoms. Photosynthetic. Some flagellated, others not. *Mischococcus, Synura, Vaucheria*.

PHYLUM RHODOPHYTA.   Red algae. Mostly photosynthetic, some parasitic. Nearly all marine, some in freshwater habitats. *Porphyra. Bonnemaisonia, Euchema*.

PHYLUM PHAEOPHYTA.   Brown algae. Photosynthetic, nearly all in temperate or marine waters. *Macrocystis, Fucus, Sargassum, Ectocarpus, Postelsia*.

PHYLUM CHLOROPHYTA.   Green algae. Mostly photosynthetic, some parasitic. Most freshwater, some marine or terrestrial. *Chlamydomonas, Spirogyra, Ulva, Volvox, Codium, Halimeda*.

PHYLUM ZYGOMYCOTA. Zygomycetes. Zygosporangia (zygote inside thick wall) formed by sexual reproduction. Bread molds, related forms. *Rhizopus, Philobolus.*

PHYLUM ASCOMYCOTA. Ascomycetes. Sac fungi. Sac-shaped cells form sexual spores (ascospores). Most yeasts and molds, morels, truffles. *Saccharomycetes, Morchella, Neurospora, Sarcoscypha, Claviceps, Ophiostoma, Candida, Aspergillus, Penicillium.*

PHYLUM BASIDIOMYCOTA. Basidiomycetes. Club fungi. Most diverse group. Produce basidiospores inside club-shaped structures. Mushrooms, shelf fungi, stinkhorns. *Agaricus, Amanita, Craterellus, Gymnophilus, Puccinia, Ustilago.*

IMPERFECT FUNGI. Sexual spores absent or undetected. The group has no formal taxonomic status. If better understood, a given species might be grouped with sac fungi or club fungi. *Arthobotrys, Histoplasma, Microsporum, Verticillium.*

LICHENS. Mutualistic interactions between fungal species and a cyanobacterium, green alga, or both. *Lobaria, Usnea, Cladonia.*

PHYLUM RHYNIOPHYTA. Earliest known vascular plants; muddy habitats. Extinct. *Cooksonia, Rhynia.*

PHYLUM PROGYMNOSPERMOPHYTA. Progymnosperms. Ancestral to early seed-bearing plants; extinct. *Archaeopteris.*

PHYLUM PTERIDOSPERMOPHYTA. Seed ferns. Fernlike gymnosperms; extinct. *Medullosa.*

PHYLUM CHAROPHYTA. Stoneworts.

PHYLUM BRYOPHYTA. Bryophytes: mosses, liverworts, hornworts. Seedless, nonvascular, haploid dominance. *Marchantia, Polytrichum, Sphagnum.*

PHYLUM PSILOPHYTA. Whisk ferns. Seedless, vascular. No obvious roots, leaves on sporophyte. *Psilotum.*

PHYLUM LYCOPHYTA. Lycophytes, club mosses. Seedless, vascular. Leaves, branching rhizomes, vascularized roots and stems. *Lepidodendron* (extinct), *Lycopodium, Selaginella.*

PHYLUM SPHENOPHYTA. Horsetails. Seedless, vascular. Some sporophyte stems photosynthetic, others nonphotosynthetic, spore-producing. *Calamites* (extinct), *Equisetum.*

PHYLUM PTEROPHYTA. Ferns. Largest group of seedless vascular plants (12,000 species), mainly tropical, temperate habitats. *Pteris, Trichomanes, Cyathea* (tree ferns), *Polystichum.*

PHYLUM CYCADOPHYTA. Cycads. Gymnosperm group (vascular, bears "naked" seeds). Tropical, subtropical. Palm-shaped leaves, simple cones on male and female plants. *Zamia.*

PHYLUM GINKGOPHYTA. Ginkgo (maidenhair tree). Type of gymnosperm. Seeds with fleshy outer layer. *Ginkgo.*

PHYLUM GNETOPHYTA. Gnetophytes. Only gymnosperms with vessels in xylem and double fertilization (but endosperm does not form). *Ephedra, Welwitchia.*

PHYLUM CONIFEROPHYTA. Conifers. Most common and familiar gymnosperms. Generally cone-bearing species with needle-like or scale-like leaves.
Family Pinaceae. Pines, firs, spruces, hemlock, larches, Douglas firs, true cedars. *Abies, Cedrus, Pinus, Pseudotsuga.*
Family Cupressaceae. Junipers, cypresses. *Cupressus, Juniperus.*
Family Taxodiaceae. Bald cypress, redwoods, bigtree, dawn redwood. *Metasequoia, Sequoia, Sequoiadendron, Taxodium.*
Family Taxaceae. Yews. *Taxus.*

PHYLUM ANTHOPHYTA. Angiosperms (the flowering plants). Largest, most diverse group of vascular seed-bearing plants. Only organisms that produce flowers, fruits.
Class Dicotyledonae. Dicotyledons (dicots). Some families of some representative orders are listed:
Family Nymphaeaceae. Water lilies.
Family Papaveraceae. Poppies.
Family Brassicaceae. Mustards, cabbages, radishes.
Family Malvaceae. Mallows, cotton, okra, hibiscus.
Family Solanaceae. Potatoes, eggplant, petunias.
Family Salicaceae. Willows, poplars.
Family Rosaceae. Roses, apples, almonds, strawberries.
Family Fabaceae. Peas, beans, lupines, mesquite.
Family Cactaceae. Cacti.
Family Euphorbiaceae. Spurges, poinsettia.
Family Cucurbitaceae. Gourds, melons, cucumbers, squashes.
Family Apiaceae. Parsleys, carrots, poison hemlock.
Family Aceraceae. Maples.
Family Asteraceae. Composites. Chrysanthemums, sunflowers, lettuces, dandelions.
Class Monocotyledonae. Monocotyledons (monocots). Some families of several different orders are listed:
Family Liliaceae. Lilies, hyacinths, tulips, onions, garlic.
Family Iridaceae. Irises, gladioli, crocuses.
Family Orchidaceae. Orchids.
Family Arecaceae. Date palms, coconut palms.
Family Cyperaceae. Sedges.
Family Poaceae. Grasses, bamboos, corn, wheat, sugarcane.
Family Bromeliaceae. Bromeliads, pineapples, Spanish moss.

PHYLUM PLACOZOA. Marine. Simplest known animal. Two cell layers, no mouth, no organs. *Trichoplax.*

PHYLUM MESOZOA. Ciliated, wormlike parasites, about the same level of complexity as *Trichoplax.*

PHYLUM PORIFERA. Sponges. No symmetry, tissues. *Euplectella.*

PHYLUM CNIDARIA. Radial symmetry, tissues, nematocysts.
Class Hydrozoa. Hydrozoans. *Hydra, Obelia, Physalia, Prya.*
Class Scyphozoa. Jellyfishes. *Aurelia.*
Class Anthozoa. Sea anemones, corals. *Telesto.*

PHYLUM CTENOPHORA. Comb jellies. Modified radial symmetry.

PHYLUM PLATYHELMINTHES. Flatworms. Bilateral, cephalized; simplest animals with organ systems. Saclike gut.
Class Turbellaria. Triclads (planarians), polyclads. *Dugesia.*
Class Trematoda. Flukes. *Clonorchis, Schistosoma.*
Class Cestoda. Tapeworms. *Diphyllobothrium, Taenia.*

PHYLUM NEMERTEA. Ribbon worms. *Tubulanus.*

PHYLUM NEMATODA. Roundworms. *Ascaris, Caenorhabditis elegans, Necator* (hookworms), *Trichinella.*

PHYLUM ROTIFERA. Rotifers. *Asplancha, Philodina.*

PHYLUM MOLLUSCA. Mollusks.
Class Polyplacophora. Chitons. *Cryptochiton, Tonicella.*

Class Gastropoda. Snails (periwinkles, whelks, limpets, abalones, cowries, conches, nudibranchs, tree snails, garden snails), sea slugs, land slugs. *Aplysia, Ariolimax, Cypraea, Haliotis, Helix, Liguus, Limax, Littorina, Patella.*

Class Bivalvia. Clams, mussels, scallops, cockles, oysters, shipworms. *Ensis, Chlamys, Mytelus, Patinopectin.*

Class Cephalopoda. Squids, octopuses, cuttlefish, nautiluses. *Dosidiscus, Loligo, Nautilus, Octopus, Sepia.*

PHYLUM BRYOZOA.   Bryozoans (moss animals).

PHYLUM BRACHIOPODA.   Lampshells.

PHYLUM ANNELIDA.   Segmented worms.

Class Polychaeta. Mostly marine worms. *Eunice, Neanthes.*

Class Oligochaeta. Mostly freshwater and terrestrial worms, but many marine. *Lumbricus* (earthworms), *Tubifex.*

Class Hirudinea. Leeches. *Hirudo, Placobdella.*

PHYLUM TARDIGRADA.   Water bears.

PHYLUM ONYCHOPHORA.   Onychophorans. *Peripatus.*

PHYLUM ARTHROPODA.

Subphylum Trilobita. Trilobites; extinct.

Subphylum Chelicerata. Chelicerates. Horseshoe crabs, spiders, scorpions, ticks, mites.

Subphylum Crustacea. Shrimps, crayfishes, lobsters, crabs, barnacles, copepods, isopods (sowbugs).

Subphylum Uniramia.
Superclass Myriapoda. Centipedes, millipedes.
Superclass Insecta.
Order Ephemeroptera. Mayflies.
Order Odonata. Dragonflies, damselflies.
Order Orthoptera. Grasshoppers, crickets, katydids.
Order Dermaptera. Earwigs.
Order Blattodea. Cockroaches.
Order Mantodea. Mantids.
Order Isoptera. Termites.
Order Mallophaga. Biting lice.
Order Anoplura. Sucking lice.
Order Homoptera. Cicadas, aphids, leafhoppers, spittlebugs.
Order Hemiptera. Bugs.
Order Coleoptera. Beetles.
Order Diptera. Flies.
Order Mecoptera. Scorpion flies. *Harpobittacus.*
Order Siphonaptera. Fleas.
Order Lepidoptera. Butterflies, moths.
Order Hymenoptera. Wasps, bees, ants.

PHYLUM ECHINODERMATA.   Echinoderms.
Class Asteroidea. Sea stars. *Asterias.*
Class Ophiuroidea. Brittle stars.
Class Echinoidea. Sea urchins, heart urchins, sand dollars.
Class Holothuroidea. Sea cucumbers.
Class Crinoidea. Feather stars, sea lilies.
Class Concentricycloidea. Sea daisies.

PHYLUM HEMICHORDATA.   Acorn worms.

PHYLUM CHORDATA.   Chordates.

Subphylum Urochordata. Tunicates, related forms.

Subphylum Cephalochordata. Lancelets.

Subphylum Vertebrata. Vertebrates.
Class Agnatha. Jawless vertebrates (lampreys, hagfishes).
Class Placodermi. Jawed, heavily armored fishes; extinct.
Class Chondrichthyes. Cartilaginous fishes (sharks, rays, skates, chimaeras).
Class Osteichthyes. Bony fishes.
Subclass Dipnoi. Lungfishes.
Subclass Crossopterygii. Coelacanths, related forms.
Subclass Actinopterygii. Ray-finned fishes.
Order Acipenseriformes. Sturgeons, paddlefishes.
Order Salmoniformes. Salmon, trout.
Order Atheriniformes. Killifishes, guppies.
Order Gasterosteiformes. Seahorses.

Order Perciformes. Perches, wrasses, barracudas, tunas, freshwater bass, mackerels.
Order Lophiiformes. Angler fishes.
Class Amphibia. Mostly tetrapods; embryo in amnion.
Order Caudata. Salamanders.
Order Anura. Frogs, toads.
Order Apoda. Apodans (caecilians).
Class Reptilia. Skin with scales, embryo enclosed in amnion.
Subclass Anapsida. Turtles, tortoises.
Subclass Lepidosaura. *Sphenodon,* lizards, snakes.
Subclass Archosaura. Dinosaurs (extinct), crocodiles, alligators.
Class Aves. Birds. (In some of the more recent schemes, dinosaurs, crocodilians, and birds are grouped in the same category.)
Order Struthioniformes. Ostriches.
Order Sphenisciformes. Penguins.
Order Procellariiformes. Albatrosses, petrels.
Order Ciconiiformes. Herons, bitterns, storks, flamingoes.
Order Anseriformes. Swans, geese, ducks.
Order Falconiformes. Eagles, hawks, vultures, falcons.
Order Galliformes. Ptarmigan, turkeys, domestic fowl.
Order Columbiformes. Pigeons, doves.
Order Strigiformes. Owls.
Order Apodiformes. Swifts, hummingbirds.
Order Passeriformes. Sparrows, jays, finches, crows, robins, starlings, wrens.
Class Mammalia. Skin with hair; young nourished by milk-secreting glands of adult.
Subclass Prototheria. Egg-laying mammals (duckbilled platypus, spiny anteaters).
Subclass Metatheria. Pouched mammals or marsupials (opossums, kangaroos, wombats, Tasmanian devil).
Subclass Eutheria. Placental mammals.
Order Insectivora. Tree shrews, moles, hedgehogs.
Order Scandentia. Insectivorous tree shrews.
Order Chiroptera. Bats.
Order Primates.
Suborder Strepsirhini (prosimians). Lemurs, lorises.
Suborder Haplorhini (tarsioids and anthropoids).
Infraorder Tarsiiformes. Tarsiers.
Infraorder Platyrrhini (New World monkeys).
Family Cebidae. Spider monkeys, howler monkeys, capuchin.
Infraorder Catarrhini (Old World monkeys and hominoids).
Superfamily Cercopithecoidea. Baboons, macaques, langurs.
Superfamily Hominoidea. Apes and humans.
Family Hylobatidae. Gibbon.
Family Pongidae. Chimpanzees, gorillas, orangutans.
Family Hominidae. Existing and extinct human species (*Homo*) and australopiths.
Order Lagomorpha. Rabbits, hares, pikas.
Order Rodentia. Most gnawing animals (squirrels, rats, mice, guinea pigs, porcupines, beavers, etc.).
Order Cetacea. Whales, porpoises.
Order Carnivora. Carnivores.
Suborder Feloidea. Cats, mongooses, hyenas.
Suborder Canoidea. Dogs, weasels, skunks, otters, raccoons, pandas, bears.
Order Pinnipedia. Seals, walruses, sea lions.
Order Proboscidea. Elephants; mammoths (extinct).
Order Sirenia. Sea cows (manatees, dugongs).
Order Perissodactyla. Odd-toed ungulates (horses, tapirs, rhinos).
Order Artiodactyla. Even-toed ungulates (camels, deer, bison, sheep, goats, antelopes, giraffes, etc.).
Order Edentata. Anteaters, tree sloths, armadillos.
Order Tubulidentata. African aardvarks.

## Metric-English Conversions

### Length

| English | | Metric |
|---|---|---|
| inch | = | 2.54 centimeters |
| foot | = | 0.30 meter |
| yard | = | 0.91 meter |
| mile (5,280 feet) | = | 1.61 kilometer |

| To convert | multiply by | to obtain |
|---|---|---|
| inches | 2.54 | centimeters |
| feet | 30.00 | centimeters |
| centimeters | 0.39 | inches |
| millimeters | 0.039 | inches |

### Weight

| English | | Metric |
|---|---|---|
| grain | = | 64.80 milligrams |
| ounce | = | 28.35 grams |
| pound | = | 453.60 grams |
| ton (short) (2,000 pounds) | = | 0.91 metric ton |

| To convert | multiply by | to obtain |
|---|---|---|
| ounces | 28.3 | grams |
| pounds | 453.6 | grams |
| pounds | 0.45 | kilograms |
| grams | 0.035 | ounces |
| kilograms | 2.2 | pounds |

### Volume

| English | | Metric |
|---|---|---|
| cubic inch | = | 16.39 cubic centimeters |
| cubic foot | = | 0.03 cubic meter |
| cubic yard | = | 0.765 cubic meters |
| ounce | = | 0.03 liter |
| pint | = | 0.47 liter |
| quart | = | 0.95 liter |
| gallon | = | 3.79 liters |

| To convert | multiply by | to obtain |
|---|---|---|
| fluid ounces | 30.00 | milliliters |
| quart | 0.95 | liters |
| milliliters | 0.03 | fluid ounces |
| liters | 1.06 | quarts |

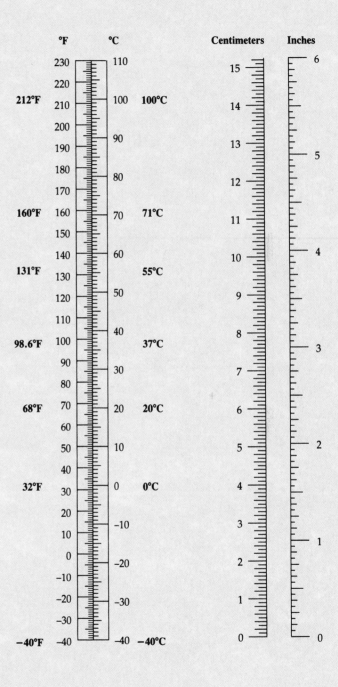

**21**
1. c *340*
2. e *334–349*
3. b *337*
4. d *336*
5. d *341*
6. d *341*
7. b *340–341*
   d *341*
   e *344–345*
   a *346*
   c *349*

**22**
1. c *358–359*
2. c *360*
3. b *362*
4. c *358*
5. d *357*
6. d *362*
7. e *364–365*
8. d *364*
9. d *360*
   e *362*
   b *364–365*
   f *359*
   g *358*
   c *360*
   a *367*

**23**
1. b *374*
2. d *375*
3. a *375*
4. c *376–377*
5. a *380*
6. a *383, 384*
7. d *383*
8. b *387*

**24**
1. c *394*
2. c *395*
3. b *CI, 399*
4. a *394*
5. a *394*
6. e *396*
   c *397*
   d *394*
   a *394–395*
   b *396*
   f *396–397*

**25**
1. d *404*
2. b *406*
3. c *404*
4. c *405*
5. b *411*
6. c *412*
   e *404–405*
   g *408*
   h *416*
   f *404*
   a *404*
   b *416–417*
   d *416–417*

**26**
1. b *424*
2. c *424*
3. c *425*
4. c *428–429*
5. b *432–435*
6. a *442*
7. a *442*
8. d *426*
   f *428*
   c *432*
   g *434–435*
   e *436*
   i *437*
   j *440*
   h *442*
   a *450*
   b *424 (Table 26.1)*

**27**
1. d *456–457*
2. a *458*
3. d *457–458*
4. b *459*
5. b *460*
6. d *462*
7. f *464*
8. e *465*
9. d *465, 469*
10. d *468, 469*
11. e *472*
12. a *478–479*
13. d *459*
    e *460*
    c *461*
    b *462*
    g *464*
    f *468–469*
    a *470*

**28**
1. b *486*
2. d *487*
3. d *492–493*
4. d *494*

# GLOSSARY

**absorption** Uptake of water and solutes from the environment by cell or multicelled organism; e.g., movement of nutrients, fluid, and ions across gut lining and into internal environment.

**acid** [L. *acidus*, sour] Any dissolved substance that donates hydrogen ions to other solutes or to water molecules.

**acoelomate** (ay-SEE-luh-mate) Absence of a fluid-filled cavity between gut and body wall.

**actin** (AK-tin) Cytoskeletal protein; subunit of microfilaments.

**adaptation** [L. *adaptare*, to fit] Of evolution, being adapted (or becoming more adapted) to a set of environmental conditions. Of a sensory neuron, a decrease or cessation in the frequency of action potentials when a stimulus is maintained at constant strength.

**adaptive radiation** Macroevolutionary pattern; a burst of genetic divergences from a lineage that gives rise to many new species, each adapted to using a novel resource or a new (or recently vacated) habitat.

**adaptive trait** Any aspect of form, function, or behavior that helps the individual survive and reproduce under prevailing conditions.

**adaptive zone** A way of life available for organisms that are physically, ecologically, and evolutionarily equipped to live it, such as "catching insects in the air at night."

**aerobic respiration** (air-OH-bik) [Gk. *aer*, air, + *bios*, life] Main ATP-forming pathway; proceeds from glycolysis through Krebs cycle and then electron transport phosphorylation. Final electron acceptor is oxygen. Typical net energy yield: 36 ATP per glucose molecule.

**aging** Of any multicelled organism showing extensive cell differentiation, a gradual and expected deterioration of the body over time.

**allele** (uh-LEEL) For a given gene locus, one of two or more slightly different molecular forms of a gene that arise through mutation and that code for different versions of the same trait.

**allele frequency** For a given locus, the relative abundance of each kind of allele among all the individuals of a population.

**allergy** Hypersensitivity to an allergen.

**allopatric speciation** [Gk. *allos*, different, + L. *patria*, native land]. Model of what may be the most common speciation route. A physical barrier arises and separates populations or subpopulations of a species, stops gene flow, and favors divergences that end in speciation.

**amino acid** (uh-MEE-no) An organic molecule with a hydrogen atom, an amino group, an acid group, and an R group, all covalently bonded to a carbon atom. Twenty kinds are the subunits of polypeptide chains.

**ammonification** (uh-moan-ih-fih-KAY-shun) Process by which some soil bacteria and fungi break down nitrogenous wastes and organic

remains; part of nitrogen cycle.

**amnion** (AM-nee-on) An extraembryonic membrane; the boundary layer of a fluid-filled sac (amniotic cavity) in which the embryos of some vertebrate embryos grow and develop, move freely, and are protected from sudden impacts and temperature shifts.

**amniote egg** Egg that has extraembryonic membranes and often a shell. Contributed to successful invasion of land by vertebrates.

**amphibian** Only type of vertebrate making the transition from water to land (evolutionarily and in their embryonic development). Existing gropus are salamanders, frogs, toads, and caecilians.

**analogous structures** (ann-AL-uh-gus) [Gk. *analogos*, similar to one another] Body parts that once differed in evolutionarily distant lineages but converged in their structure and function in response to similar environmental pressures.

**anaphase** (AN-uh-faze) Of mitosis, stage when sister chromatids of each chromosome move apart to opposite spindle poles. In anaphase I of meiosis, each duplicated chromosome and its homologue move to opposite spindle poles. In anaphase II of meiosis, sister chromatids of each chromosome move to opposite poles.

**animal** Multicelled heterotroph that feeds on other organisms, is motile for at least part of life cycle, develops through embryonic stages, has tissues (except for *Trichoplax* and sponges), and most often organs and organ systems.

**Animalia** Kingdom of animals.

**annelid** Type of invertebrate; a segmented worm (e.g., oligochaete, leech, or polychaete).

**antibiotic** Metabolic product of soil microbes that kills bacterial competitors for nutrients.

**antibody** [Gk. *anti*, against] Antigen-binding receptor. Only B cells make antibodies, then position them at their surface or secrete them.

**antigen** (AN-tih-jen) [Gk. *anti*, against, + *genos*, race, kind] Any molecular configuration that certain lymphocytes recognize as nonself and that triggers an immune response.

**apical dominance** Growth-inhibiting effect of a terminal bud on growth of lateral buds.

**Archaebacteria** Kingdom of prokaryotes more like eukaryotic cells than eubacteria; includes methanogens, halophiles, and thermophiles.

**archipelago** An island chain some distance away from a continent.

**area effect** Idea that larger islands support more species than smaller ones at equivalent distances from sources of colonizing species.

**arthropod** Invertebrate having a hardened exoskeleton, specialized body segments, and jointed appendages. Insects are examples.

**artificial selection** Selection of traits among individuals of a population in an artificial environment, under contrived conditions.

**asexual reproduction** Any of a number of modes of reproduction by which offspring arise from a single parent and inherit the genes of that parent only.

**atom** Smallest particle unique to an element; has one or more positively charged protons, electrons, and (except for hydrogen), neutrons.

**atomic number** Number of protons in the nucleus of each atom of an element.

**australopith** (OHSS-trah-low-pith) [L. *australis*, southern, + Gk. *pithekos*, ape] One of earliest known hominids; a primate that may be on or near evolutionary road to modern humans.

**autoimmune response** Misdirected immune response in which lymphocytes mount an attack against normal body cells.

**autosome** Type of chromosome that is the same in males and in females of the species.

**autotroph** (AH-toe-trofe) [Gk. *autos*, self, + *trophos*, feeder] Organism that makes its own organic compounds using an environmental energy source (e.g., sunlight) along with carbon dioxide as its carbon source.

**axon** Cylindrical extension of neuron cell body, specialized for the rapid propagation of action potentials.

**bacterial conjugation** The transfer of plasmid DNA from one bacterial cell to another.

**bacteriophage** (bak-TEER-ee-oh-fahj) [Gk. *baktērion*, small staff, rod, + *phagein*, to eat] Category of viruses that infect bacterial cells.

**Barr body** In body cells of female mammals, one of either of the two X chromosomes that was condensed to inactivate its genes.

**basal body** A centriole which, after giving rise to microtubules of a flagellum or cilium, remains attached to its base in the cytoplasm.

**behavior** Response to external and internal stimuli based on sensory, neural, endocrine, and effector components. Has a genetic basis, can evolve, and can be modified by learning.

**bilateral symmetry** Body plan in which left and right halves generally are mirror images.

**binary fission** Asexual reproductive mode of protozoans and some other animals. The body divides into two parts of the same or different sizes. *Compare* prokaryotic fission.

**binomial system** Of taxonomy, assigning a generic and a specific name to each species.

**biogeographic realm** [Gk. *bios*, life, + *geographein*, to describe the Earth's surface] One of six vast land areas, each with distinctive kinds and numbers of plants and animals.

**biological clock** Internal time-measuring mechanism that helps adjust an organism's daily activities, seasonal activities, or both in response to environmental cues.

**biological magnification** The increasing concentration of a nondegradable or slowly degradable substance in body tissues as it is passed along food chains.

**biological species concept** Defines a species as one or more populations of individuals that are interbreeding under natural conditions, that are producing fertile offspring, and that are isolated reproductively from other such populations. Applies to sexually reproducing species only.

**bioluminescence** Any organism flashing with fluorescent light by way of an ATP-driven reaction involving enzymes (luciferases).

**biosphere** [Gk. *bios*, life, + *sphaira*, globe] All regions of the Earth's waters, crust, and atmosphere in which organisms live.

**bipedalism** Habitually walking upright on two feet, as by ostriches and hominids.

**bird** Only vertebrate that produces feathers; evolutionarily linked with dinosaurs.

**blood** Fluid connective tissue of water, solutes, and formed elements (blood cells and platelets). Blood transports substances to and from cells, and helps maintain internal environment.

**bone** Vertebrate organ with mineral-hardened connective tissue (bone tissue); helps move body, protects other organs, stores minerals. Some (e.g., breastbone) produce blood cells.

**brain** Of most nervous systems, integrating center that receives and processes sensory input and issues coordinated commands for responses by muscles and glands.

**brown alga** Photoautotrophic protistan with chlorophylls $a$, $c_1$, and $c_2$, and carotenoids (e.g., fucoxanthin). Mostly marine. Range in size from microscopic to giant multicelled kelps.

**bulk flow** In response to a pressure gradient, a movement of more than one kind of molecule in the same direction in the same medium, as in blood, sap, or air.

**Calvin–Benson cycle** Cyclic, light-independent reactions; "synthesis" part of photosynthesis. Uses ATP and NADPH from light-dependent reactions. RuBP or some compound to which carbon has been affixed is rearranged and regenerated, and a sugar phosphate forms.

**camouflage** Adaptation in coloration, form, patterning, or behavior that helps predators or prey hide in the open (blend with their surroundings) and escape detection.

**cancer** Malignant tumor; mass of cells that have grossly altered plasma membrane and cytoplasm, grow and divide abnormally, and adhere weakly to home tissue (which leads to metastasis). Lethal unless eradicated.

**carbohydrate** [L. *carbo*, charcoal, + *hydro*, water] Molecule of carbon, hydrogen, and oxygen mostly in a 1:2:1 ratio. Carbohydrates are structural materials, energy stores, and transportable energy forms. Monosaccharide, oligosaccharide, or polysaccharide.

**carbon cycle** An atmospheric cycle. Carbon moves from its largest reservoirs (sediments, rocks, and the ocean), through the atmosphere (mostly as $CO_2$), through food webs, and back to the reservoirs.

**carcinogen** (kar-SIN-uh-jen) Any substance or agent that can trigger cancer.

**cardiovascular system** Organ system that has blood, one or more hearts, and blood vessels.

**carnivore** [L. *caro*, *carnis*, flesh, + *vovare*, to devour] Animal that eats other animals.

**carotenoid** (kare-OTT-en-oyd) An accessory pigment. Different kinds absorb blue-violet and blue-green wavelengths, the energy of which is transferred to chlorophylls. They reflect yellow, orange, and red wavelengths.

**carrying capacity** The maximum number of individuals in a population (or species) that a given environment can sustain indefinitely.

**cartilage** Connective tissue with solid, pliable intercellular material that resists compression.

**cDNA** DNA molecule copied from a mature mRNA transcript by reverse transcription.

**cell** [L. *cella*, small room] Smallest living unit; organized unit with a capacity to survive and reproduce on its own, given DNA instructions, energy sources, and raw materials.

**cell cycle** Events by which a cell increases in mass, roughly doubles its cytoplasmic components, duplicates its DNA, then divides its nucleus and cytoplasm. Extends from the time a cell forms until it completes division.

**cell differentiation** Key development process. Different cell lineages become specialized in their composition, structure, and function by activating and suppressing some fraction of the genome in different ways.

**cell junction** Site that joins cells physically, functionally, or both (e.g., tight junction in animals; plasmodesma in plants).

**cell plate** Disklike structure that forms in a plant cell after nuclear division; becomes a crosswall, with new plasma membrane on both surfaces, that divides the cytoplasm.

**cell theory** Theory stating that all organisms consist of one or more cells, the cell is the smallest unit with a capacity for independent life, and all cells arise from preexisting cells.

**cell wall** Of most bacteria, many protistan and fungal cells, and plant cells, the outermost, semirigid, permeable structure that helps the cell retain its shape and resist rupturing when the internal fluid pressure increases.

**central nervous system** Brain and spinal cord.

**central vacuole** Large, fluid-filled organelle of living, mature plant cell. Stores amino acids, sugars, ions, and toxic wastes. As it enlarges during growth, it forces the primary cell wall to expand and cell surface area to increase.

**cephalization** (sef-ah-lah-ZAY-shun) [Gk. *kephalikos*, head] The concentration of sensory structures and nerve cells in a head; occurred during evolution of bilateral animals.

**chemical bond** A union between the electron structures of two or more atoms or ions.

**chemical energy** Potential energy of molecules.

**chemical synapse** (SIN-aps) [Gk. *synapsis*, union] Narrow cleft between a presynaptic neuron and a postsynaptic cell. Molecules of neurotransmitter diffuse across it.

**chemoautotroph** (KEE-moe-AH-toe-trofe) Type of bacterium that can synthesize its own organic compounds using only carbon dioxide as the carbon source and an inorganic substance as the energy source.

**chemoreceptor** Sensory receptor that detects chemical energy (ions or molecules dissolved in the fluid bathing it).

**chlorofluorocarbon** (KLORE-oh-FLOOR-oh-car-bun) Compound of chlorine, fluorine, and carbon that has been contributing to ozone thinning.

**chlorophyll** (KLOR-uh-fill) [Gk. *chloros*, green, + *phyllon*, leaf] Main photosynthetic pigment; absorbs violet-to-blue and red wavelengths but transmits green.

**chloroplast** (KLOR-uh-plast) The organelle of photosynthesis in plants and many protistans.

**chordate** Animal with a notochord, dorsal hollow nerve cord, pharynx, and gill slits in pharynx wall during at least part of life cycle.

**chromatid** (CROW-mah-tid) Of a duplicated eukaryotic chromosome, one of two DNA molecules (with associated proteins) attached at centromere. Mitosis or meiosis separates them; each becomes a separate chromosome.

**chromosome** (CROW-moe-some) [Gk. *chroma*, color, + *soma*, body] Of eukaryotic cells, a DNA molecule, duplicated or unduplicated, with a profusion of associated proteins. Of prokaryotic cells (bacteria), a circular DNA molecule with few, if any, proteins attached.

**chromosome number** All chromosomes in a given type of cell. *See* haploidy; diploidy.

**cilium** (SILL-ee-um), plural **cilia** Short motile or sensory structure projecting from surface of certain eukaryotic cells; its core is a 9 + 2 array of microtubules.

**circadian rhythm** (ser-KAYD-ee-un) [L. *circa*, about, + *dies*, day] Cycle of physiological events completed every twenty-four hours or so independently of environmental change.

**circulatory system** Organ system that moves substances to and from cells, and often helps stabilize body temperature and pH. Typically consists of a heart, blood vessels, and blood.

**classification scheme** A way of organizing and retrieving information about species.

**cloaca** (kloe-AY-kuh) Chamber or duct in last portion of gut of some animals that also serves in excretion, reproduction, and sometimes respiration.

**club fungus** Fungus with club-shaped cells that produce and bear spores.

**cnidarian** Radial invertebrate at tissue level of organization; the only nematocyst producer. Medusae and polyps are typical body forms.

**codominance** In heterozygotes, simultaneous expression of a pair of nonidentical alleles that specify different phenotypes.

**coelom** (SEE-lum) [Gk. *koilos*, hollow] Cavity lined with peritoneum between the gut and body wall of most animals.

**coevolution** Joint evolution of two species that interacting closely. When one evolves, the change affects selection pressures that are operating between the two, so that the other also evolves.

**cofactor** Metal ion or coenzyme; it helps an enzyme catalyze a reaction or it transfers electrons, atoms, or functional groups to a different substrate.

**colon** (CO-lun) Large intestine.

**commensalism** [L. *com*, together, + *mensa*, table] Ecological interaction between two (or more) species in which one benefits directly and the other is affected little, if at all.

**community** All populations in a habitat. Also, a group of organisms with similar life-styles in a habitat, such as a bird community.

**comparative morphology** [Gk. *morph*, form] Scientific study of comparable body parts of adults or embryonic stages of major lineages.

**competitive exclusion** Theory that two or more species that require identical resources cannot coexist indefinitely.

**complement system** Set of proteins circulating in inactive form in vertebrate blood. Different kinds promote inflammation, induce lysis of pathogens, and stimulate phagocytes during nonspecific defenses and immune responses.

**compound** Substance consisting of two or more elements in unvarying proportions.

**concentration gradient** Between two adjoining regions, a difference in the number of molecules (or ions) of a substance. All molecules are in constant motion. As they collide, they career outward to an adjoining region where they are less concentrated. Barring other forces, all substances diffuse down such a gradient.

**condensation reaction** Two molecules become covalently bonded into a larger molecule, and water often forms as a by-product.

**conifer** Type of gymnosperm; generally, an evergreen woody tree or shrub with needle-like or scale-like leaves.

**connective tissue proper** Animal tissue with a characteristic proportion of fibroblasts and other cells, fibers (e.g., collagen, elastin), and a ground substance consisting of modified polysaccharides.

**consumer** [L. *consumere*, to take completely] A heterotroph that feeds on cells or tissues of other organisms for carbon and energy (e.g., herbivores, carnivores, and parasites).

**continuous variation** Of a population, a more or less continuous range of small differences in a given trait among its individuals.

**contractile vacuole** (kun-TRAK-till VAK-you-ohl) [L. *contractus*, to draw together] Organelle that takes up excess water in cell body and contracts to expel water through a pore.

**control group** A group used as a standard for comparison with an experimental group. Ideally, a control group is identical with the experimental group in all respects except for the one variable being studied.

**cortex** [L. *cortex*, bark] Generally, a rindlike layer such as kidney cortex or cell cortex. In vascular plants, a ground tissue that supports plant parts and stores food.

**crossing over** At prophase I of meiosis, an interaction in which nonsister chromatids (of a pair of homologous chromosomes) break at corresponding sites and exchange segments; genetic recombination is the result.

**culture** Sum of behavior patterns of a social group, passed between generations by way of learning and symbolic behavior, especially language.

**cuticle** (KEW-tih-kull) Body covering. Of plants, a transparent covering of waxes and cutin on outer epidermal cell walls. Of annelids, a thin, flexible coat. Of arthropods, a lightweight exoskeleton hardened with protein and chitin.

**cyst** Of many microorganisms, a resting stage with thick outer layers that typically forms under adverse conditions. Of skin, abnormal, fluid-filled sac without an external opening.

**cytokinesis** (SIGH-toe-kih-NEE-sis) [Gk. *kinesis*, motion] Cytoplasmic division; splitting of a parent cell into daughter cells.

**cytoplasm** (SIGH-toe-plaz-um) All cell parts, particles, and semifluid substances between the plasma membrane and the nucleus (or nucleoid, in bacteria).

**cytoskeleton** Dynamic internal framework of eukaryotic cells. Its microtubules and other components structurally support the cell and organize and move its internal parts. It helps free-living cells move in their environment.

**decomposer** [partly L. *dis-*, to pieces, + *companere*, arrange] Fungal or bacterial heterotroph. Obtains carbon and energy from remains, products, or wastes of organisms. Collectively, decomposers help cycle nutrients to producers in ecosystems.

**degradative pathway** A stepwise series of metabolic reactions that break down organic compounds to products of lower energy.

**dentition** (den-TIH-shun) Collectively, the type, size, and number of an animal's teeth.

**dermal tissue system** Tissues that cover and protect all exposed surfaces of a plant.

**dermis** Skin layer beneath the epidermis that consists primarily of dense connective tissue.

**deuterostome** (DUE-ter-oh-stome) [Gk. *deuteros*, second, + *stoma*, mouth] Category of bilateral animals in which the first indentation to form in the early embryo becomes an anus (e.g., an echinoderm or chordate).

**development** Of multicelled species, emergence of specialized, morphologically distinct body parts according to a genetic program.

**dicot** (DIE-kot) [Gk. *di*, two, + *kotylēdōn*, cup-shaped vessel] Dicotyledon. Flowering plant generally characterized by embryos with two cotyledons; net-veined leaves; and floral parts arranged in fours, fives, or multiples of these.

**diffusion** Net movement of like molecules or ions down their concentration gradient. In the absence of other forces, they collide constantly and randomly owing to their inherent energy. It occurs most frequently where they are most crowded, so the net movement is outward from regions of higher to lower concentrations.

**digestive system** Body sac or tube having one or two openings and often specialized regions for ingesting, digesting, and absorbing food, then eliminating undigested residues.

**directional selection** Mode of natural selection by which allele frequencies underlying a range of phenotypic variation shift in a consistent direction, in response to directional change or to new conditions in the environment.

**disruptive selection** Mode of natural selection by which forms of a trait at both ends of range of variation are favored and the intermediate forms are selected against.

**distance effect** Idea that only species adapted for long-distance dispersal can be potential colonists of islands far from their home range.

**diversity of life** Sum total of all variations in form, function, and behavior that accumulated in different lineages. Such variations generally are adaptive to prevailing conditions or were adaptive to conditions in the past.

**DNA** Deoxyribonucleic acid (dee-OX-ee-RYE-bow-new-CLAY-ik). For cells and many viruses, a nucleic acid that is the molecule of inheritance. Hydrogen bonds hold its two helically twisted nucleotide strands together. DNA's nucleotide sequence encodes instructions for synthesizing proteins, hence new individuals of a species.

**DNA clone** Many identical copies of foreign DNA that was inserted into plasmids and later replicated repeatedly after being taken up by population of host cells (typically, bacteria).

**DNA fingerprint** DNA fragments, inherited in a Mendelian pattern from each parent, that give each individual a unique identity. For humans, the most informative fragments are from tandem repeats (short regions of repeated DNA) that differ greatly among individuals.

**dominance hierarchy** Social organization in which some individuals of the group have adopted a subordinate status to others.

**dominant allele** Of diploid cells, an allele that masks phenotypic effect of any recessive allele paired with it.

**dormancy** [L. *dormire*, to sleep] A predictable time of metabolic inactivity for many spores, cysts, seeds, perennials, and some animals.

**double fertilization** Of flowering plants only, fusion of a sperm nucleus with an egg nucleus (a zygote forms), and fusion of another sperm nucleus with nuclei of endosperm mother cell, which gives rise to a nutritive tissue.

**doubling time** Time it takes for a population to double in size.

**duplication** Gene sequence repeated several to many hundreds or thousands of times. Even normal chromosomes have such sequences.

**ecdysone** Hormone that has major influence over early development of many insects.

**echinoderm** Type of invertebrate with calcified spines, needles, or plates on body wall. Radially symmetrical, but with some bilateral features. Sea stars and sea urchins are examples.

**ecology** [Gk. *oikos*, home, + *logos*, reason] Scientific study of how organisms interact with one another and with their physical and chemical environment.

**ecosystem** Array of organisms and their physical environment, all interacting by a flow of energy and a cycling of materials.

**ectoderm** [Gk. *ecto*, outside, + *derma*, skin] The first-formed, outermost primary tissue layer of animal embryos; gives rise to nervous system tissues and integument's outer layer.

**egg** Mature female gamete; an ovum.

**electron** Negatively charged unit of matter, with particulate and wavelike properties, that occupies one of the orbitals around the atomic nucleus. Atoms gain, lose, or share electrons.

**element** Substance that cannot be degraded by ordinary means into a substance having different properties.

**embryo** (EM-bree-oh) [Gk. *en*, in, + *bryein*, to swell] Of animals, a multicelled body formed by way of cleavage, gastrulation, and other early developmental events. Of plants, a young sporophyte, from the time of the earliest cell divisions after fertilization until germination.

**emerging pathogen** Deadly pathogen, either a newly mutated strain of an existing type or one that evolved long ago and is now exploiting an increased presence of human hosts.

**end product** Substance present at the end of a metabolic pathway.

**endangered species** A species that is highly vulnerable to extinction by natural events (e.g., genetic drift after a severe bottleneck) or by human activities (e.g., accidental introduction of competitive, exotic species).

**endocrine gland** Ductless gland that secretes hormones, which later enter the bloodstream.

**endocrine system** Integrative system of cells, tissues, and organs, functionally linked to the nervous system, that exerts control by way of its hormones and other chemical secretions.

**endocytosis** (EN-doe-sigh-TOE-sis) Cell uptake of substances when part of plasma membrane forms a vesicle around them. Three routes are receptor-mediated endocytosis, bulk transport of extracellular fluid, and phagocytosis.

**endoderm** [Gk. *endon*, within, + *derma*, skin] Inner primary tissue layer of animal embryos; source of inner gut lining and derived organs.

**endodermis** Sheetlike wrapping of single cells around root vascular cylinder; helps control uptake of water and dissolved nutrients.

**endometrium** (EN-doe-MEET-ree-um) [Gk. *metrios*, of the womb] Inner lining of uterus.

**endoplasmic reticulum** or **ER** (EN-doe-PLAZ-mik reh-TIK-yoo-lum) Organelle that starts at nucleus and curves through cytoplasm. New polypeptide chains acquire side chains inside rough ER (with ribosomes on its cytoplasmic side); smooth ER (with no ribosomes) is a site of lipid synthesis.

**endoskeleton** [Gk. *endon*, within, + *skleros*, hard] Of chordates, an internal framework of cartilage, bone, or both that works with skeletal muscle to support and move body, and to maintain posture.

**endosperm** (EN-doe-sperm) Nutritive tissue that surrounds an embryo sporophyte inside the seed of a flowering plant.

**endospore** Resting structure formed by some bacteria; encloses a duplicate of the bacterial chromosome and a portion of cytoplasm.

**endosymbiosis** Continuing physical contact between one species and another species that lives in its body.

**energy** Capacity to do work.

**enhancer** A short DNA base sequence that is a binding site for an activator protein.

**enzyme** (EN-zime) A protein or one of a few RNAs that greatly speed (catalyze) reactions between substances, most often at functional groups.

**epidermis** Outermost tissue layer of plants and all animals above sponge level of organization.

**epithelium** (EP-ih-THEE-lee-um) Tissue that covers the animal body's external surfaces and lines its internal cavities and tubes. It has one free surface and one resting on a basement membrane that is next to a connective tissue.

**equilibrium, dynamic** [Gk. *aequus*, equal, + *libra*, balance] Point at which a reaction runs forward as fast as in reverse, so no net change in reactant or product concentrations.

**estrogen** (ESS-trow-jen) Female sex hormone that helps oocytes mature, induces changes in the uterine lining during menstrual cycle and pregnancy, helps maintain secondary sexual traits, and affects growth and development.

**estrus** (ESS-truss) [Gk. *oistrus*, frenzy] Of most mammals, a cyclic period during which the female is sexually receptive to the male.

**ethylene** (ETH-il-een) Plant hormone that stimulates fruit ripening and abscission.

**Eubacteria** Kingdom of all prokaryotic cells except archaebacteria.

**eukaryotic cell** (yoo-CARE-EE-oh-tic) [Gk. *eu*, good, + *karyon*, kernel] Cell having a nucleus and other membrane-bound organelles.

**eutrophication** The enrichment of a body of water with nutrients; typically leads to reduced transparency and a phytoplankton-dominated community.

**evaporation** [L. *e-*, out, + *vapor*, steam] The conversion of a substance from liquid state to gaseous state under input of heat energy.

**evolution, biological** [L. *evolutio*, unrolling] Genetic change in a line of descent over time, brought about by microevolutionary events (gene mutation, natural selection, genetic drift, and gene flow).

**evolutionary tree** Treelike diagram in which each branch signifies a separate line of descent from a common ancestor and each branch point signifies a time of divergence.

**excitatory postsynaptic potential** (EPSP) A graded potential that arises at an input zone of an excitable cell and that drives the cell membrane closer to threshold.

**excretion** Removal of excess water, solutes, and wastes, and some harmful substances from body by way of a urinary system or glands.

**exocrine gland** (EK-suh-krin) [Gk. *es*, out of, + *krinein*, to separate] Glandular structure that secretes products, usually through ducts or tubes, to a free epithelial surface.

**exocytosis** (EK-so-sigh-TOE-sis) Release of the contents of a vesicle at the cell surface, where the vesicle's membrane fuses with and becomes part of the plasma membrane.

**exodermis** Cylindrical sheet of cells just inside the root epidermis of most flowering plants; helps control uptake of water and solutes.

**exoskeleton** [Gk. *skleros*, hard, stiff] External skeleton, as in arthropods.

**exotic species** Species that left its established home range, deliberately or accidentally, and successfully took up residence elsewhere.

**experiment** Test that simplifies observation in nature or in the laboratory by manipulating and controlling conditions under which the observations are made.

**extinction** Irrevocable loss of a species.

**extracellular fluid** Of most animals, all fluid not inside cells; plasma (the liquid portion of blood) plus interstitial fluid (the liquid that occupies spaces between cells and tissues).

**family pedigree** Chart of genetic relationship of family individuals through the generations.

**fat** Lipid with a glycerol head and one, two, or three fatty acid tails. Tryglycerides have three. The carbon backbone of unsaturated tails has single covalent bonds; that of saturated tails has one or more double bonds.

**fate map** Surface diagram of certain early embryos (e.g., of *Drosophila*) showing where the differentiated cells of the adult originate.

**fertilization** [L. *fertilis*, to carry, to bear] The Fusion of a sperm nucleus with the nucleus of an egg, which thus becomes a zygote.

**filter feeder** Animal that filters food from a current of water directed through a body part (e.g, through a sea squirt's pharynx).

**filtration** First step in urine formation; the pressure of heart contractions filters blood by forcing water and all solutes except proteins from glomerular capillaries into Bowman's capsule of nephrons.

**fin** Of fishes generally, appendage that helps stabilize, propel, and guide body in water.

**fish** Aquatic animal of the most ancient, diverse vertebrate lineage; a jawless, cartilaginous, or bony fish.

**fitness** Increase in adaptation to environment brought about by genetic change.

**fixation** Loss of all but one kind of allele at a gene locus; all individuals in a population are homozygous for it.

**fixed action pattern** Program of coordinated, stereotyped muscle activity that is completed independently of feedback from environment.

**flagellum** (fluh-JELL-um), plural **flagella** Motile structure of many free-living eukaryotic cells. Its core has a 9 + 2 array of microtubules.

**flower** Of angiosperms only, a reproductive structure with nonfertile parts (sepals, petals) and fertile parts (stamens, carpels) attached to a receptacle (modified base of a floral shoot).

**follicle** (FOLL-ih-kul) Small sac, pit, or cavity, as around a hair; also a mammalian oocyte with its surrounding layer of cells.

**food chain** Straight-line sequence of who eats whom in an ecosystem.

**food web** Cross-connecting, interlinked food chains consisting of producers, consumers, and decomposers, detritivores, or both.

**forebrain** Most complex portion of vertebrate brain; includes cerebrum (and cerebral cortex), olfactory lobes, and hypothalamus.

**fossil** Recognizable, physical evidence of an organism that lived in the distant past.

**fossil fuel** Coal, petroleum, or natural gas; a nonrenewable energy source that formed long ago from remains of swamp forests.

**fossilization** How fossils form. An organism or traces of it become buried in sediments or volcanic ash. Water and dissolved inorganic compounds infiltrate remains. Accumulating sediments exert pressure above the burial site. Over time, the pressure and chemical changes transform the remains to stony hardness.

**fruit** [L. after *frui*, to enjoy] Flowering plant's mature ovary, often with accessory structures.

**Fungi** Kingdom of fungi which, as a group, are major decomposers. Also includes diverse pathogens and parasites.

**fungus** Eukaryotic heterotroph that secretes digestive enzymes to break down food outside the body into molecules that its cells absorb (e.g., extracellular digestion and absorption). Saprobes feed on nonliving organic matter, and parasites feed on cells or tissues of living organisms.

**gamete** (GAM-eet) [Gk. *gametēs*, husband, and *gametē*, wife] Haploid cell, formed by meiotic cell division of a germ cell; required for sexual reproduction. Eggs and sperm are examples.

**gamete formation** Formation of sex cells (e.g., sperm and eggs); occurs in reproductive tissues or organs in most eukaryotic species.

**gametophyte** (gam-EET-oh-fite) [Gk. *phyton*, plant] Haploid gamete-producing body that forms during plant life cycles.

**gastrulation** (gas-tru-LAY-shun) Stage of animal development; major reorganization of new cells into two or three primary tissue layers.

**gene** [short for German *pangan*, after Gk. *pan*, all, + *genes*, to be born] Unit of information about a heritable trait, passed from parents to offspring. Each gene has a specific location on a chromosome (e.g., its locus).

**gene flow** Microevolutionary process; alleles enter and leave a population as an outcome of immigration and emigration, respectively.

**gene frequency** Abundance of a given allele relative to other alleles at same locus in a population.

**gene library** Mixed collection of bacteria that house many different cloned DNA fragments.

**gene locus** A gene's chromosomal location.

**gene pair** Two alleles at the same gene locus on a pair of homologous chromosomes.

**gene pool** All genotypes in a population.

**gene therapy** Generally, the transfer of one or more normal genes into an organism to correct or lessen adverse effects of a genetic disorder.

**genetic code** [After L. *genesis*, to be born] The correspondence between nucleotide triplets in DNA (then mRNA) and specific sequences of amino acids in resulting polypeptide chains; the basic language of protein synthesis in cells.

**genetic disease** Illness in which expression of one or more genes increases susceptibility to an infection or weakens immune response to it.

**genetic disorder** Inherited condition that causes mild to severe medical problems.

**genetic divergence** Gradual accumulation of differences in gene pools of populations or subpopulations of a species after a geographic barrier arises and separates them; mutation, natural selection, and genetic drift thereafter are operating independently in each one.

**genetic drift** Change in allele frequencies over the generations, as brought about by chance alone. Population size influences its effect on genetic and phenotypic diversity, because small populations are more vulnerable to losing alleles entirely.

**genetic engineering** Altering the information content of DNa molecules with recombinant DNA technology.

**genetic equilibrium** In theory, a state in which a population is not evolving. These conditions must be met: no mutation, the population very large in size and isolated from others of same species, and no natural selection (all members reproduce equally by random mating).

**genome** All of the DNA in a haploid number of chromosomes for a given species.

**genotype** (JEEN-oh-type) Genetic constitution of an individual; a single gene pair or the sum total of an individual's genes.

**genus**, plural **genera** (JEEN-US, JEN-er-ah) [L. *genus*, race or origin] A grouping of all species perceived to be more closely related to one another in their morphology, ecology, and evolutionary history than to other species at the same taxonomic level.

**geologic time scale** Time scale for the Earth's history. Its major subdivisions correspond to mass extinctions. Its absolute dates have been refined by radiometric dating.

**germ cell** Animal cell of a lineage set aside for sexual reproduction; gives rise to gametes.

**germination** (jur-min-AY-shun) Of seeds and spores, resumption of growth after dispersal, dormancy, or both.

**gibberellin** (JIB-er-ELL-un) Type of plant hormone that promotes elongation of stems, helps seeds and buds break dormancy, and contributes to flowering.

**gill** Organ of respiration. Most have a thin, moist, vascularized layer for gas exchange.

**gland** Secretory cell or structure derived from epithelium and often connected to it.

**glycocalyx** Sticky mesh of polysaccharides, polypeptides, or both around the cell wall of many bacteria.

**Golgi body** (GOHL-gee) Organelle of lipid assembly, polypeptide chain modification, and packaging of both in vesicles for export or for transport to locations in cytoplasm.

**gonad** (GO-nad) Primary reproductive organ in which animal gametes are produced.

**granum**, plural **grana** In many chloroplasts, a stack of flattened, membranous compartments that have chlorophyll and other light-trapping pigments and reaction sites for ATP formation.

**gravitropism** (GRAV-ih-TROPE-izm) [L. *gravis*, heavy, + Gk. *trepein*, to turn] Tendency of a plant to grow directionally in response to the Earth's gravitational force.

**green alga** Type of protistan evolutionarily, structurally, and biochemically most similar to plants (e.g., nearly all are photoautotrophs with starch grains and chlorophylls *a* and *b* in chloroplasts; and some have cell walls of cellulose, pectin, and other polysaccharides typical of plants).

**greenhouse effect** Overall warming of lower atmosphere. Gaseous molecules (e.g., carbon dioxide and methane) impede the escape of infrared wavelengths (heat) from the Earth's sunlight-warmed surface. The gases continually absorb those wavelengths and radiate much of their energy back toward Earth.

**ground substance** Intercellular material in some animal tissues; made of cell secretions and other noncellular components.

**ground tissue system** Tissues (parenchyma, especially) making up most of the plant body.

**growth** Of multicelled species, increases in the number, size, and volume of cells. Of bacteria, increase in the number of cells in a population.

**gut** Generally, a sac or tube from which food is absorbed into internal environment. Also, a gastrointestinal tract (from stomach onward).

**gymnosperm** (JIM-noe-sperm) [Gk. *gymnos*, naked, + *sperma*, seed] Vascular plant that bears seeds at exposed surfaces of its reproductive structures (e.g., on cone scales).

**habitat** [L. *habitare*, to live in] Type of place where an organism or species normally lives; characterized by physical and chemical features and by its array of species.

**halophile** Archaebacterium of saline habitats.

**heart** Muscular pump; its contractions keep blood circulating through the animal body.

**heat** Thermal energy; a form of kinetic energy.

**hemoglobin** (HEEM-oh-glow-bin) [Gk. *haima*, blood, + L. *globus*, ball] Iron-containing, oxygen-transporting protein of red blood cells.

**herbivore** [L. *herba*, grass, + *vovare*, to devour] Plant-eating animal (e.g., snail, deer, manatee).

**hermaphrodite** Individual having both male and female gonads.

**heterocyst** (HET-er-oh-sist) Cyanobacterial cell that modifies itself and makes a nitrogen-fixing enzyme when nitrogen supplies dwindle.

**heterotroph** (HET-er-oh-trofe) [Gk. *heteros*, other, + *trophos*, feeder] Organism unable to make its own organic compounds; feeds on autotrophs, other heterotrophs, organic wastes.

**heterozygous condition** (HET-er-oh-ZYE-guss) [Gk. *zygoun*, join together] For a given trait, having a pair of nonidentical alleles at a gene locus (that is, on a pair of homologous chromosomes).

**higher taxon** (plural, **taxa**) One of ever more inclusive groupings that reflect relationships among species. Family, order, class, phylum, and kingdom are examples.

**homeostasis** (HOE-me-oh-STAY-sis) [Gk. *homo*, same, + *stasis*, standing] State in which physical and chemical aspects of internal environment (blood, interstitial fluid) are being maintained within ranges suitable for cell activities.

**homeotic gene** A master gene governing the development of specific body parts.

**hominid** [L. *homo*, man] All species on or near evolutionary road leading to modern humans.

**hominoid** Apes, humans, and recent ancestors.

**homologous chromosome** (huh-MOLL-uh-gus) [Gk. *homologia*, correspondence] Of cells with a diploid chromosome number, one of a pair of chromosomes that are identical in size, shape, and gene sequence, and that interact at meiosis. Nonidentical sex chromosomes (e.g., X and Y) also interact as homologues during meiosis.

**homologous structures** The same body parts that became modified differently, in different lines of descent from a common ancestor.

**homology** Similarity in one or more body parts in different species; attributable to descent from a common ancestor.

**homozygous condition** (HOE-moe-ZYE-guss) For a specified trait, having a pair of identical alleles at a gene locus (on a pair of homologous chromosomes).

**homozygous dominant condition** Having a pair of dominant alleles at a gene locus (on a pair of homologous chromosomes).

**homozygous recessive condition** Having a pair of recessive alleles at a gene locus (on a pair of homologous chromosomes).

**hormone** [Gk. *hormon*, stir up, set in motion] Signaling molecule secreted by one cell that stimulates or inhibits activities of any other cell having receptors for it. Animal hormones are picked up by bloodstream, which delivers them to cells some distance away. In plants, hormones do not travel far from source cells.

**horsetail** Seedless vascular plant with ancient, tree-size ancestors. The sporophytes of the one existing genus have rhizomes, scale-shaped leaves, and hollow photosynthetic stems with silica-reinforced ribs.

**humus** Decomposing organic matter in soil. Amount varies in soils of different types.

**hybrid offspring** Of a genetic cross, offspring with a pair of nonidentical alleles for a trait.

**hydrologic cycle** A biogeochemical cycle, driven by solar energy, in which water moves through the atmosphere, on or through land, to the ocean, and back to the atmosphere.

**hydrolysis** (high-DRAWL-ih-sis) [L. *hydro*, water, + Gk. *lysis*, loosening] Cleavage reaction that breaks covalent bonds and splits a molecule into two or more parts. Commonly, $H^+$ and $OH^-$ (derived from a water molecule) become attached to the exposed bonding sites.

**hydrostatic pressure** Pressure exerted by a volume of fluid against a wall, membrane, or some other structure that encloses the fluid.

**hydrothermal vent ecosystem** Ecosystem near a steaming fissure in the ocean floor. Chemoautotrophic bacteria are the basis of its food webs.

**hypha** (HIGH-fuh), plural **hyphae** [Gk. *hyphe*, web] Fungal filament with chitin-reinforced walls; component of a mycelium.

**hypodermis** Subcutaneous layer with stored fat that helps insulate the body; anchors skin but still allows it to move somewhat.

**hypothesis** In science, a possible explanation of a phenomenon, one that has the potential to be proved false by experimental tests.

**immune response** Events by which B cells and T cells recognize antigen and give rise to antigen-sensitized populations of effector cells and memory cells.

**immunoglobulin** (Ig) One of five classes of antibodies, each with antigen-binding sites as well as other sites with specialized functions.

**imprinting** Time-dependent form of learning, usually during a sensitive period for a young animal, triggered by exposure to sign stimuli.

**in vitro fertilization** Conception outside the body ("in glass" petri dishes or test tubes).

**inbreeding** Nonrandom mating among close relatives that share many identical alleles; a form of genetic drift in a small group of relatives that are preferentially interbreeding.

**incomplete dominance** Condition in which one allele of a pair is not fully dominant over the other; a heterozygous phenotype in between both homozygous phenotypes emerges.

**indirect selection theory** Idea that altruistic individuals can pass on their genes indirectly by helping relatives survive and reproduce.

**inductive logic** Pattern of thinking by which an individual derives a general statement from specific observations.

**infection** Invasion and multiplication of a pathogen in a host. Disease follows if defenses are not mobilized fast enough; the pathogen's activities interfere with normal body functions.

**inflammation, acute** Rapid response to tissue injury by phagocytes and diverse proteins (e.g., histamine, complement, clotting factors). Signs include localized redness, heat, swelling, pain.

**inheritance** The transmission, from parents to offspring, of genes that specify structures and functions characteristic of the species.

**inhibiting hormone** Hypothalmic signaling molecule that suppresses a particular secretion by the anterior lobe of the pituitary gland.

**inhibitor** Substance able to bind with a specific molecule and interfere with its functioning.

**instinctive behavior** A behavior performed without having been learned by experience.

**integration, neural** [L. *integrare*, coordinate] Moment-by-moment summation of excitatory and inhibitory synapses acting on a neuron.

**integrator** A control center (e.g., brain) that receives, processes, and stores sensory input, then puts together and issues commands for coordinated responses to it.

**integument** Of animals, protective body cover (e.g., skin). Of seed-bearing plants, one or more layers around an ovule; becomes a seed coat.

**integumentary exchange** (in-teg-you-MEN-tuh-ree) Respiration across a thin, moist, and often vascularized surface layer of animal body.

**intermediate** Substance that forms between the start and end of a metabolic pathway.

**intermediate filament** One of the ropelike cytoskeletal elements that impart mechanical strength to animal cells and tissues.

**internode** In vascular plants, the stem region between two successive nodes.

**interphase** Of a cell cycle, interval between nuclear divisions when a cell increases in mass and roughly doubles the number of its cytoplasmic components. It also duplicates its chromosomes (replicates its DNA) during interphase, but *not* between meiosis I and II.

**interspecific competition** Ecological interaction in which individuals that belong to different species compete for a share of resources in the same habitat.

**interstitial fluid** (IN-ter-STISH-ul) [L. *interstitus*, to stand in the middle of something] The portion of extracellular fluid that occupies the spaces between animal cells and tissues.

**intraspecific competition** Ecological interaction in which individuals that belong to the same population compete for a share of resources in their habitat.

**intron** One of the noncoding portions of a pre-mRNA transcript. All introns are excised before translation.

**inversion** A linear stretch of DNA within a chromosome that has become oriented in the reverse direction, with no molecular loss.

**invertebrate** Any animal without a backbone. Of the 2 million named species in the animal kingdom, all but 50,000 are invertebrates.

**juvenile** Of some animals, a post-embryonic stage that changes only in size and proportion to become the adult (no metamorphosis).

**karyotype** (CARE-ee-oh-type) For an individual or a species, a preparation of metaphase chromosomes sorted by length, centromere location, and other defining features.

**keratin** Tough, water-insoluble protein made by vertebrate epidermal cells, which die and accumulate as keratinized bags at the surface of skin to form a barrier against dehydration, bacteria, and many toxins.

**key innovation** A structural or functional modification to the body that, by chance, gives a lineage the opportunity to exploit the environment in more efficient or novel ways.

**keystone species** A species that dominates a community and dictates its structure.

**kidney** One of a pair of vertebrate organs that filter ions and other substances from blood; it controls the amounts returned and so helps maintain the solute concentrations and volume of extracellular fluid.

**kilocalorie** 1,000 calories of heat energy (the amount required to raise the temperature of 1 kilogram of water by 1°C). Used as the unit of measure for the caloric content of foods.

**kinetic energy** Energy of motion.

**lactate fermentation** An anaerobic pathway of ATP formation. Pyruvate from glycolysis is converted to three-carbon lactate, and NAD$^+$ is regenerated. Net energy yield: two ATP.

**lactation** Production and secr etion of milk by hormone-primed mammary glands.

**large intestine** Colon; gut region that receives unabsorbed residues from small intestine, and concentrates and stores feces until expulsion.

**larva**, plural **larvae** An immature stage that develops between the embryo and adult in many animal life cycles.

**lateral root** Outward branching from the first (primary) root of a taproot system.

**leaching** Removal of some nutrients from soil as water percolates through it.

**leaf** Chlorophyll-rich plant part adapted for sunlight interception and photosynthesis.

**learned behavior** Lasting modification in behavior as a result of experience or practice.

**lethal mutation** Mutation with drastic effects on phenotype; usually causes death.

**lichen** (LY-kun) Symbiotic interaction between a fungus and a photoautotroph.

**life cycle** Recurring pattern of genetically programmed events from the time individuals are produced until they themselves reproduce.

**ligament** A strap of dense connective tissue that bridges a joint.

**light-dependent reactions** The first stage of photosynthesis. Sunlight energy is trapped and converted to chemical energy of ATP, NADPH, or both, depending on the pathway.

**light-independent reactions** Second stage of photosynthesis. ATP makes phosphate-group transfers required to build sugar phosphates. NADPH delivers electrons and hydrogen atoms for the synthesis reactions, which also require carbon from carbon dioxide. Sugar phosphates enter other reactions by which starch, cellulose, and other end products are assembled.

**lignification** Deposition of lignin in secondary wall of many plant cells. Lignin imparts strength and rigidity, stabilizes and protects other wall components, and forms a waterproof barrier. Important in vascular plant evolution.

**limiting factor** Any essential resource that, in short supply, limits population growth.

**lineage** (LIN-ee-age) Line of descent.

**linkage group** All the genes on a chromosome.

**lipid** Mainly a greasy or oily hydrocarbon. Lipid molecules strongly resist dissolving in water but quickly dissolve in nonpolar substances. Some types serve as the main reservoirs of stored energy in all cells; others are structural materials (as in membranes) and cell products (e.g., surface coatings).

**lipid bilayer** Structural basis of cell membranes. Two layers of mostly phospholipid molecules. Hydrophobic tails of the lipids are sandwiched between the hydrophilic heads, and heads are dissolved in intracellular or extracellular fluid.

**lipoprotein** Molecule that forms when proteins circulating in blood combine with cholesterol, triglycerides, and phospholipids absorbed from the small intestine.

**liver** In vertebrates and many invertebrates, a large gland that stores, converts, and helps maintain blood levels of organic compounds; inactivates most hormone molecules that have completed their tasks; inactivates compounds that can be toxic at high concentrations.

**local signaling molecule** One of the secretions by many cell types that alter chemical conditions only in localized tissue regions. Prostaglandin is an example.

**logic** Thought patterns by which an individual draws a conclusion that does not contradict evidence used to support that conclusion.

**lung** Internal sac-shaped respiratory surface that evolved in oxygen-poor habitats as an adaptation that increases the surface area for gas exchange. A pair occur in a few fishes and in amphibians, birds, reptiles, and mammals.

**lycophyte** Seedless vascular plant with tree-size ancestors of ancient swamp forests. They require free water to complete life cycle. Most have leaves, branching rhizome, vascularized roots and stems (e.g., club mosses).

**lymphatic system** Supplement to vertebrate circulatory system. Its vessels deliver fluid and solutes from interstitial fluid to blood; and its lymphoid organs have roles in body defenses.

**lymphocyte** A T cell or B cell.

**lysis** [Gk. *lysis*, a loosening] Gross damage to a plasma membrane, cell wall, or both that lets the cytoplasm to leak out; causes cell death.

**lysogenic pathway** Latent period that extends many viral replication cycles. Viral genes get integrated into host chromosome and may stay inactivated through many host cell divisions but eventually are replicated in host progeny.

**lysosome** (LYE-so-sohm) Important organelle of intracellular digestion.

**lysozyme** Infection-fighting enzyme present in mucous membranes (e.g., of mouth, vagina).

**lytic pathway** Of viruses, a rapid replication pathway that ends with lysis of host cell.

**macroevolution** Large-scale patterns, trends, and rates of change among higher taxa.

**macrophage** Phagocytic white blood cell; roles in nonspecific defenses and immune responses. One of the key antigen-presenting cells.

**mammal** Only vertebrate whose females nourish offspring with milk from mammary glands.

**mass extinction** Catastrophic event or phase in geologic time when entire families or other major groups are irrevocably lost.

**mast cell** A basophil-like cell that releases histamine during tissue inflammation.

**mechanoreceptor** Sensory cell or nearby cell that detects mechanical energy (changes in pressure, position, or acceleration).

**medusa** (meh-DOO-sah) [Gk. *Medousa*, one of three sisters in Greek mythology having snake-entwined hair] Of cnidarian life cycles, a free-swimming, bell-shaped stage, often with oral lobes and tentacles extending below the bell.

**megaspore** Haploid spore that forms by way of meiosis in the ovary of seed-bearing plants; one of its cellular descendants develops into an egg.

**meiosis** (my-OH-sis) [Gk. *meioun*, to diminish] Two-stage nuclear division process that halves the chromosome number of a parental germ cell (to haploid number). Each daughter nucleus receives one of each type of chromosome. Basis of gamete formation. Also, basis of formation of spores that give rise to gamete-producing bodies (gametophytes).

**memory** The capacity to store and retrieve information about past sensory experience.

**menstrual cycle** A recurring cycle, lasting twenty-eight days on average in adult human females. A secondary oocyte is released from one of a pair of ovaries, and the lining of the uterus is primed for pregnancy. The hormones estrogen, progesterone, FSH, and LH control the cyclic activity.

**menstruation** Sloughing of a blood-enriched endometrium when pregnancy does not occur.

**mesoderm** (MEH-zoe-derm) [Gk. *mesos*, middle, + *derm*, skin] Primary tissue layer important in evolution of all large, complex animals; gives rise to many internal organs and part of the integument.

**mesophyll** (MEH-zoe-fill) A photosynthetic parenchyma with abundant air spaces for gas exchange between its cells.

**messenger RNA** (mRNA) A single strand of ribonucleotides transcribed from DNA, then translated into a polypeptide chain. The only RNA encoding protein-building instructions.

**metabolic pathway** (MEH-tuh-BALL-ik) Orderly sequence of enzyme-mediated reactions by which cells maintain, increase, or decrease the concentrations of particular substances.

**metabolism** (meh-TAB-oh-lizm) [Gk. *meta*, change] All the controlled, enzyme-mediated chemical reactions by which cells acquire and use energy to synthesize, store, degrade, and eliminate substances in ways that contribute to growth, survival, and reproduction.

**metamorphosis** (me-tuh-MOR-foe-sis) [Gk. *meta*, change, + *morphe*, form] Major changes in body form during the transition from the embryo to the adult; involves hormonally controlled size increases, reorganization of tissues, and remodeling of body parts.

**metaphase** Of meiosis I, a stage when all pairs of homologous chromosomes have become positioned at the spindle equator. Of mitosis or meiosis II, a stage when all the duplicated chromosomes have become positioned at the equator of the microtubular spindle.

**metazoan** Any multicelled animal.

**methanogen** Anaerobic archaebacterium that produces methane gas as by-product.

**micelle** (my-SELL) Of fat digestion, tiny droplet of bile salts, fatty acids, and monoglycerides; role in fat absorption from small intestine.

**microevolution** Of a population, any change in allele frequencies resulting from mutation, genetic drift, gene flow, natural selection, or some combination of these.

**micrograph** Photograph of an image that came into view with the aid of a microscope.

**microorganism** Organism, usually single celled, too small to be observed without a microscope.

**microspore** Walled haploid spore; becomes a pollen grain in gymnosperms and angiosperms.

**migration** Recurring pattern of movement between two or more locations in response to environmental rhythms; e. g., circadian rhythms and seasonal changes in daylength. It requires activation or suppression of internal timing mechanisms that govern physiological and behavioral functions of migratory animals.

**mimicry** (MIM-ik-ree) Close resemblance in form, behavior, or both between one species (the mimic) and another (its model). Serves in deception, as when an orchid mimics a female insect and so attracts males that pollinate it.

**mineral** Any element or inorganic compound that formed by natural geologic processes and is required for normal cell functioning.

**mitochondrion** (MY-toe-KON-dree-on) Double-membrane organelle of ATP formation. Only site of aerobic respiration's second and third stages. May have endosymbiotic origins.

**mitosis** (my-TOE-sis) [Gk. *mitos*, thread] Type of nuclear division that maintains the parental chromosome number for daughter cells. The basis of growth in size, tissue repair, and often asexual reproduction for eukaryotes.

**mixture** Two or more elements intermingled in proportions that can and usually do vary.

**model** Theoretical, detailed description or analogy that helps people visualize something that has not yet been directly observed.

**molar** Tooth with a platform having cusps (surface bumps) that help crush, grind, and shear food; one of the cheek teeth.

**molecular clock** Model used to calculate the time of origin of one lineage or species relative to others. The underlying assumption is that neutral mutations accumulate in a lineage at predictable rates that can be measured as a series of ticks back through time.

**molecule** A unit of matter in which chemical bonds hold together two or more atoms of the same or different elements.

**mollusk** Only invertebrate with a tissue fold (mantle) draped over a soft, fleshy body; most have an external or internal shell. Enormous diversity in body plans and sizes, as in chitons, gastropods (e.g, snails), bivalves (e.g., clams), and cephalopods (e.g., octopuses, squids).

**molting** Periodic shedding of body structures that are too small, worn out, or both. Permits certain animals to grow in size or renew some parts (e.g., exoskeletons, shells, hairs, feathers, and horns). Especially characteristic of insects and other arthropods.

**Monera** In earlier classification schemes, a prokaryotic kingdom that encompasses both archaebacteria and eubacteria.

**monocot** (MON-oh-kot) Monocotyledon; a flowering plant with one cotyledon in seeds, floral parts generally in threes (or multiples of three), and often parallel-veined leaves.

**monomer** Small molecule used as a subunit of polymers, such as sugar monomers of starch.

**monosomy** Presence of a chromosome that has no homologue in a diploid cell.

**morphogenesis** (MORE-foe-JEN-ih-sis) [Gk. *morphe*, form, + *genesis*, origin] Inherited program of orderly changes in size, shape, and proportions of an animal embryo, leading to specialized tissues and early organs.

**morphological convergence** Macroevolutionary pattern. In response to similar environmental pressures over time, evolutionarily distant lineages evolve in similar ways and end up being alike in appearance, functions, or both.

**morphological divergence** Macroevolutionary pattern. Genetically diverging lineages slowly undergo change from the body form of their common ancestor.

**motor neuron** Type of neuron specialized to swiftly relay commands from the brain or spinal cord to muscle cells, gland cells, or both.

**multicelled organism** Organism composed of many cells with coordinated metabolic activity;

most show extensive cell differentiation into tissues, organs, and organ systems.

**multiple allele system** Three or more slightly different molecular forms of a gene that occur among individuals of a population.

**muscle tissue** Tissue with arrays of cells able to contract under stimulation, then passively lengthen and return to their resting position.

**mutagen** (MEW-tuh-jen) Any environmental agent, such as a virus or ultraviolet radiation, that can alter DNA's molecular structure.

**mutation** [L. *mutatus*, a change, + -*ion*, an act, a result, or a process] Heritable change in the molecular structure of DNA. Original source of all new alleles and, ultimately, the diversity of life.

**mutation frequency** Of a population, the number of times that a mutation at a particular locus has arisen.

**mutation rate** Of a gene locus, the probability that a spontaneous mutation will occur during or between DNA replication cycles.

**mutualism** [L. *mutuus*, reciprocal] Symbiotic interaction that benefits both participants.

**mycelium** (my-SEE-lee-um), plural **mycelia** [Gk. *mykes*, fungus, mushroom, + *helos*, callus] Mesh of tiny, branching filaments (hyphae); the food-absorbing portion of most fungi.

**mycorrhiza** (MY-coe-RIZE-uh) "Fungus-root." A form of mutualism between fungal hyphae and young plant roots. The plant gives up some carbohydrates and the fungus gives up some of its absorbed mineral ions.

**myofibril** (MY-oh-FY-brill) One of the many internal threadlike structures of a muscle cell, each divided into sarcomeres.

**natural killer cell** Cytotoxic lymphocyte that reconnoiters for tumor cells and virus-infected cells, then touch-kills them.

**natural selection** Microevolutionary process; the outcome of differences in survival and reproduction among individuals that vary in details of heritable traits. Over generations, it typically leads to increased fitness.

**necrosis** (neh-CROW-sis) Passive death of many cells that results from severe tissue damage.

**negative feedback mechanism** A homeostatic mechanism by which a condition that has changed as a result of some activity triggers a response that reverses the change.

**nematocyst** (NEM-at-uh-sist) [Gk. *nema*, thread, + *kystis*, pouch] Cnidarian capsule that has a dischargeable, tube-shaped thread, sometimes barbed; releases a toxin or sticky substance.

**nerve** Cordlike bundle of the axons of sensory neurons, motor neurons, or both sheathed in connective tissue.

**nerve cord** A prominent longitudinal nerve. Most animals have one, two, or three. The nervous system of chordate embryos develops from a tubular, dorsal nerve cord.

**nerve net** Simple nervous system in epidermis of cnidarians and some other invertebrates; a diffuse mesh of simple, branching nerve cells interacts with contractile cells, sensory cells, or both.

**nervous system** Integrative organ system with nerve cells interacting in signal-conducting and information-processing pathways. Detects and processes stimuli, and elicits responses from effectors (e.g., muscles and glands).

**neural tube** Embryonic and evolutionary forerunner of brain and spinal cord.

**neuroglia** (NUR-oh-GLEE-uh) Collectively, cells that structurally and metabolically support neurons. They make up about half the volume of nervous tissue in vertebrates.

**neuromuscular junction** A chemical synapse between axonal endings of a motor neuron and a muscle cell.

**neuron** (NUR-on) Type of nerve cell; basic communication unit in most nervous systems.

**neurotransmitter** Any of a class of signaling molecules secreted by neurons. It acts on cell next to it, then is rapidly degraded or recycled.

**neutral mutation** Mutation that has little or no effect on phenotype. Natural selection cannot change its frequency in a population because it does not affect survival or reproduction.

**neutron** Unit of matter, one or more of which occupies the atomic nucleus and has mass but no electric charge.

**niche** (NITCH) [L. *nidas*, nest] Sum total of all activities and relationships in which individuals of a species engage as they secure and use the resources required to survive and reproduce.

**nitrification** (nye-trih-fih-KAY-shun) Process by which certain bacteria in soil break down ammonia or ammonium to nitrite, then other bacteria break down nitrite to nitrate (which is a form that plants can take up). Key part of the nitrogen cycle.

**nitrogen cycle** Atmospheric cycle. Nitrogen moves from its largest reservoir (atmosphere), through the ocean, ocean sediments, soils, and food webs, then back to the atmosphere.

**nitrogen fixation** Process by which certain bacterial species convert gaseous nitrogen to ammonia, which swiftly dissolves in their cytoplasm to form ammonium. Ammonium can be used in biosynthesis.

**node** Stem site where one or more leaves form.

**notochord** (KNOW-toe-kord) Of chordates, a rod of stiffened tissue (not cartilage or bone) that is a supporting structure for the body.

**nuclear envelope** Outermost portion of a cell nucleus; composed of a double membrane (two lipid bilayers and associated proteins).

**nucleus** (NEW-klee-us) [L. *nucleus*, a kernel] Of atoms, a central core of one or more protons and (in all but hydrogen atoms) neutrons. In a eukaryotic cell, the organelle that physically separates DNA from cytoplasmic machinery.

**numerical taxonomy** Study of the degree of relatedness between an unidentified organism and a known group through comparisons of traits. Used to classify prokaryotes, which are poorly represented in the fossil record.

**nutrient** Element with a direct or indirect role in metabolism that no other element fulfills.

**nutrition** Processes by which food is selectively ingested, digested, absorbed, and converted to the body's own organic compounds.

**oligosaccharide** (oh-LIG-oh-SAC-uh-RID) Short-chain carbohydrate of two or more covalently bonded sugar monomers. Disaccharides (two monomers) are examples.

**omnivore** [L. *omnis*, all, + *vovare*, to devour] An animal that feeds at more than one trophic level.

**oocyte** Immature egg of all animals and some protistans.

**oogenesis** (oo-oh-JEN-uh-sis) Process by which a germ cell develops into a mature oocyte.

**orbital** One of the volumes of space around the atomic nucleus in which one or at most two electrons are likely to be at any instant.

**organ** Body structure having definite form and function that consists of more than one tissue.

**organ formation** Developmental stage in which primary tissue layers give rise to cell lineages unique in structure and function. Descendants of those lineages give rise to all the different tissues and organs of the adult.

**organ system** Two or more organs that are interacting chemically, physically, or both in a common task.

**organelle** (or-GUN-ell) Membrane-bound sac or compartment in the cytoplasm having one or more specialized metabolic functions. Most eukaryotic cells have a profusion of them.

**organic compound** Molecule of one or more elements covalently bonded to some number of carbon atoms.

**osmosis** (oss-MOE-sis) [Gk. *osmos*, pushing] The diffusion of water in response to water concentration gradient between two regions that are separated by a selectively permeable membrane. The greater the number of ions and molecules dissolved in a solution, the lower its water concentration.

**osmotic pressure** Force that operates after hydrostatic pressure develops in a cell or in an enclosed body region; the amount of force that stops further increases in fluid volume by countering the inward diffusion of water.

**ovary** (OH-vuh-ree) In most animals, a female gonad. In flowering plants, the enlarged base of one or more carpels. A fruit is a mature ovary often combined with other floral parts.

**oviduct** (OH-vih-dukt) One of a pair of ducts through which eggs travel from an ovary to the uterus. Formerly called a fallopian tube.

**ovulation** (OHV-you-LAY-shun) Release of a secondary oocyte from an ovary during one menstrual cycle.

**ovule** (OHV-youl) [L. *ovum*, egg] Tissue mass in a plant ovary that develops into a seed. Consists of female gametophyte with egg cell, nutrient-rich tissue, and a jacket (cell layers) that will become a seed coat.

**ovum** (OH-vum) Mature secondary oocyte.

**ozone thinning** Pronounced seasonal thinning of the atmosphere's ozone layer, especially above the Earth's polar regions.

**pancreas** (PAN-cree-us) Gland with roles in digestion and organic metabolism. Secretes enzymes and bicarbonate into small intestine; also secretes insulin and glucagon, hormones that travel the bloodstream to target cells.

**parasite** [Gk. *para*, alongside, + *sitos*, food] Organism that lives in or on a host organism for at least part of its life cycle. It feeds on specific tissues and usually does not kill its host outright.

**parasitism** Symbiotic interaction in which one species (a parasite) benefits and the other (its host) is harmed. The parasite lives inside or on a host and feeds on its cells or tissues.

**parasitoid** Type of insect larva that grows and develops in a host organism (usually another insect), consumes its soft tissues, and kills it.

**parasympathetic nerve** (PARE-uh-SIM-pu-THET-ik) An autonomic nerve. Signals carried by such nerves tend to slow overall body activities and divert energy to basic tasks, and to help make small adjustments in internal organ activity by acting continually in opposition to sympathetic nerve signals.

**parenchyma** (par-ENG-kih-mah) Simple tissue that makes up the bulk of a plant; has roles in photosynthesis, storage, secretion, other tasks.

**parthenogenesis** (par-THEN-oh-GEN-uh-sis) An unfertilized egg giving rise to an embryo.

**passive transport** Process by which a transport protein that spans a cell membrane passively permits a solute to diffuse through its interior. Also called facilitated diffusion.

**pathogen** (PATH-oh-jen) [Gk. *pathos*, suffering, + *genēs*, origin] Any virus, bacterium, fungus, protistan, or parasitic worm that can infect an organism, multiply in it, and cause disease.

**peat bog** Compressed, soggy, highly acidic mat of accumulated remains of peat mosses.

**penis** Male copulatory organ by which sperm is deposited in a female reproductive tract.

**perennial** [L. *per-*, throughout, + *annus*, year] Flowering plant that lives for three or more growing seasons.

**pericycle** (PARE-ih-sigh-kul) [Gk. *peri-*, around, + *kyklos*, circle] One or more cell layers just inside the endodermis that give rise to lateral roots and contribute to secondary growth.

**periderm** Protective cover that replaces plant epidermis during extensive secondary growth.

**peripheral nervous system** (per-IF-ur-uhl) [Gk. *peripherein*, to carry around] All nerves leading into and out from the spinal cord and brain. Includes ganglia of those nerves.

**peritoneum** (pare-ih-tuh-NEE-um) Membrane that lines the coelom and helps maintain the positions of soft organs inside it.

**pest resurgence** Directional selection for an insecticide-resistant strain of a pest species.

**phagocyte** (FAG-uh-sight) [Gk. *phagein*, to eat, + *kytos*, hollow vessel] Cell that captures prey by phagocytosis (e.g., amoebas); also cells that use same process for defense and day-to-day tissue housekeeping (e.g., macrophages).

**phagocytosis** (FAG-uh-sigh-TOE-sis) [Gk. *phagein*, to eat, + *kytos*, hollow vessel] Engulfment of foreign cells or particles by way of pseudopod formation and endocytosis.

**pharynx** (FARE-inks) Among invertebrates, a muscular tube to the gut. In some chordates, a gas exchange organ. In land vertebrates, a dual entrance to the esophagus and trachea.

**phenotype** (FEE-no-type) [Gk. *phainein*, to show, + *typos*, image] Observable trait or traits of an individual that arise from gene interactions and gene–environment interactions.

**pheromone** (FARE-oh-moan) [Gk. *phero*, to carry, + *-mone*, as in hormone] Hormone-like, nearly odorless exocrine gland secretion. A signaling molecule between individuals of the same species that integrates social behavior.

**phloem** (FLOW-um) Plant vascular tissue that conducts sugars and other solutes. Includes living cells (sieve tubes) that connect to form conducting tubes, and adjoining companion cells that help load solutes into the tubes.

**phospholipid** Organic compound that has a glycerol backbone, two fatty acid tails, and a hydrophilic head of two polar groups (one being phosphate). Phospholipids are the main structural material of cell membranes.

**phosphorus cycle** Movement of phosphorus (mainly phosphate ions) from land, through food webs, to ocean sediments, then back to land. As for other minerals, Earth's crust is the largest reservoir in this sedimentary cycle.

**photoautotroph** Photosynthetic autotroph; any organism that synthesizes its own organic compounds using carbon dioxide as the source of carbon atoms and sunlight as the energy source. Nearly all plants, some protistans, and a few bacteria are photoautotrophs.

**photoperiodism** Biological response to change in relative lengths of daylight and darkness.

**photoreceptor** Light-sensitive sensory cell.

**photosynthesis** Trapping of sunlight energy, followed by its conversion to chemical energy (ATP, NADPH, or both) and then synthesis of sugar phosphates, which become converted into sucrose, cellulose, starch, and other end products. The main pathway by which energy and carbon enter the web of life.

**phototropism** [Gk. *photos*, light, + *trope*, a turning, direction] Change in the direction of cell movement or growth in response to light (e.g., as when differences in cell elongation cause a stem to bend toward light).

**phylogeny** Evolutionary relationships among species, starting with an ancestral form and including branches leading to descendants.

**phytochrome** A light-sensitive pigment. Its controlled activation and inactivation take part in plant hormone activities that govern leaf expansion, stem branching, stem lengthening and often seed germination and flowering.

**phytoplankton** (FIE-toe-PLANK-tun) [Gk. *phyton*, plant, + *planktos*, wandering] Aquatic community of floating or weakly swimming photoautotrophs (e.g., "pastures of the seas").

**pigment** Any light-absorbing molecule.

**pioneer species** Any opportunistic colonizer of barren or disturbed habitats. Adapted for rapid growth and dispersal.

**placenta** (play-SEN-tuh) Blood-engorged organ of pregnant female placental mammals; made of some endometrial tissue and extraembryonic membranes. Allows exchanges between the mother and fetus without an intermingling of their bloodstreams, thus sustaining the new individual and allowing its blood vessels to develop apart from the mother's.

**plankton** [Gk. *planktos*, wandering] Of aquatic habitats, a community of suspended or weakly swimming organisms, mostly microscopic.

**plant** Generally, a multicelled photoautotroph with well-developed root and shoot systems; photosynthetic cells that include starch grains as well as chlorophylls *a* and *b*; and cellulose, pectin, and other polysaccharides in cell walls.

**Plantae** Kingdom of plants.

**plasma** (PLAZ-muh) Liquid portion of blood; mainly water in which ions, proteins, sugars, gases, and other substances are dissolved.

**plasma membrane** Outermost cell membrane; structural and functional boundary between cytoplasm and the fluid outside the cell.

**plasmid** Of many bacteria, a small, circular molecule of extra DNA that carries only a few genes and that is replicated independently of the bacterial chromosome.

**plate tectonics** Theory that great slabs (plates) of the Earth's outer layer (lithosphere) float on the hot, plastic, underlying mantle. All plates are in motion and have rafted continents to new positions over time. The geologic changes have profoundly affected the evolution of life.

**pollen grain** [L. *pollen*, fine dust] Immature or mature, sperm-bearing male gametophyte of gymnosperms and angiosperms.

**pollination** Arrival of a pollen grain on the landing platform (stigma) of a carpel.

**pollutant** Natural or synthetic substance with which an ecosystem has no prior evolutionary experience, in terms of kinds or amounts; it accumulates to disruptive or harmful levels.

**polymorphism** (poly-MORE-fizz-um) [Gk. *polus*, many, + *morphe*, form] The persistence of two or more qualitatively different forms of a trait (morphs) in a population.

**polyp** (POH-lip) Vase-shaped, sedentary stage of cnidarian life cycles.

**population** All individuals of the same species occupying the same area.

**population density** Count of individuals of a population occupying a specified area or specified volume of a habitat.

**population distribution** Dispersal pattern for individuals of a population through a habitat.

**population size** The number of individuals that make up the gene pool of a population.

**positive feedback mechanism** A homeostatic control mechanism. It sets in motion a chain of events that intensifies change from an original condition; after a limited time, intensification reverses the change.

**predation** Ecological interaction in which a predator feeds on a prey organism.

**predator** [L. *prehendere*, to grasp, seize] A heterotroph that feeds on other living organisms (its prey), that lives neither in or nor on them (as parasites do), and that may or may not end up killing them.

**pressure flow theory** Explanation of how organic compounds move through phloem of vascular plant. The compounds follow solute concentration and pressure gradients between sources (e.g., photosynthetically active leaves where they form) and sinks (e.g., growing parts where they are being used or stored).

**primary growth** Plant growth originating at root tips and shoot tips.

**primary immune response** Defensive actions by white blood cells and their secretions, as elicited by first-time recognition of antigen. Includes antibody- and cell-mediated responses.

**primary productivity, gross** Of ecosystems, the rate at which primary producers capture and store a given amount of energy in their cells and tissues during a specified interval.

**primary productivity, net** Of ecosystems, the rate of energy storage in primary producer cells and tissues in excess of rate of aerobic respiration during a specified interval.

**primary wall** A wall of polysaccharides, glycoproteins, and cellulose that is flexible and thin enough to allow new plant cells to divide or change shape during growth and development.

**primate** Mammalian lineage dating from the Eocene; includes prosimians, tarsioids, and anthropoids (monkeys, apes, and humans).

**probability** The chance that each outcome of a given event will occur is proportional to the number of ways the outcome can be reached.

**producer** Autotroph (self-feeder); it nourishes itself using sources of energy and carbon from its physical environment. Photoautotrophs and chemoautotrophs are examples.

**progesterone** (pro-JESS-tuh-rown) A sex hormone secreted by ovaries and the corpus luteum of female mammals.

**prokaryotic cell** (pro-CARE-EE-oh-tic) [L. *pro*, before, + Gk. *karyon*, kernel] Archaebacterium or eubacterium; single-celled organism, most often walled; lacks the profusion of membrane-bound organelles observed in eukaryotic cells.

**prokaryotic fission** Cell division mechanism by which a bacterial cell reproduces.

**prophase** Of mitosis, a stage when duplicated chromosomes start to condense, microtubules form a spindle, and the nuclear envelope starts to break up. Duplicated pairs of centrioles (if present) are moved to opposite spindle poles.

**prophase I** The first stage of meiosis I. Each duplicated chromosome starts to condense. It pairs with its homologue; nonsister chromatids usually undergo crossing over. Each becomes attached to microtubular spindle. One of the duplicated pairs of centrioles (if present) is moved to opposite spindle pole.

**prophase II** First stage of meiosis II. In each daughter cell, spindle microtubules attach to kinetochores of each chromosome and move them toward spindle's equator. One centriole pair (if present) is already at each spindle pole.

**protein** Organic compound composed of one or more polypeptide chains.

**Protista** Kingdom of protistans. Chytrids; water molds; slime molds; protozoans; sporozoans; euglenoids; chrysophytes; dinoflagellates; and red, brown, and green algae are major groups.

**protistan** (pro-TISS-tun) [Gk. *prōtistos*, primal, very first] Diverse species, ranging from single cells to giant kelps, that are photoautotrophs, heterotrophs, or both. Some are thought to be most like the earliest eukaryotic cells. All are unlike bacteria in having a nucleus, large ribosomes, mitochondria, ER, Golgi bodies, chromosomes with many proteins attached, and cytoskeletal microtubules.

**proton** Positively charged particle; one or more reside in nucleus of each atom. An unbound (free) proton is called a hydrogen ion ($H^+$).

**protostome** (PRO-toe-stome) [Gk. *proto*, first, + *stoma*, mouth] Lineage of coelomate, bilateral animals that includes mollusks, annelids, and arthropods. The first indentation to form in protostome embryos becomes the mouth.

**protozoan** Type of protistan that may resemble the single-celled heterotrophs that gave rise to animals. Amoeboid, animal-like, and ciliated protozoans are major categories.

**pulmonary circuit** Vertebrate cardiovascular route in which oxygen-poor blood flows from the heart to the lungs, where it is oxygenated before flowing back to the heart.

**radial symmetry** Animal body plan having four or more roughly equivalent parts around a central axis (e.g., sea anemone).

**radioisotope** Unstable atom (uneven number of protons and neutrons). It spontaneously emits particles and energy; over a predictable time span, it decays into a different atom.

**radiometric dating** Method of measuring the proportions of (1) a radioisotope in a mineral trapped long ago in newly formed rock and (2) a daughter isotope that formed from it by radioactive decay in the same rock. Used to assign absolute dates to fossil-containing rocks and to the geologic time scale.

**rearrangement, molecular** Conversion of one organic compound to another through changes in its internal bonds.

**receptor, molecular** Type of membrane protein that binds an extracellular substance (e.g., hormone).

**receptor, sensory** Sensory cell or specialized cell adjacent to it that can detect a stimulus.

**recessive allele** [L. *recedere*, to recede] In heterozygotes, an allele whose expression is fully or partially masked by expression of its partner. Fully expressed only in homozygous recessives.

**recombinant chromosome** Of eukaryotes, a chromosome that emerges from meiosis with a combination of alleles that differs from a parental combination of alleles.

**recombinant DNA** Any molecule of DNA that incorporates one or more nonparental nucleotide sequences. Outcome of microbial gene transfer in nature or recombinant DNA technology.

**recombination** Any enzyme-mediated reaction that inserts one DNA sequence into another. "Generalized" recombination uses any pair of homologous sequences between chromosomes as substrates, as during crossing over. Site-specific recombination uses only a short stretch of homology between viral and bacterial DNA. A different reaction can insert transposable elements at new, random sites in bacterial or eukaryotic genomes; no homology is required.

**red alga** Type of protistan. Most are multicelled, aquatic photoautotrophs with an abundance of phycobilins that mask chlorophyll *a*.

**red blood cell** Erythrocyte. Cell that serves in rapid transport of oxygen in blood.

**reflex** [L. *reflectere*, to bend back] Stereotyped, simple movement in response to stimuli. In simple reflex arcs, sensory neurons synapse directly on motor neurons.

**regulatory protein** Component of mechanisms that control transcription, translation, and gene products by interacting with DNA, RNA, new polypeptide chains, or proteins (e.g., enzymes).

**reproduction** Any process by which a parental cell or organism produces offspring. Among eukaryotes, asexual modes (e.g., binary fission, budding, vegetative propagation) and sexual modes. Bacteria employ prokaryotic fission. Viruses do not reproduce themselves; host organisms execute their replication cycle.

**reproductive isolating mechanism** A heritable feature of body form, functioning, or behavior that prevents interbreeding between two or more genetically divergent populations.

**reproductive success** Production of viable offspring by the individual.

**reptile** Carnivorous species belonging to the first vertebrate lineage to escape dependency on free water, by way of internal fertilization, efficient kidneys, amniote eggs, and other adaptations. Examples are dinosaurs (extinct), crocodilians, snakes, lizards, and tuataras.

**resource partitioning** Of two or more species that compete for the same resource, a sharing of the resource in different ways or at different times, which allows them to coexist.

**respiration** [L. *respirare*, to breathe] Of all animals, exchange of environmental oxygen with carbon dioxide from cells (e.g., through integumentary exchange or a respiratory system).

**respiratory surface** Thin, moist epithelium that functions in gas exchange in animals.

**resting membrane potential** Of a neuron and other excitable cells, a steady voltage difference across the plasma membrane in the absence of outside stimulation.

**restriction enzyme** One of a class of bacterial enzymes that cut apart foreign DNA injected into the cell body, as by viruses. Important tool of recombinant DNA technology.

**reticular formation** Low-level pathway of information flow through vertebrate nervous system. Mesh of interneurons that extends from the upper spinal cord, through the brain stem, and into the cerebral cortex.

**rhizoid** Simple rootlike absorptive structure of some fungi and nonvascular plants.

**ribosomal RNA** (rRNA) Type of RNA that combines with proteins to form ribosomes, on which polypeptide chains are assembled.

**ribosome** Structure composed of two subunits of rRNA and proteins. Has binding sites for mRNA and tRNAs, which interact to produce a polypeptide chain in translation stage of protein synthesis.

**RNA** Ribonucleic acid. Any of a class of single-stranded nucleic acids that function in transcribing and translating the genetic instructions encoded in DNA into proteins. A molecule of mRNA, rRNA, or tRNA.

**root** Plant part, typically belowground, that absorbs water and dissolved minerals, anchors aboveground parts, and often stores food.

**root hair** Threadlike extension of a specialized epidermal cell of a young root. Increases root surface area for absorbing water and minerals.

**root nodule** Localized swelling on a root of certain legumes and other plants. Develops when nitrogen-fixing bacteria infect the plant, multiply, and interact symbiotically with it.

**salinization** Salt buildup in soil by evaporation, poor drainage, and heavy irrigation.

**saliva** Glandular secretion that is mixed with food and starts starch breakdown in the mouth.

**salt** Compound that releases ions other than $H^+$ and $OH^-$ in solution.

**saltatory conduction** Of myelinated neurons, a rapid form of action potential propagation. Excitation hops to nodes between jellyrolled membranes of cells making up myelin sheath.

**saprobe** Heterotroph that obtains energy and carbon from nonliving organic matter and so causes its decay (e.g., many fungal species).

**sclerenchyma** (skler-ENG-kih-mah) Simple plant tissue that supports mature plant parts and commonly protects seeds. Most of its cells have thick, lignin-impregnated walls.

**sea-floor spreading** Ongoing event in which molten rock erupts from immense, continuous ridges on the ocean floor, flows laterally in both directions, and hardens to form new crust. Elsewhere, it forces older crust down into vast trenches in the seafloor.

**second messenger** Molecule within a cell that mediates a hormonal signal by initiating the cellular response to it.

**secondary immune response** Immune action against previously encountered antigen, more rapid and prolonged than a primary response owing to swift participation of memory cells.

**secondary sexual trait** A trait associated with maleness or femaleness but with no direct role in reproduction (e.g., distribution of body hair and body fat). The primary sexual trait is the presence of male or female gonads.

**secondary wall** A wall on the inner surface of the primary wall of an older plant cell that stopped growing but needs structural support. Contains lignin in older cells of woody plants.

**secretion** A cell acting on its own or as part of glandular tissue releases a substance across its plasma membrane, to the surroundings.

**sedimentary cycle** Biogeochemical cycle. An element having no gaseous phase moves from land, through food webs, to the seafloor, then returns to land through long-term uplifting.

**seed** Mature ovule with an embryo sporophyte inside and integuments that form a seed coat.

**segmentation** Of animal body plans, a series of units that may or may not be similar to one another in appearance. Of tubular organs, an oscillating movement produced by rings of circular muscle in the tube wall.

**segregation, theory of** [L. *se-*, apart, + *grex*, herd] Mendelian theory. Sexually reproducing organisms inherit pairs of genes (on pairs of homologous chromosomes), the two genes of each pair are separated from each other at meiosis, and they end up in separate gametes.

**selective gene expression** Control of which gene products a cell makes or activates during a specified interval. Depends on the type of cell, its adjustments to changing chemical conditions, which external signals it is receiving, and its built-in control systems.

**selective permeability** Of a cell membrane, a capacity to let some substances but not others cross it at certain sites, at certain times. The capacity arises as an outcome of its lipid bilayer structure and its transport proteins.

**selfish behavior** An individual protects or increases its own chance to produce offspring regardless of consequences to its social group.

**selfish herd** Social group held together simply by reproductive self-interest.

**semen** (SEE-mun) Sperm-bearing fluid expelled from a penis during male orgasm.

**senescence** (sen-ESS-cents) [L. *senescere*, to grow old] Processes leading to the natural death of an organism or to parts of it (e.g., leaves).

**sensation** Conscious awareness of a stimulus.

**sensory neuron** Type of neuron that detects a stimulus and relays information about it toward an integrating center (e.g., a brain).

**sensory system** The "front door" of a nervous system; it detects external and internal stimuli and relays information to integrating centers that issue commands for responses.

**sessile animal** (SESS-ihl) Animal that remains attached to a substrate during some stage (often the adult stage) of its life cycle.

**sex chromosome** A chromosome with genes that influence primary sex determination (whether male or female gonads will develop in the new individual). Depending on the species, somatic cells have one or two sex chromosomes, of the same or different type. In mammals, females are XX and males XY.

**sexual dimorphism** Occurrence of female and male phenotypes among the individuals of a sexually reproducing species.

**sexual reproduction** Production of offspring by meiosis, gamete formation, and fertilization.

**sexual selection** A microevolutionary process. Natural selection favors a trait that gives the individual a competitive edge in attracting or keeping a mate, hence in reproductive success.

**shell model** Model of electron distribution in which all orbitals available to electrons of atoms occupy a nested series of shells.

**shifting cultivation** Cutting and burning trees in a plot of land, followed by tilling ashes into soil. Once called slash-and-burn agriculture.

**shoot system** Aboveground plant parts (e.g., stems, leaves, and flowers).

**sieve-tube member** One of the cells that join together as phloem's sugar-conducting tubes.

**sign stimulus** Simple environmental cue that triggers a response to a stimulus that the nervous system is prewired to recognize.

**sink** Any region of a plant where cells are storing or using food (e.g., roots).

**sister chromatid** Of a duplicated chromosome, one of two DNA molecules (and associated proteins) attached at the centromere until they are separated from each other during mitosis or meiosis. After separation, each is then called a chromosome in its own right.

**six-kingdom classification scheme** A recent phylogenetic scheme that groups all organisms into the kingdoms Eubacteria, Archaebacteria, Protista, Fungi, Plantae, and Animalia.

**skeletal muscle** An organ with hundreds to many thousands of muscle cells bundled inside a sheath of connective tissue, which extends past the muscle as tendons.

**small intestine** Part of the vertebrate digestive system in which digestion is completed and most dietary nutrients are absorbed.

**social behavior** Diverse interactions among individuals of a species, which display, send, and respond to shared forms of communication that have genetic and learned components.

**social parasite** Animal that exploits the social behavior of another species to assure its own survival and reproduction.

**soil** Mixture of mineral particles of variable sizes and decomposing organic material; air and water occupy spaces between particles.

**solute** (SOL-yoot) [L. *solvere*, to loosen] Any substance dissolved in a solution. Spheres of hydration around charged parts of its ions and molecules keep them dispersed.

**solvent** Any fluid (e.g., water) in which one or more substances are dissolved.

**somatic cell** (so-MAT-ik) [Gk. *soma*, body] Any body cell that is not a germ cell. (Germ cells are the forerunners of gametes.)

**somatic nervous system** Nerves leading from a central nervous system to skeletal muscles.

**somite** One of many paired segments in a vertebrate embryo that give rise to most bones, skeletal muscles of head and trunk, and dermis.

**source** Any plant part where photosynthetic cells are making organic compounds.

**speciation** (spee-see-AY-shun) The formation of a daughter species from a population or subpopulation of a parent species by way of microevolutionary processes. Routes vary in their details and in length of time before the required reproductive isolation is completed.

**species** (SPEE-sheez) [L. *species*, a kind] One kind of organism. Of sexually reproducing organisms, one or more natural populations in which individuals are interbreeding and are reproductively isolated from other such groups.

**sperm** [Gk. *sperma*, seed] Mature male gamete.

**spermatogenesis** (sper-MAT-oh-JEN-ih-sis) Formation of mature sperm from a germ cell.

**sphere of hydration** A clustering of water molecules around individual molecules or ions of a substance placed in water owing to positive and negative interactions among them.

**spinal cord** The part of the central nervous system in a canal inside the vertebral column; site of direct reflex connections between sensory and motor neurons; also has tracts to and from the brain.

**spleen** A lymphoid organ that is a filtering station for blood, a reservoir of red blood cells, and a reservoir of macrophages.

**spore** Reproductive or resting structure of one or a few cells, often walled or coated and adapted for resisting adverse conditions, for dispersal, or both. May be nonsexual or sexual (formed by way of meiosis). Sporozoans, fungi, plants, and some bacteria form spores.

**sporophyte** [Gk. *phyton*, plant] A vegetative body that grows, by mitotic cell divisions, from a plant zygote and that produces spore-bearing structures.

**stabilizing selection** Mode of natural selection by which intermediate phenotypes in the range of variation are favored and extremes at both ends are eliminated.

**stamen** (STAY-mun) A male reproductive part of a flower; usually a pollen-bearing structure (anther) on a single stalk (filament).

**stem cell** Self-perpetuating animal cell that stays unspecialized. Some of its daughter cells also are self-perpetuating; others differentiate into specialized cells (e.g., red blood cells that arise from stem cells in bone marrow).

**steroid hormone** Lipid-soluble hormone made from cholesterol that acts on a target cell's DNA by entering the nucleus alone or bound to intracellular receptor. Some act by binding to a receptor on a target's plasma membrane.

**stigma** Sticky or hairy surface tissue on upper part of a carpel (or fused carpels) that captures pollen grains and favors their germination.

**stimulus** [L. *stimulus*, goad] A specific form of energy (e.g., pressure, light, and heat) that activates a sensory receptor able to detect it.

**stomach** A muscular, stretchable sac that mixes and stores ingested food, helps break it apart mechanically and chemically, and controls its expulsion (e.g., into the small intestine).

**strain** One of two organisms with differences that are too minor to classify it as a separate species (e.g., *Escherichia coli* strain 018:K1:H).

**stratification** Stacked layers of sedimentary rock that resulted from a gradual deposition of volcanic ash, silt, and other materials over time.

**stream** A flowing-water ecosystem that starts out as a freshwater spring or seep.

**stromatolite** Fossilized mats of shallow-water, microbial communities (mainly cyanobacteria) of Archean to Precambrian times. Their tacky gel secretions blocked out ultraviolet radiation but trapped sediments; so new mats had to grow over older ones, like cake layers. Stromatolites are found on all continents; some are over a half mile thick and hundreds of miles across.

**succession, primary** (suk-SESH-un) [L. *succedere*, to follow after] Ecological pattern by which a community develops in orderly progression, from the time pioneer species colonize a barren habitat to the climax community.

**succession, secondary** Ecological pattern by which a disturbed area of a community recovers and moves back toward the climax state. It is typical of abandoned croplands, forest burns, and storm-battered intertidal zones.

**surface-to-volume ratio** Mathematical relation in which volume increases with the cube of the diameter, but surface area increases only with the square. If a growing cell were simply to expand in diameter, its volume of cytoplasm would increase faster than the surface area of the plasma membrane required to service it. In general, this constraint keeps cells small, elongated, or with infoldings or outfoldings of its plasma membrane.

**swim bladder** Adjustable flotation device that changes in volume as it exchanges gases with blood; it helps many fishes maintain neutral buoyancy in water.

**symbiosis** (sim-by-OH-sis) [Gk. *sym*, together, + *bios*, life, mode of life] Individuals of one species live near, in, or on those of another species for at least part of life cycle (e.g., in commensalism, mutualism, and parasitism).

**sympathetic nerve** An autonomic nerve that deals mainly with increasing overall body activities at times of heightened awareness, excitement, or danger; also works continually in opposition with parasympathetic nerves to make minor adjustments in internal organ activities.

**sympatric speciation** [Gk. *sym*, together, + *patria*, native land] A speciation event within the home range of an existing species, in the absence of a physical barrier. Such species may form instantaneously, as by polyploidy.

**systemic circuit** (sis-TEM-ik) Of vertebrates, a cardiovascular route in which oxygen-enriched blood flows from the heart through the rest of the body (where it gives up oxygen and takes up carbon dioxide), then back to the heart.

**taproot system** A primary root together with all of its lateral branchings.

**target cell** Any cell with molecular receptors that can bind with a particular hormone or some other signaling molecule.

**taxonomy** Field of biology that deals with identifying, naming, and classifying species.

**telophase** (TEE-low-faze) Of meiosis I, the stage when one member of each pair of homologous chromosomes has arrived at a spindle pole. Of mitosis and of meiosis II, the stage when chromosomes decondense into threadlike structures and two daughter nuclei form.

**temperature** A measure of the kinetic energy of ions or molecules in a specified region.

**tendon** A cord or strap of dense connective tissue that attaches a muscle to bone.

**territory** An area that an animal is defending against competitors for mates, food, water, living space, other resources.

**test** A means to determine the accuracy of a prediction, as by conducting experimental or observational tests and by developing models. Scientific tests are made under controlled conditions in nature or the laboratory.

**testis**, plural **testes** A primary reproductive organ (gonad) of some male animals; it produces male gametes and sex hormones.

**testosterone** (tess-TOSS-tuh-rown) A type of sex hormone necessary for the development and functioning of the male reproductive system of vertebrates.

**theory, scientific** A testable explanation of a broad range of related phenomena, one that has been subjected to extensive experimental testing and can be used with a high degree of confidence. A scientific theory remains open to tests, revision, and tentative acceptance or rejection.

**thermophile** A type of archaebacterium that is adapted to unusually hot aquatic habitats, such as hot springs and hydrothermal vents.

**thermoreceptor** Sensory cell or specialized cell next to it that detects radiant energy (heat).

**thigmotropism** (thig-MOE-truh-pizm) [Gk. *thigm*, touch] An orientation of the direction of growth in response to physical contact with a solid object, as when a vine curls around a fencepost.

**threshold** Of an excitable cell (e.g., a neuron or muscle cell), the minimum amount of change in the resting membrane potential that will trigger an action potential.

**thymus gland** A lymphoid organ that has endocrine functions. Lymphocytes of the immune system multiply, differentiate, and mature in its tissues; its hormone secretions affect their functioning.

**thyroid gland** Endocrine gland located in front of the trachea; its hormones have widespread effects on growth and development, and on the overall metabolic rates of warm-blooded animals.

**tissue** Of multicelled organisms, a group of cells and intercellular substances that function together in one or more specialized tasks.

**touch-killing** Mechanism by which cytotoxic T cells directly release perforins and toxins onto a target cell and cause its destruction.

**toxin** A normal metabolic product of one species with chemical effects that can hurt or kill individuals of a different species.

**trace element** Any element that represents less than 0.01 percent of body weight.

**trachea** (TRAY-kee-uh), plural **tracheae** An air-conducting tube of respiratory systems. Of land vertebrates, the windpipe through which air passes between the larynx and bronchi.

**tracheal respiration** Of certain invertebrates (e.g., insects), respiration by way of finely branching tracheae that start at openings in the integument and dead-end in body tissues.

**tracheid** (TRAY-kid) One of two types of cells in xylem that conduct water and minerals.

**tract** A cordlike bundle of axons of sensory neurons, motor neurons, or both inside the brain or spinal cord. Comparable to a nerve.

**transpiration** Evaporative water loss from a plant's aboveground parts, leaves especially.

**transport protein** One of many kinds of membrane proteins involved in active or passive transport of water-soluble substances across the lipid bilayer of a cell membrane. Solutes on one side of the membrane pass through the protein's interior to the other side.

**transposable element** A stretch of DNA that can move at random from one location to another in the individual's genome. Often it inactivates the genes into which it becomes inserted and causes changes in phenotype.

**triglyceride** (neutral fat) A type of lipid that has three fatty acid tails attached to a glycerol backbone. Triglycerides are the body's most abundant lipids and its richest energy source.

**trisomy** (TRY-so-mee) The presence of three chromosomes of a given type in a cell rather than the two characteristic of a parental diploid chromosome number.

**trophic level** (TROE-fik) [Gk. *trophos*, feeder] Of an ecosystem, all organisms that are the same number of transfer steps away from the energy input into the system.

**tropical rain forest** A biome characterized by regular, heavy rainfall, an annual mean temperature of 25°C, humidity of 80 percent or more, and stunning biodiversity. Presently being obliterated in regions with fast-growing human populations but limited food, fuel, and lumber; projections are that most will disappear by 2035.

**tropism** (TROE-pizm) Of plants, a directional growth response to an environmental factor (e.g., growth toward light).

**true breeding** Of a sexually reproducing species, a lineage in which one version only of a trait shows up through the generations in all parents and their offspring.

**turgor pressure** (TUR-gore) [L. *turgere*, to swell] Internal fluid pressure on a cell wall when water moves into the cell by osmosis.

**uniformity theory** Early theory that the Earth's surface changes in gradual, uniformly repetitive ways (major floods, earthquakes, and other infrequent annual catastrophes were not considered unusual). Helped change Darwin's view of evolution. Has since been replaced by plate tectonics theory.

**uracil** (YUR-uh-sill) Nitrogen-containing base of a nucleotide in RNA but not DNA. Like thymine, uracil can base-pair with adenine.

**urinary bladder** The distensible sac in which urine is stored before being excreted.

**urinary excretion** Mechanism by which excess water and solutes are removed from the body through the urinary system.

**urinary system** Organ system that adjusts the volume and composition of blood, and thereby helps maintain extracellular fluid.

**urine** Fluid consisting of any excess water, wastes, and solutes; it forms in kidneys by filtration, reabsorption, and secretion.

**uterus** (YOU-tur-us) [L. *uterus*, womb] Of a female placental mammal, a muscular, pear-shaped organ in which embryos are contained and nurtured during pregnancy.

**vaccination** Immunization procedure against a specific pathogen.

**vaccine** An antigen-containing preparation, swallowed or injected, designed to increase immunity to certain diseases by inducing formation of armies of effector and memory B and T cells.

**vagina** The part of the reproductive system of mammalian females that receives sperm, forms part of the birth canal, and channels menstrual flow to the exterior.

**variable** Of an experimental test, a specific aspect of an object or event that may differ over time and among individuals. A single variable is directly manipulated in an attempt to support or disprove a prediction; any other variables that might influence the results are identical (ideally) in both the experimental group and one or more control groups.

**vascular bundle** An array of primary xylem and phloem in multistranded, sheathed cords that thread lengthwise throughout the ground tissue system.

**vascular cambium** A lateral meristem that increases stem or root diameter.

**vascular cylinder** Arrangement of vascular tissues as a central cylinder in roots.

**vascular plant** Plant with xylem and phloem, and usually with well-developed roots, stems, and leaves.

**vascular tissue system** Xylem and phloem, the conducting tissues that distribute water and solutes through a vascular plant.

**vein** Of a cardiovascular system, any of the large-diameter vessels that lead back to the heart. Of leaves, one of the vascular bundles that thread through photosynthetic tissues.

**vernalization** Environmental stimulation of flowering, by exposure to low temperatures.

**vertebra**, plural **vertebrae** One of a series of hard bones, with cushioning intervertebral disks stacked between them, that serve as a backbone and that protect the spinal cord.

**vertebrate** Animal with a backbone.

**vesicle** (VESS-ih-kul) [L. *vesicula*, little bladder] One of a variety of small, membrane-bound sacs in the cytoplasm that function in the transport, storage, or digestion of substances.

**vessel member** Type of cell in xylem; dead at maturity, but its wall becomes part of a water-conducting pipeline (a vessel).

**vestigial** (ves-TIDJ-ul) Applies to a small body part, tissue, or organ abnormally developed or degenerated and unable to function like its normal counterpart (e.g., vestigial wings of mutant fruit flies; "tail bones" of humans).

**villus** (VIL-us), plural **villi** Any of several fingerlike absorptive structures projecting from the free surface of an epithelium.

**viroid** An infectious particle consisting only of very short, tightly folded strands or circles of RNA. Viroids might have evolved from introns, which they resemble.

**virus** A noncellular infectious agent that is composed of DNA or RNA and a protein coat; it can become replicated only after its genetic material enters a host cell and subverts the host's metabolic machinery.

**viscera** All soft organs inside an animal body (e.g., heart, lungs, and stomach).

**vision** Perception of visual stimuli. Requires a focusing of light precisely onto a layer of photoreceptive cells that is dense enough to sample details of a light stimulus, followed by image formation in a brain.

**visual signal** An observable action or cue that functions as a communication signal.

**vitamin** Any of more than a dozen organic substances that an organism requires in small amounts for metabolism but that it generally cannot synthesize for itself.

**vocal cord** Of certain animals, one of the thickened, muscular folds of the larynx that help produce sound waves for vocalization.

**water potential** Sum of two opposing forces (osmosis and turgor pressure) that can cause a directional movement of water into or out of an enclosed volume (e.g., a walled cell).

**wavelength** A wavelike form of energy in motion. The horizontal distance between the crests of every two successive waves.

**wax** Molecule having long-chain fatty acids packed together and linked to long-chain alcohols or to carbon rings. Waxes have a firm consistency and repel water.

**wild-type allele** Allele that occurs normally or with the greatest frequency at a given gene locus among individuals of a population.

**wing** A body part that serves in flight, as among birds, bats, and many insects. A bird wing is a forelimb having feathers, strong muscles, and extremely lightweight bones. A bat wing is a modified forelimb; four thin, elongated digits are the framework for a thin integumentary membrane. An insect wing develops as a lateral fold of the exoskeleton.

**X chromosome** A type of sex chromosome. In mammals, an XX pairing causes an embryo to develop into a female; an XY pairing causes it to develop into a male.

**X-linked gene** Any gene that is located on an X chromosome.

**X-linked recessive inheritance** Recessive condition in which the responsible, mutated gene is located on the X chromosome.

**xylem** (ZYE-lum) [Gk. *xylon*, wood] A tissue having pipelines that conduct water and solutes through vascular plants. The pipelines are the interconnecting walls of cells that are dead at maturity.

**Y chromosome** Distinctive chromosome in males or females of many species, but not both (e.g., human males are XY and human females, XX).

**Y-linked gene** Gene on a Y chromosome.

**yolk sac** An extraembryonic membrane. In most shelled eggs, it holds nutritive yolk; in humans, part of the yolk sac becomes a site of blood cell formation and some cells give rise to forerunners of gametes.

**zooplankton** A community of suspended or weak-swimming heterotrophs of freshwater or marine habitats. Most of its species are microscopic; commonly, rotifers and copepods are among the most abundant.

**zygote** (ZYE-goat) The first cell of a new individual, formed by fusion of a sperm nucleus with egg nucleus at fertilization; a fertilized egg.

# CREDITS AND ACKNOWLEDGMENTS

**Diversity of Life**
This page constitutes an extension of the copyright page. We have made every effort to trace the ownership of all copyrighted material and to secure permission from copyright holders. In the event of any question arising as to the use of any material, we will be pleased to make the necessary corrections in future printings. Thanks are due to the following authors, publishers, and agents for permission to use the material indicated.

ART BY LISA STARR 23.9 (b); 23.19; 27.16 (a); 27.19; 27.30; 27.33; 27.37; 28.2; 28.4; 28.5 (a, below); 28.8; 28.9; 28.11

CHAPTER 21 21.1 Jeff Hester and Paul Scowen, Arizona State University, and NASA / 21.2 Painting by William K. Hartmann / 21.3 (a) Painting by Chesley Bonestell / 21.4 Art by Precision Graphics / 21.5 (a) Sidney W. Fox; (b) W. Hargreaves and D. Deamer / 21.7 Art by Precision Graphics / 21.8 Bill Bachman/Photo Researchers / 21.9 (a) Stanley M. Awramik; (b–f) Andrew H. Knoll, Harvard University / 21.10 P. L. Walne and J. H. Arnott, *Planta*, 77: 325–354, 1967 / 21.11 Art by Raychel Ciemma / 21.12 Robert K. Trench / 21.13 (a, b) Neville Pledge/South Australian Museum; (c, d) Chip Clark; / 21.14 (a, b, d) Art by Raychel Ciemma; (c) Patricia G. Gensel / 21.15 (above) Painting by Megan Rohn courtesy David Dilcher; (below) © John Gurche 1989 / 21.16 © John Gurche 1989 / 21.17 NASA Galileo Imaging Team / 21.18 Art by Raychel Ciemma / 21.19 (a, b) Paintings © Ely Kish; (c) Field Museum of Natural History, Chicago, and artist Charles R. Knight (Neg. No. CK8T) / 21.20 (left) Maps by L. K. Townsend after A. M. Ziegler, C. R. Scotese, and S. F. Barrett, "Mesozoic and Cenozoic Paleogeographic Maps" and J. Krohn and J. Sundermann, "Paleotides Before the Permian" in F. Brosche and J. Sundermann (Eds.) *Tidal Friction and the Earth's Rotation II*, Springer-Verlag, 1983. /

CHAPTER 22 22.1 (a–c) Tony Brain and David Parker/SPL/Photo Researchers / 22.2 Lee D. Simon/Photo Researchers / 22.3 Art by Raychel Ciemma / 22.4 (a) Stanley Flegler/Visuals Unlimited; (b) P. Hawtin, University of Southhampton/SPL/Photo Researchers; (c) CNRI/SPL/Photo Researchers / 22.5 Courtesy of K. Amako and A. Umeda, Kyshu University, Japan / 22.6 L. J. LeBeau, University of Illinois Hospital/BPS / 22.7 Art by Raychel Ciemma; Micrograph L. Santo. / 22.10 (a) R. Robinson/Visuals Unlimited; (b) James Jones, *Archives of Microbiology*, 136: 254–261, 1983. Reprinted by permission of Springer-Verlag. / 22.11 (a) © 2000 PhotoDisc, Inc.; (b) Barry Rokeach; (c) courtesy of NASA; (d) © Alan L. Detrick, Science Source/Photo Researchers, Inc. / 22.12 (a) John D. Cunningham/Visuals Unlimited; (b) Tony Brain/SPL/Photo Researchers; (c) P. W. Johnson and J. MeN. Sieburth, Univ. Rhode Island/BPS / 22.13 T. J. Beveridge, University of Guelph/BPS / 22.14 © Ken Greer/Visuals Unlimited / 22.15 (a) Richard Blakemore; (b) Hans Reichenbach, Gesellschaft fur Biotechnologische Forschung, Braunschweig, Germany / 22.17 (a) K. G. Murti/Visuals Unlimited; (b) George Musil/Visuals Unlimited; (c, d) Kenneth M. Corbett / 22.18, 22.19 Art by Palay/Beaubois and Precision Graphics / 22.20 R. Regnery, Centers for Disease Control and Prevention, Atlanta / 22.21 E. A. Zottola, University of Minnesota /

CHAPTER 23 23.1 (a) Steven C. Wilson/Entheos; (b) Ronald W. Hoham, Dept. of Biology, Colgate University; (c) Edward S. Ross; (d) Gary W. Grimes and Steven L'Hernault / 23.2 (a) M. S. Fuller, *Zoosporic Fungi in Teaching and Research*, M. S. Fuller and A. Jaworski (eds.), 1987, Southeastern Publishing Company, Athens, GA; (b) Claude

Taylor and the University of Wisconsin Dept. of Botany; (c) Heather Angel; (d) W. Merrill / 23.3 (a) Art by Leonard Morgan; (b) M. Claviez, G. Gerish, and R. Guggenheim; (c) London Scientific Films; (d–f) Carolina Biological Supply Company; (g) courtesy Robert R. Kay from R. R. Kay, et al., *Development*, 1989 Supplement, pp. 81–90, © The Company of Biologists Ltd., 1989. / 23.4 M. Abbey/Visuals Unlimited / 23.5 (a) Neal Ericsson; (b) John Clegg/Ardea, London; (c) Courtesy of Allan W. H. Bé and David A. Caron; (d) Dr. Howard J. Spero; (e, g) art redrawn from V. & J. Pearse and M. & R. Buchsbaum, *Living Invertebrates*, The Boxwood Press, 1987. Used by permission; (f) G. Shih and R. Kessel/Visuals Unlimited; (h) T. E. Adams/Visuals Unlimited / 23.6 (a) Frieder Sauer/Bruce Coleman Ltd.; (b) Art by Raychel Ciemma redrawn from V. & J. Pearse and M. & R. Buchsbaum, *Living Invertebrates*, The Boxwood Press, 1987. Used by permission; (c) Art by Raychel Ciemma; (d) Sidney L. Tamm; (e) Ralph Buchsbaum and sketch from V. & J. Pearse and M. & R. Buchsbaum, *Living Invertebrates*, The Boxwood Press, 1987. Used by permission / 23.7 Art by Raychel Ciemma / 23.8 (a, b) Dr. Stan Erlandsen, University of Minnesota / 23.9 (a) © Oliver Meckes/Photo Researchers, Inc.; (b) After Prescott, *Microbiology*, 3rd Ed. / 23.10 David M. Phillips/Visuals Unlimited / 23.11 courtesy of Saul Tzipori, Div. of Infectious Diseases, Tufts University School of Veterinary Medicine / 23.12 © A. M. Siegelman/Visuals Unlimited 23.13 © Gary Head / 23.14 Art by L. Morgan; micrograph Steven L'Hernault / 23.15 Art by Palay/Beaubois / 23.16 (a, b) Ronald W. Hoham, Dept. of Biology, Colgate University; (c) Greta Fryxell, University of Texas, Austin; (d) *Emiliania huxleyi*. Photograph by Vita Pariente. Scanning electron micrograph taken on a Jeol T330A instrument at the Texas A&M University Electron Microscopy Center / 23.17 (a) S. Carty, Heidelberg College; (b, left) © S. Berry/Visuals Unlimited; (b, right) C. C. Lockwood; (c, d) North Carolina State University, Aquatic Botany Lab / 23.18 (a) D. P. Wilson/Eric & David Hosking; (b) Douglas Faulkner/Sally Faulkner Collection / 23.19 Art by Raychel Ciemma / 23.20 (a) J. R. Waaland, University of Washington/BPS; (b) Lewis Trusty/Animals Animals; (below) from T. Garrison, *Oceanography: An Invitation to Marine Science*, Brooks/Cole, 1993. / 23.21 (a) Herve Chaumeton/Agence Nature; (b) © Linda Sims/Visuals Unlimited; (c) Brian Parker/Tom Stack and Associates; (d, e) D. S. Littler; (f) Manfred Kage/Peter Arnold, Inc.; (g, h) Ronald W. Hoham, Dept. Biology, Colgate University / 23.22 Photograph D. J. Patterson/Seaphot Limited: Planet Earth Pictures; Art by Raychel Ciemma / 23.23 Carolina Biological Supply Company /

CHAPTER 24 24.1 © Sherry K. Pittam / 24.2 (a–f) Robert C. Simpson/Nature Stock / 24.3 (a, b) Robert C. Simpson/Nature Stock / 24.4 Art by Raychel Ciemma; Micrograph Garry T. Cole, University of Texas, Austin/BPS / 24.5 (a) Jane Burton/Bruce Coleman, Ltd.; (b) Thomas J. Duffy / p. 396 Micrograph J. D. Cunningham/Visuals Unlimited / 24.6 Micrographs Ed Reschke; Art by Raychel Ciemma / 24.7 (a) After T. Rost, et al., *Botany*, Wiley, 1979; (b) M. Eichelberger/Visuals Unlimited; (c) Robert C. Simpson/Nature Stock; (d)

G. L. Barron, University of Guelph; (e) Garry T. Cole, University of Texas, Austin/BPS / 24.8 N. Allin and G. L. Barron / 24.9 (a ) After Raven, Evert, and Eichhorn, *Biology of Plants*, 4th Ed., Worth Publishers, New York, 1986; (b) © Mark E. Gibson/Visuals Unlimited; (c) Gary Head; (d) Mark Mattock/Planet Earth Pictures; (e) Edward S. Ross / 24.10 (a) Prof. DJ Read, University of Sheffield; (b) © 1990 Gary Braasch; (c) F. B. Reeves / 24.11 (a) Dr. P. Marazzi/SPL/Photo Researchers; (b) Eric Crichton/Bruce Coleman, Ltd. / 24.12 John Hodgin /

CHAPTER 25 25.1 (a) Pat & Tom Leeson/Photo Researchers; (b, c) Edward S. Ross / 25.2 Art by Raychel Ciemma after E. O. Dodson and P. Dodson, *Evolution Process and Product*, 3rd Ed., p. 401, PWS / 25.4 (a) Craig Wood/Visuals Unlimited; (b) Jane Burton/Bruce Coleman Ltd.; Art by Raychel Ciemma / 25.5 (a) Fred Bavendam/Peter Arnold; (b) John D. Cunningham/Visuals Unlimited / 25.6 (a, b) Kingsley R. Stern; (c) John D. Cunningham/Visuals Unlimited / 25.7 (b) Kingsley R. Stern; (c) Ed Reschke/Peter Arnold; (d) William Ferguson; (e) W. H. Hodge; (f) Kratz/ZEFA / 25.8 Art by Raychel Ciemma; Photograph (inset) A. & E. Bomford, Ardea, London; Photograph Lee Casebere / 25.9 Art by Raychel Ciemma; Brian Parker/Tom Stack & Associates; (below) Field Museum of Natural History, Chicago (Neg. #7500C) / 25.10 (a) Kathleen B. Pigg, Arizona State University; (b) George J. Wilder/Visuals Unlimited / 25.11 (a) Stan Elams/Visuals Unlimited; (b) Gary Head / 25.12 (a) Jeff Gnass Photography; (b) R. J. Erwin/Photo Researchers; (c) Robert & Linda Mitchell; (d) Doug Sokell/Visuals Unlimited / 25.13 (a) Robert & Linda Mitchell; (b) Ed Reschke / 25.14 (a) Joyce Photographics/Photo Researchers; (b) Runk/Schoenberger/Grant Heilman, Inc.; (c) William Ferguson; (d) Kingsley R. Stern / 25.15 (a) Edward S. Ross; (b) William Ferguson; (c) Robert & Linda Mitchell Photography; (d) William Ferguson; (e) Biological Photo Service / 25.16 (left) Edward S. Ross; (top) Robert & Linda Mitchell Photography; (bottom) R. J. Erwin/Photo Researchers; Art by Raychel Ciemma / 25.17 (a) David Hiser, Photographers/Aspen, Inc.; (b) Terry Livingstone; (c) © 1993 Trygve Steen; (d) © 1994 Robert Glenn Ketchum / 25.18 (a) Art by Jennifer Wardrip; (b) M. P. L. Fogden/Bruce Coleman Ltd.; (c) Heather Angel; (d) L. Mellichamp/Visuals Unlimited; (e) Peter F. Zika/Visuals Unlimited / 25.19 Art by Raychel Ciemma / 25.20 (a) John Mason/Ardea, London; (b) Earl Roberge/Photo Researchers; (c) ZEFA-Rein; (d) George Loun/Visuals Unlimited; (e) Dick Davis/Photo Researchers / 25.21 Bob Cerasoli / 25.22 (left, center) © 1989, 1991 Clinton Webb; (right) Gary Head /

CHAPTER 26 26.1 Courtesy of Department of Library Services, American Museum of Natural History (Neg. #K10273) / 26.2 Photograph by Lisa Starr, Jack Carey / 26.3 Art by D. & V. Hennings / 26.4 Art by Raychel Ciemma / 26.5 (left) After Laszlo Meszoly in L. Margulis, *Early Life*, Jones and Bartlett, 1982 / 26.6 Bruce Hall / 26.7 (a–c) Art by Raychel Ciemma; (d, left) Art by Raychel Ciemma, after Bayer and Owre, *The Free-Living Lower Invertebrates*, © 1968 Macmillan; (d, right) Don W. Fawcett/Visuals Unlimited; (e) Marty